高等学校信息技术
人才能力培养系列教材

BigData

BigData Technology
Foundation and Practices

大数据技术基础与实战

薛志东 张双双 卢璟祥 等 编著

人民邮电出版社
北京

图书在版编目（CIP）数据

大数据技术基础与实战 / 薛志东等编著. -- 北京 ：
人民邮电出版社，2021.9
高等学校信息技术人才能力培养系列教材
ISBN 978-7-115-56719-2

Ⅰ. ①大… Ⅱ. ①薛… Ⅲ. ①数据处理－高等学校－
教材 Ⅳ. ①TP274

中国版本图书馆CIP数据核字(2021)第122319号

内 容 提 要

本书从实战的角度出发，带领读者一步一步学习大数据技术的相关技能。本书不仅提供相应命令、配置文件，还提供模拟环境演示等，并尽可能通过案例和实操降低大数据技术的学习门槛，力求让读者学以致用。

本书主要内容包括大数据技术概述、Linux 基础与集群搭建、Hadoop 集群配置、HDFS、MapReduce 分布式编程、Hive 大数据仓库、HBase 数据库部署与操作、数据获取与 Flume 应用、基于 Spark 的内存计算，以及利用大数据平台处理图像。

本书可作为数据科学与大数据、大数据技术与应用、软件工程、计算机科学与技术等专业的大数据实训课程的教材，也可供初学者自学使用。

◆ 编　　著　薛志东　张双双　卢璟祥　等
　责任编辑　邹文波
　责任印制　王　郁　马振武

◆ 人民邮电出版社出版发行　　北京市丰台区成寿寺路 11 号
　邮编　100164　　电子邮件　315@ptpress.com.cn
　网址　https://www.ptpress.com.cn
　固安县铭成印刷有限公司印刷

◆ 开本：787×1092　1/16
　印张：15.25　　　　　　　　2021 年 9 月第 1 版
　字数：381 千字　　　　　　 2024 年 9 月河北第 5 次印刷

定价：59.80 元

读者服务热线：(010)81055256　印装质量热线：(010)81055316
反盗版热线：(010)81055315
广告经营许可证：京东市监广登字 20170147 号

编 委 会

前言

大数据与我们的生活息息相关，为我们的出行、购物、就业、医疗等带来巨大便利。大数据存在巨大的潜在价值，学习和应用大数据技术成为社会各界的共识。当我们意识到应该应用好大数据的时候，新的概念又开始影响我们的生活，如5G、虚拟现实、人工智能等。好像还没来得及深入学习大数据，我们就又落伍了。其实，大可不必惊慌，5G、虚拟现实、人工智能等技术不仅会对大数据技术提出更高的要求，而且会给学习者提供更好的机遇。例如，通过5G网络产生的海量物联网数据、虚拟现实内容数据等均需要大数据技术支持，而人工智能技术则从我们建设的大数据平台源源不断地消费数据，并通过学习达到某种“智能”。因而，可以预见，在未来几十年中，新技术只会促进大数据技术的发展和增加社会对大数据人才的需求。

党的二十大报告特别强调了推动战略新兴产业、加快物联网发展、加快数字经济发展等对我国建设现代化产业体系的重要作用。大数据技术，作为一种新兴的信息技术，不仅仅是数据科学、计算机科学与技术、软件工程、网络空间安全、管理科学与工程等相关领域大数据工作者需要掌握的技术，而且是高校和企业响应党的号召、服务全面建设社会主义现代化国家要大力发展的技术。

我们编写本书的初衷是，希望能借助本书引导读者积极、努力地学习、应用大数据技术，服务于我国现代化产业体系建设。

本书重在基础，重在实践，采用循序渐进的方式带领读者进入大数据的世界，帮助读者在感受大数据魅力的同时掌握大数据相关技能。本书在实践部分不仅提供相应命令、配置文件，还提供模拟环境演示等，让读者轻松地一步一步跟着教程进行学习。本书主要内容如下。

第1章大数据技术概述。本章首先介绍大数据的概念与特性、大数据处理流程，接着介绍Hadoop大数据技术，包括HDFS、MapReduce、HBase、YARN、Hive、Flume、Spark等，最后介绍如何配置基本实践环境，为学习后文做准备。

第2章Linux基础与集群搭建。本章重点介绍Linux基础知识，包括常用命令、网络配置、集群配置等。本章旨在为读者打牢基础，帮助读者轻松应对Linux环境中的常见操作，为后续大数据实践环境中的操作做铺垫。

第3章Hadoop集群配置。本章以伪分布式集群为例，详细介绍Hadoop集群配置时需要的基本环境准备、集群安装与启动过程、日志查看方法等内容。通过本章的学习，读者就可以搭建属于自己的Hadoop大数据环境。

第4章HDFS。本章首先介绍HDFS的基本概念、文件的读取、文件的写入、数据备份等理论知识，接着介绍HDFS基本命令、数据平衡优化，最后介绍HDFS API的使用方法。

第5章MapReduce分布式编程。本章首先以词频统计为例，介绍MapReduce分布式计算框架，并对Shuffle过程做简单讲解；接着介绍性能优化方法及YARN数据处理框架；最后介绍一个MapReduce实战案例，即绘制频度分布。

第6章Hive大数据仓库。本章从Hive安装及配置开始，分别介绍创建数据库、创建表、数据查询及自定义函数运算、自定义函数编程等内容，最后以实战的形式进一步介绍Hive。

第7章HBase数据库部署与操作。本章依次介绍HBase基本知识、HBase的安装、HBase Shell

操作、HBase 客户端 API、数据过滤、HBase 客户端选择及配置优化、HBase 与 MapReduce 集成、HBase 集群监控等内容，最后以公有云网盘为例介绍 HBase 的一个具体使用场景。

第 8 章 数据获取与 Flume 应用。本章首先介绍一些公开数据源网站，然后详细介绍两种数据获取方式——使用网络爬虫获取数据和使用 Flume 获取数据，最后以案例的形式对前面的内容进行整合。

第 9 章 基于 Spark 的内存计算。本章立足于实战，重点介绍 Spark 程序和 RDD 编程、Spark 生态系统，最后以一个简单案例让读者加深对 Spark 的认识。

第 10 章 利用大数据平台处理图像。本章首先介绍图像的基本概念，然后指出 Hadoop 直接处理海量图像存在的问题并提出解决途径，接着介绍一个图像处理库——HIPI，最后介绍 HIPI 的 hibDownload 工具。

本书由薛志东负责策划并主持全书的编写；张双双主要编写了第 1 章～第 4 章；杜海朋编写了第 5 章；董英豪编写了第 6 章；姚益阳编写了第 7 章；卢璟祥主要编写了第 8 章～第 10 章；张萌参与了第 8 章的编写，并提供第 5、6、7、8、9 章的案例。所有编者以及张辉、王维参加了全书的校对、审定工作。

在本书的编写过程中，编者参考、引用了相关技术的官方文档和大量互联网资源，在此向有关机构、作者表示感谢，并在参考文献一一列出，若有遗漏和不妥之处，敬请相关作者指正。

感谢 Oracle 公司杨素兰女士和广州跨象乘云软件技术有限公司产品经理孟焯先生为本书编写提供的支持与帮助；感谢华中科技大学陈传波教授、肖来元教授、李国徽教授、沈刚教授，以及陈维亚博士、区士颀博士、石强博士对本书编写工作予以的支持与帮助。

由于编者水平有限，书中难免存在不足之处，敬请读者批评指正。

编者

2021 年 6 月

目 录

第 1 章 大数据技术概述

当前各行各业都尝试通过大数据技术对产业进行升级、改造，从而出现了工业大数据、金融大数据、环境大数据、医疗健康大数据、教育大数据等新概念。在实际使用大数据的过程中，人们对大数据的概念及价值有了新的认识。本章在介绍大数据的概念与基本特性的基础上，介绍大数据处理流程及 Hadoop 大数据技术，并对实践开发需要的 VirtualBox 的安装与配置进行介绍。

1.1 大数据的概念与基本特性

大数据是指在一定时间内，无法用常规软件工具对其内容进行抓取、处理、分析和管理的数据集合。大数据中的数据一般会涉及两种以上的数据形式。大数据有 4 个特性——规模性(Volume)、多样性（ Variety ）、高速性（ Velocity ）、价值性（ Value ），简称 4V 特性，如图 1-1 所示。

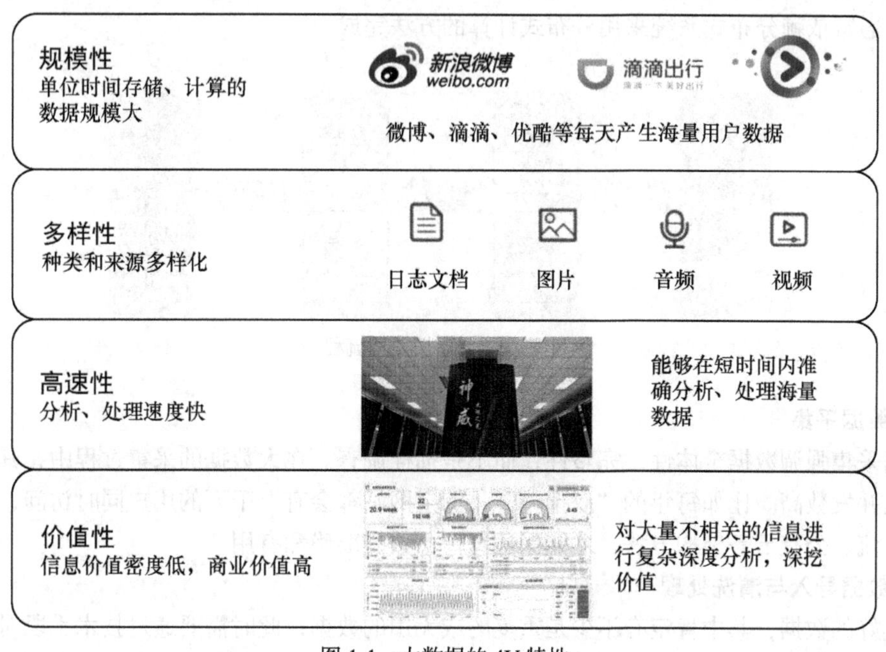

图 1-1 大数据的 4V 特性

（1）规模性：指大数据的规模大。大数据的存储单位已经从过去的 GB、TB，发展到 PB、EB。随着网络和信息技术的高速发展，数据开始爆发式增长。不仅社交网络、移动网络、各种智能终端等都成为数据的来源，企业也面临着自身数据的大规模增长。

（2）多样性：指大数据承载信息的数据形式多样、繁杂。可将大数据分为结构化、非结构化和半结构化数据。结构化数据，如财务系统数据、信息管理系统数据、医疗系统数据等，其特点是数据间因果关系强；非结构化的数据，如图片、音频、视频等，其特点是数据间没有因果关系；半结构化数据，如 HTML 文档、邮件、网页等，其特点是数据间的因果关系弱。

（3）高速性：指大数据被创建、移动、使用的速度快。为满足企业快速创建数据、处理与分析数据、实时反馈的需求，大数据越来越依赖高速处理器或服务器。

（4）价值性：指大数据的信息密度低，而价值高。大数据最大的价值在于可从大量不相关的多样数据中，挖掘出对未来趋势与模式预测分析有价值的数据，并通过机器学习方法、人工智能方法或数据挖掘方法深度分析，发现新规律和新知识并将之运用于农业、金融、医疗等各个领域，从而最终达到改善社会治理、提高生产效率、推进科学研究的目的。

因而，大数据技术可以视为从大量的、纷杂多样的数据中快速提取或发现数据价值的技术与方法的总称。大数据技术也就是应对数据规模大、数据挖掘速度慢、数据类型繁杂等带来的难以挖掘数据价值的问题而产生的技术。

1.2 大数据处理流程

一般而言，我们可以将大数据处理流程分为 4 个步骤：数据采集、数据导入与清洗处理、数据统计与数据分析、数据挖掘和应用，如图 1-2 所示。这 4 个步骤看起来与一般的数据处理分析没有太大区别，但实际上大数据的数据集更多、更大，数据相互之间的关联也更多，需要的计算量更大，通常依赖分布式系统采用分布式计算的方法完成。

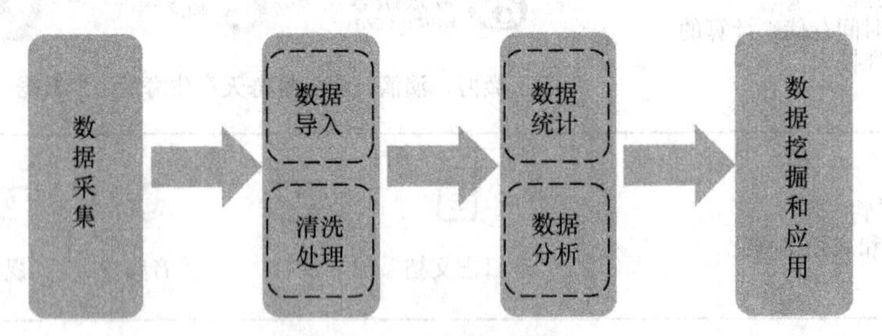

图 1-2　大数据处理流程

1. 数据采集

数据采集强调数据全体性、完整性，而不是抽样调查。在大数据的采集过程中，其主要特点和挑战是并发数高。比如每年的“双十一”，淘宝网站都会有上千万的用户同时访问，需要依靠合理的分流、公有云等架构方法，才能保证每一个数据准确和有用。

2. 数据导入与清洗处理

采集好的数据，其中肯定有不少是重复的或无用的数据，此时需要通过技术手段对数据进行处理，将这些来自前端的数据导入集中的大型分布式数据库，并进行简单的清洗和预处理工作。

而这个过程中最大的挑战是导入的数据规模十分庞大。

3. 数据统计与数据分析

数据统计与数据分析需要用工具来处理，比如可视化工具、SPSS 工具、一些结构算法模型，并进行分类、汇总以满足企业的数据分析需求。这个过程最大的特点是目的清晰，按照一定规则去分类、汇总，才能得到有效的分析结果，这也很耗费系统资源。

4. 数据挖掘和应用

采集数据的最终目的无疑是通过挖掘数据背后的联系，分析原因并找出规律，然后将之应用到实际业务中。数据挖掘是指在通过各种算法对前面几个步骤中的数据进行计算分析后，预测结果、大胆假设，使用数据验证并得出结论。数据挖掘过程的主要挑战是挖掘算法复杂，计算涉及的数据量和计算量都很大。

大数据处理的实现至少需要上述 4 个基本步骤，不过有关细节、工具的使用、数据的完整性等需要结合业务、行业特点和时代变化等不断更新。

1.3 Hadoop 大数据技术

大数据技术涉及大数据处理的各个阶段，包括采集、存储、计算处理和可视化等，而 Hadoop 则是一个集合了大数据不同阶段技术的生态系统。

1.3.1 Hadoop 简介

Hadoop 来自 Apache Lucene 搜索引擎子项目 Nutch。Google 公司为了解决其搜索引擎中大规模 Web 网页数据的处理问题，提出了 MapReduce 大规模数据并行处理技术。道・卡廷（Doug Cutting）尝试依据 Google MapReduce 的设计思想，用 Java 实现了一套新的 MapReduce 并行处理软件系统，并将其与 Nutch 分布式文件系统（Nutch Distributed File System，NDFS）结合，用以支持 Nutch 搜索引擎的数据处理。2006 年，NDFS 和 MapReduce 从 Nutch 项目中分离出来，成为一套独立的大规模数据处理软件系统，并命名为“Hadoop”。

Hadoop 生态圈不仅包含 Hadoop Common、HDFS、Hadoop YARN、Hadoop MapReduce 等组件，还包含 Ambari、Avro、Cassandra、Chukwa、HBase、Hive、Flume、Mahout、Pig、Spark、Tez、MLlib、Tachyon、ZooKeeper 等。下面简单介绍 Hadoop 大数据技术核心组件。

1. HDFS

Hadoop 分布式文件系统（Hadoop Distributed File System，HDFS）源于 Google 公司发表于 2003 年 10 月关于 Google 文件系统（Google File System，GFS）的论文，是 Hadoop 体系中数据存储管理的基础。HDFS 是一个高度容错的系统，能检测和应对硬件故障，能够运行在低成本的硬件上。HDFS 简化了文件的一致性模型，通过流式数据访问，提供高吞吐量应用程序的数据访问功能，适合带有大型数据集的应用程序。HDFS 提供了一次写入、多次读取的机制，数据以块的形式，同时分布在集群的不同物理机上，具有很高的读取效率和很强的容错性。HDFS 的架构是基于一组特定的节点构建的，这是由它自身的特点决定的。这些节点包括一个 NameNode 和若干个 DataNode。其中，NameNode 在 HDFS 内部提供元数据服务，DataNode 为 HDFS 提供存储块服务。

2. MapReduce

MapReduce 是一种用于大数据计算的分布式计算框架，源于 Google 公司在 2004 年 12 月发表的 MapReduce 论文。MapReduce 屏蔽了分布式计算框架的细节，将计算抽象成映射（Map）和规约（Reduce）两部分。其中 Map 对数据集上的独立元素进行指定的操作，生成键值（Key Value）对形式的中间结果。Reduce 则对中间结果中相同“键”的所有“值”进行规约，以得到最终结果。MapReduce 非常适合在大量计算机组成的分布式并行环境里处理数据。

3. HBase

HBase 是一个建立在 HDFS 之上，面向列并针对结构化数据的可伸缩、高可靠、高性能、分布式的动态模式数据库，源于 Google 公司在 2006 年 11 月发表的 Bigtable 论文。HBase 采用增强的稀疏排序映射（键值对）表，键由行关键字、列关键字和时间戳构成。HBase 提供了对大规模数据的随机、实时读/写访问。HBase 中保存的数据可以使用 MapReduce 来处理，它将数据存储和并行计算完美地结合在一起。HBase 是 Google Bigtable 的开源实现，将 HDFS 作为其文件存储系统，利用 MapReduce 处理 HBase 中的海量数据，将 ZooKeeper 作为协同服务。

4. YARN

另一种资源协调者（Yet Another Resource Negotiator，YARN）是由第一代经典 MapReduce 演变而来的，主要是为了解决原始 Hadoop 扩展性较差、不支持多计算框架而提出的。YARN 是一个通用的运行时框架，用户可以编写自己的计算框架，在该运行环境中运行。

5. Hive

Hive 是 Facebook 公司开源的、基于 Hadoop 的一个数据仓库，最初用于解决海量结构化的日志数据统计问题。Hive 使用类结构化查询语言（Structured Query Language，SQL）的 Hive 查询语言（Hive Query Language，HQL）实现数据查询，并将 HQL 转化为在 Hadoop 上执行 MapReduce 任务（Task）。Hive 用于离线数据分析，可让不熟悉 MapReduce 的开发人员使用 HQL 实现数据查询分析，降低了大数据处理应用的门槛。Hive 本质上是基于 HDFS 的应用程序，其数据都存储在 Hadoop 兼容的文件系统（例如，Amazon S3、HDFS）中。

6. Flume

Flume 是 Cloudera 公司开源的日志收集系统，具有分布式、高可靠、高容错、易于定制和扩展的特点。Flume 将数据从产生、传输、处理并最终写入目标路径的过程抽象为数据流。在具体的数据流中，数据源支持在 Flume 中定制数据发送方，从而支持收集各种不同协议数据。同时，Flume 不仅能对日志数据进行过滤、格式转换等简单处理，也能够将日志写入各种数据目标。

7. Spark

Spark 是一个更快、更通用的数据处理平台。最早 Spark 是加州大学伯克利分校 AMP 实验室（UC Berkeley AMPLab）开源的类 MapReduce 的通用并行计算框架。和 Hadoop 相比，Spark 可以让程序在内存中运行的速度提升约 100 倍，或者在磁盘上运行的速度提升约 10 倍。

8. Spark Streaming

Spark Streaming 支持对流数据的实时处理，以“微批”的方式对实时数据进行计算。它是构建在 Spark 上处理流数据的框架，基本原理是将流数据分成小的时间（几秒）片断，以类似批量（Batch）处理的方式来处理小部分数据，也可以用于准实时计算。

9. MLlib

机器学习库（Machine Learning Library，MLlib）提供了多种常用算法，这些算法用来在集群上处理分类、回归、聚类、协同过滤等。MLlib 是 Spark 常用的机器学习算法的实现库，同时包括相关的测试和数据生成器。Spark 的设计初衷就是为了支持一些迭代的作业（Job），这正好符合很多机器学习算法的特点。

10. Tachyon

Tachyon 是以内存为中心的分布式存储系统，诞生于加州大学伯克利分校 AMP 实验室。它拥有高性能和容错能力，并具有类 Java 的文件应用程序接口（Application Program Interface，API）、插件式的底层文件系统、兼容 MapReduce 和 Apache Spark 等特征，能够为集群框架（如 Spark、MapReduce）提供可靠的内存级速度的文件共享服务。Tachyon 充分使用内存和文件对象之间的“血统”（Lineage）信息，速度很快，官方称最高比 HDFS 吞吐量高约 300 倍。

1.3.2　Hadoop 的发行版本

Hadoop 版本包括 Apache Hadoop（又称社区版）和第三方发行版 Hadoop，其中 Apache Hadoop 是一款支持数据密集型分布式应用并以 Apache 2.0 许可协议发布的开源软件框架。第三方发行版 Hadoop 遵从 Apache 开源协议，用户可以免费地任意使用和修改。很多厂家在 Apache Hadoop 的基础上开发自己的 Hadoop 产品，比如 Cloudera 公司的 CDH、MapR 公司的 MapR、Amazon 公司的 Amazon Elastic MapReduce、华为公司的 FusionInsight 等。

1. Cloudera 公司的 CDH

在 Hadoop 生态系统中，规模最大、知名度最高的公司之一则是 Cloudera。Cloudera 公司由来自 Facebook、Google 和 Yahoo 这 3 家公司的前工程师杰夫·哈默巴切（Jeff Hammerbacher）、克里斯托弗·比塞格利亚（Christophe Bisciglia）、埃姆·阿瓦达拉（Amr Awadallah）以及 Oracle 公司前高管迈克·奥尔森（Mike Olson）在 2008 年创建。2014 年，Cloudera 公司正式进入中国；2018 年 10 月，Cloudera 与 Hortonworks 公司合并。Cloudera 公司提供的 CDH 拥有强大的 Hadoop 部署、管理和监控工具，以及众多的部署案例，Cloudera 公司开发、贡献了可实时处理大数据的 Impala 项目。

2. MapR 公司的 MapR

MapR与其竞争产品相比，使用了一些不同的概念，特别是为了获取更好的性能和易用性，支持本地 UNIX 文件系统而不是 HDFS。可以使用本地 UNIX 命令来代替 Hadoop 命令。除此之外，MapR 的产品还凭借诸如快照、镜像或有状态的故障恢复之类的高可用性特性来与其竞争产品相区别。MapR 公司也领导着 Apache Drill 项目，该项目是 Google 公司的 Dremel 开源项目的重新实现，目的是在 Hadoop 数据上执行类似 SQL 的查询以提供实时处理。

3. Amazon 公司的 Amazon Elastic MapReduce

Amazon Elastic MapReduce（EMR）是一个托管的解决方案，其运行在由 Amazon 弹性计算云（Amazon Elastic Compute Cloud，Amazon EC2）和 Amazon 简易存储服务（Amazon Simple Storage Service，Amazon S3）组成的云基础设施之上。如果需要一次性或不常见的大数据处理，选择 EMR 可能会节省大笔开支。EMR 默认只包含了 Hadoop 生态系统中的 Pig 和 Hive 项目，并优化为使用 S3 中的数据。EMR 上的文件 I/O 相比于物理机上 Hadoop 集群或私有 EC2 集群来说会慢很多，并有更大的延时。

4. 华为公司的 FusionInsight

华为公司的 FusionInsight 大数据平台，能够帮助企业快速构建海量数据信息处理系统，通过对企业内部和外部的巨量信息数据进行实时与非实时的分析挖掘，发现全新价值点和企业商机。FusionInsight 是完全开放的大数据平台，可运行在开放的 x86 架构服务器上。它以海量数据处理引擎和实时数据处理引擎为核心，并针对金融、通信等数据密集型行业的运行维护、应用开发等需求，打造了敏捷、智慧、可信的平台软件、建模中间件及运营管理系统（Operation Management System，OMS），让企业可以更快、更准、更稳地从各类繁杂、无序的海量数据中发现价值。

FusionInsight 还集成了企业知识引擎和实时决策支持中心等功能。丰富的知识库和分析套件工具、全方位企业实时知识引擎和决策中心，能够帮助运营商在瞬息万变的数字商业环境中快速决策，实现敏捷的商业成功。开发者合作伙伴可以在华为 FusionInsight 上，基于大数据的各类商业应用场景，比如增强型商务智能（Business Intelligence，BI）、客户智能和数据开放，为金融、通信等行业的客户实现数据的价值——效率提升和收入提升。FusionInsight 整体架构如图 1-3 所示。

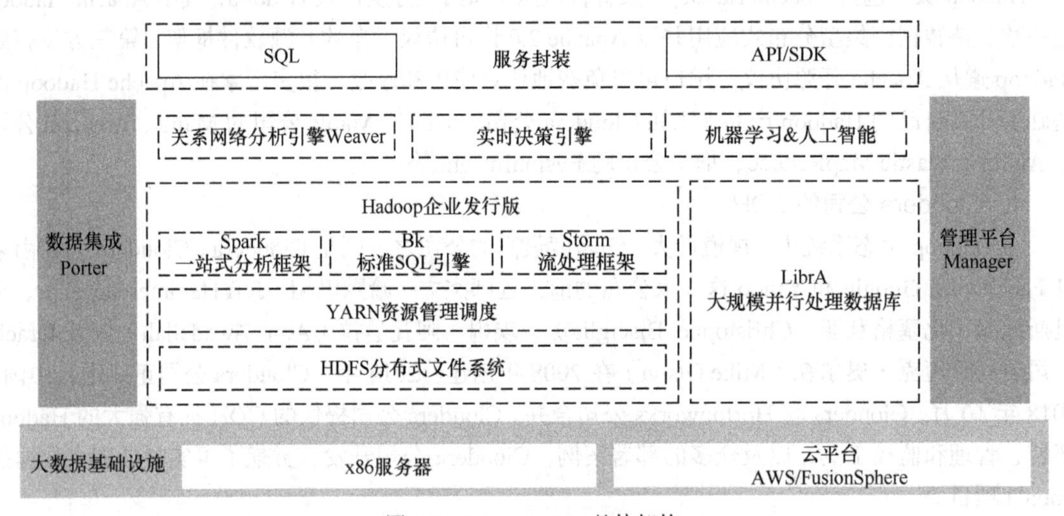

图 1-3　FusionInsight 整体架构

第三方发行版针对社区版的不足以及自身企业优势进行优化，有以下优点。

（1）基于 Apache 协议，100%开源。

（2）版本管理清晰。比如 Cloudera 的 CDH1、CDH2、CDH3、CDH4 等，以及后面加上的补丁版本，如 CDH4.1.0 patch level 923.142，表示在原生态 Apache Hadoop 0.20.2 基础上添加了 1065 个补丁。

（3）在兼容性、安全性、稳定性上有所增强。第三方发行版通常都经过了大量的测试验证，有众多部署实例，大量地运行到各种生产环境。

（4）版本更新快。通常情况，比如 CDH 每个季度会有一个更新，每一年会发布一个新版本。

（5）基于稳定版 Apache Hadoop，并应用了最新的 bug 修复或 Feature 的补丁。

（6）提供了部署、安装、配置工具，大大提高了集群部署的效率，可以在几个小时内部署好集群。

（7）运维简单。提供了管理、监控、诊断、修改配置的工具，管理配置方便，定位问题快速、准确，使运维工作简单、有效。

尽管第三方发行版有诸多优势，本书依然采用社区版作为教学环境，旨在帮助学生深入理解 Hadoop 的基本知识。

1.4　实践环境准备

通过前文的学习，相信读者已经对大数据的基本概念与特性、大数据处理流程、Hadoop 的基础知识有了初步了解。

本节对本书的实践环境进行介绍，本书的所有操作都在此环境中进行。实践操作使用 VirtualBox 搭建虚拟机，构建基础实践环境（注：VirtualBox 并不属于大数据技术，只是用于构建实践教学环境的工具）。

VirtualBox 是一个很受欢迎的开源虚拟机软件，可以在其官方网站下载最新版本，如图 1-4 所示。本书实践的物理机使用 Windows 操作系统，选择 Windows hosts 选项下载 VirtualBox 的 Windows 版本安装文件（VirtualBox-5.2.18-124319-Win.exe），并从 CentOS 官网下载 CentOS 7 的 Minimal 镜像文件（CentOS-7-x86_64-Minimal-1804.iso）。

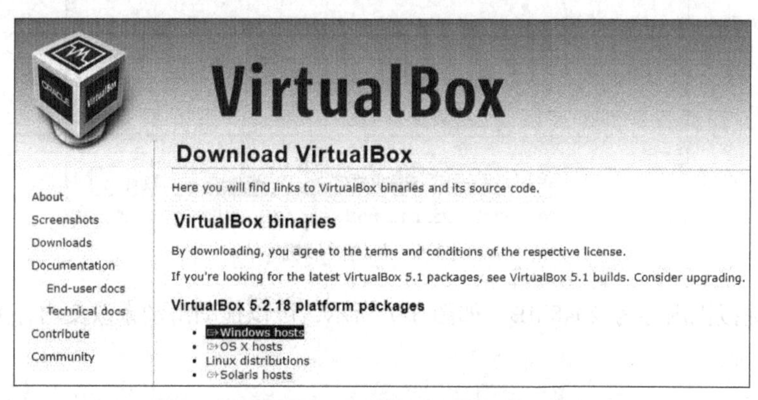

图 1-4　VirtualBox 下载界面

（1）安装 VirtualBox 之前，首先需要确认 BIOS 中的 Virtualization 选项已经开启。双击 VirtualBox 安装文件以完成安装，并启动 VirtualBox，VirtualBox 主界面如图 1-5 所示。

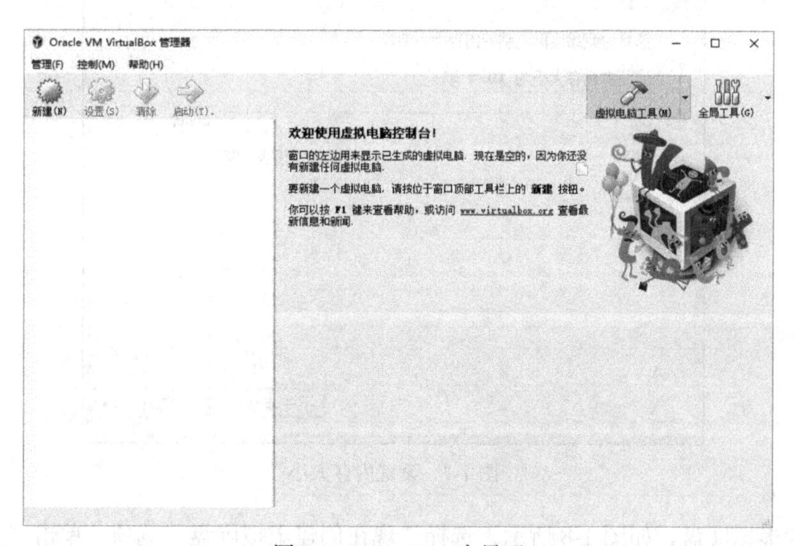

图 1-5　VirtualBox 主界面

（2）单击左上角的“新建”按钮新建虚拟机。如图 1-6 所示，在弹出的对话框中填写名称，并选择类型和版本。名称填写为 hadoop，类型选择 Linux，版本选择 Red Hat (64-bit)，单击“下一步”按钮。注意，如果版本中没有 64 位的选项，首先确定当前主机是否支持 64 位系统；如果不支持，则只能安装 32 位系统。

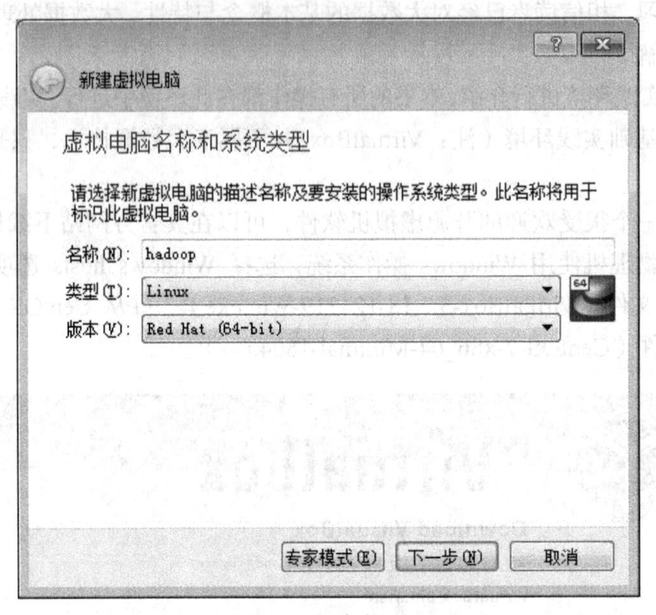

图 1-6　新建虚拟机

（3）设置虚拟机内存为 2048MB，如图 1-7 所示，可以拖动滑动条改变内存大小，单击“下一步”按钮。

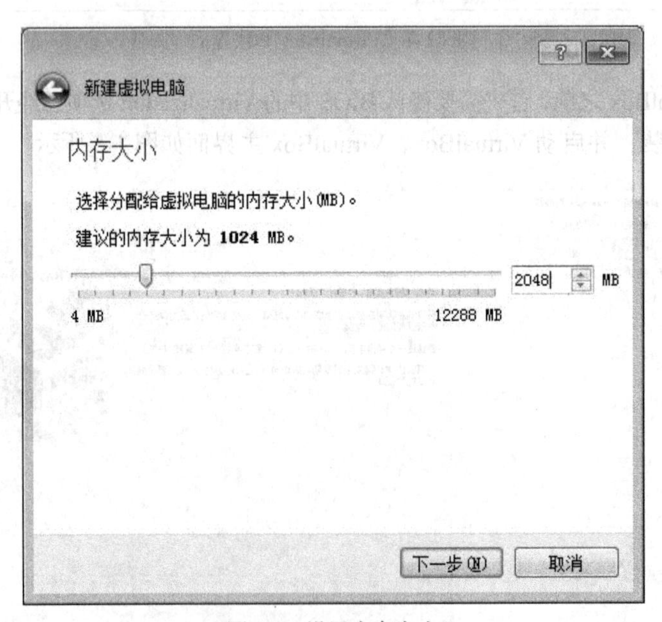

图 1-7　设置内存大小

（4）设置虚拟硬盘，如图 1-8 所示，选择“现在创建虚拟硬盘”选项，单击“创建”按钮。

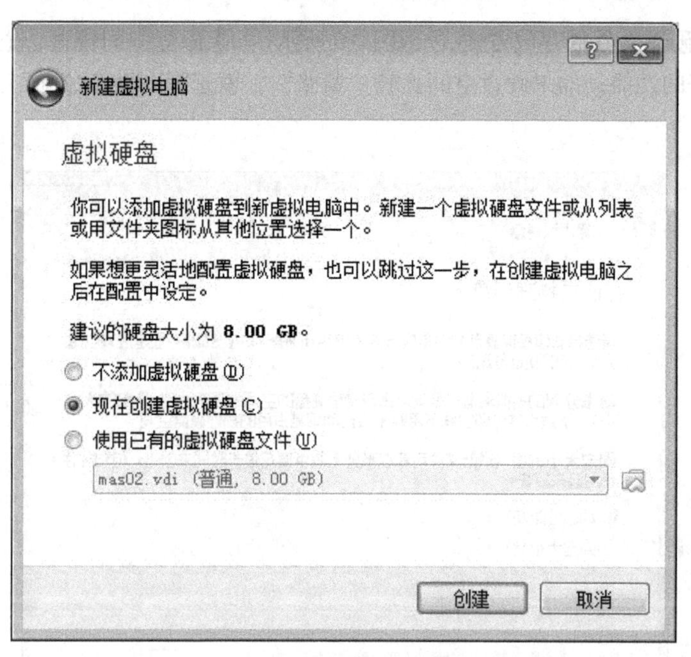

图 1-8 设置虚拟硬盘

（5）虚拟硬盘文件类型选择 VDI，如图 1-9 所示。VDI 是 VirtualBox 的基本格式，目前仅 VirtualBox 软件支持这种文件类型。VHD 是 Microsoft Virtual PC 的基本格式，此种文件类型在 Microsoft 产品中比较受欢迎。VMDK 由 VMware 软件团队开发， Sun xVM、QEMU、VirtualBox、SUSE Studio、DiscUtils 也支持这种文件类型。VMDK 具有将存储的文件分割为小于 2GB 文件的附加功能，如果文件系统的文件大小存在限制，可考虑选择 VMDK 文件类型。设置好后单击“下一步”按钮。

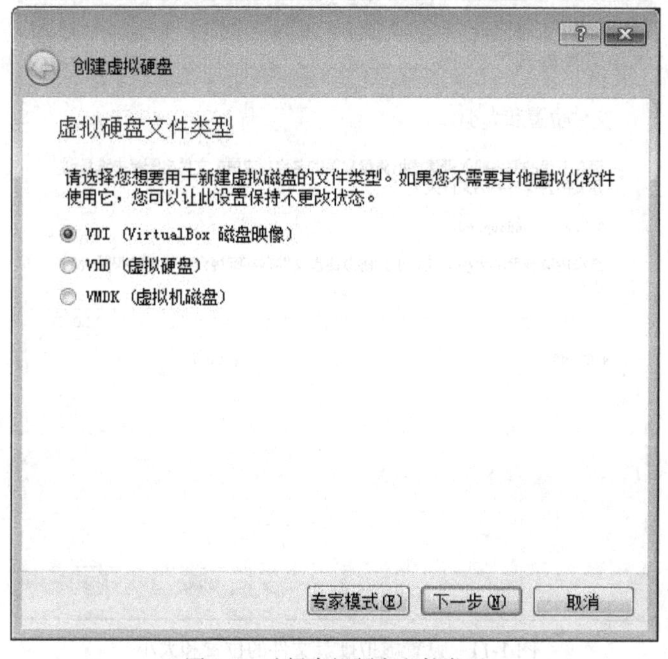

图 1-9 选择虚拟硬盘文件类型

（6）设置虚拟硬盘文件的存放方式，如图 1-10 所示。如果磁盘空间较大，就选择“固定大小”，这样可以获得较好的性能；如果硬盘空间比较“紧张”，就选择“动态分配”。单击“下一步”按钮。

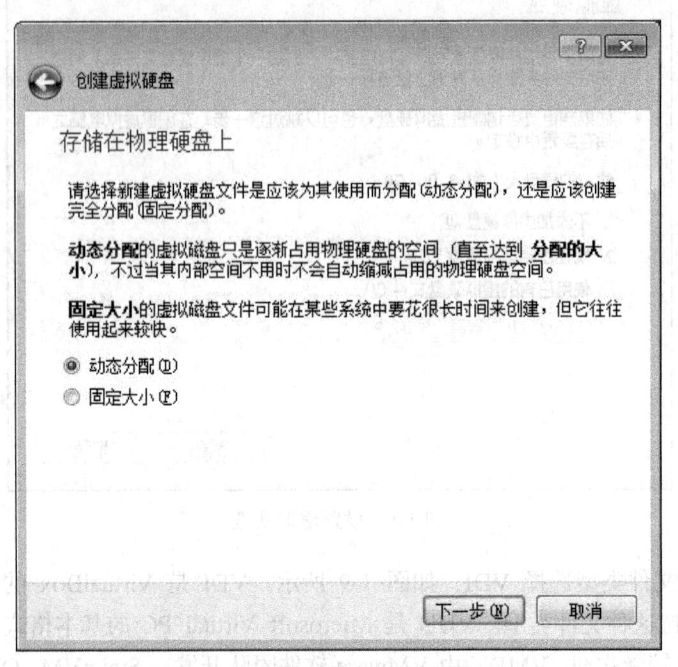

图 1-10　设置虚拟硬盘文件的存放方式

（7）设置虚拟硬盘文件的位置和大小，如图 1-11 所示。选择一个容量充足的磁盘来存放虚拟硬盘文件，因为该文件通常都比较大，然后单击“创建”按钮。

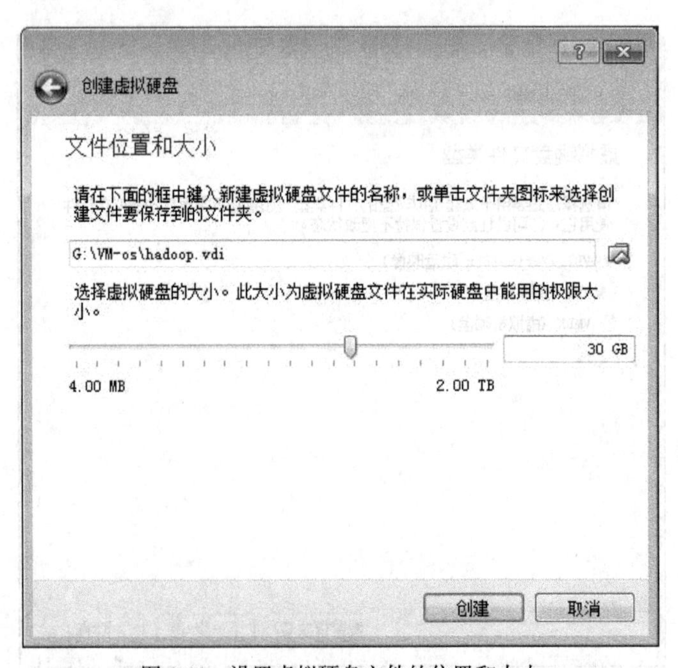

图 1-11　设置虚拟硬盘文件的位置和大小

（8）这样，一个“空壳”虚拟机就创建好了，其基本信息如图 1-12 所示。接下来开始安装 CentOS 7。单击左上角的“启动”按钮来启动虚拟机。

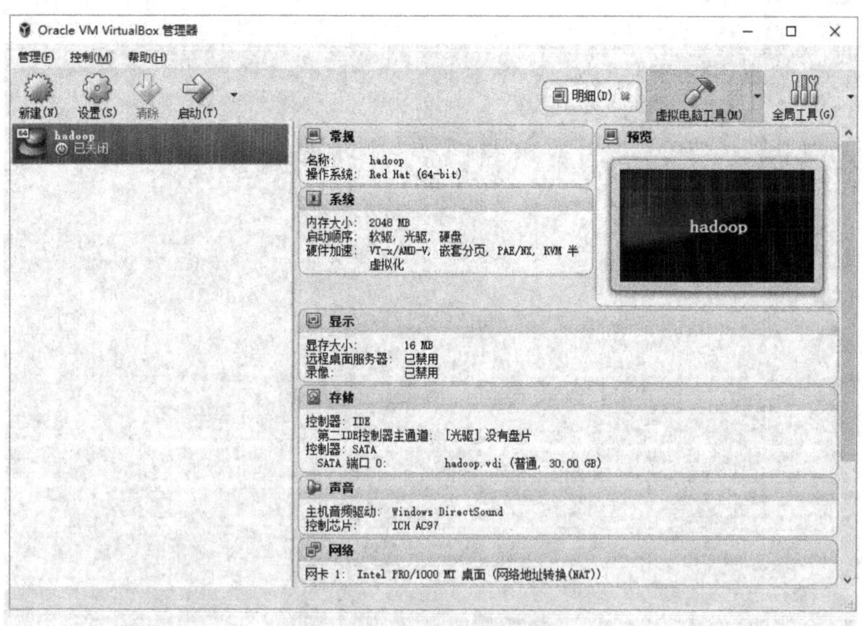

图 1-12　虚拟机的基本信息

（9）如图 1-13 所示，在“选择启动盘”对话框中选择下载好的 CentOS 7 镜像文件后，单击“启动”按钮。

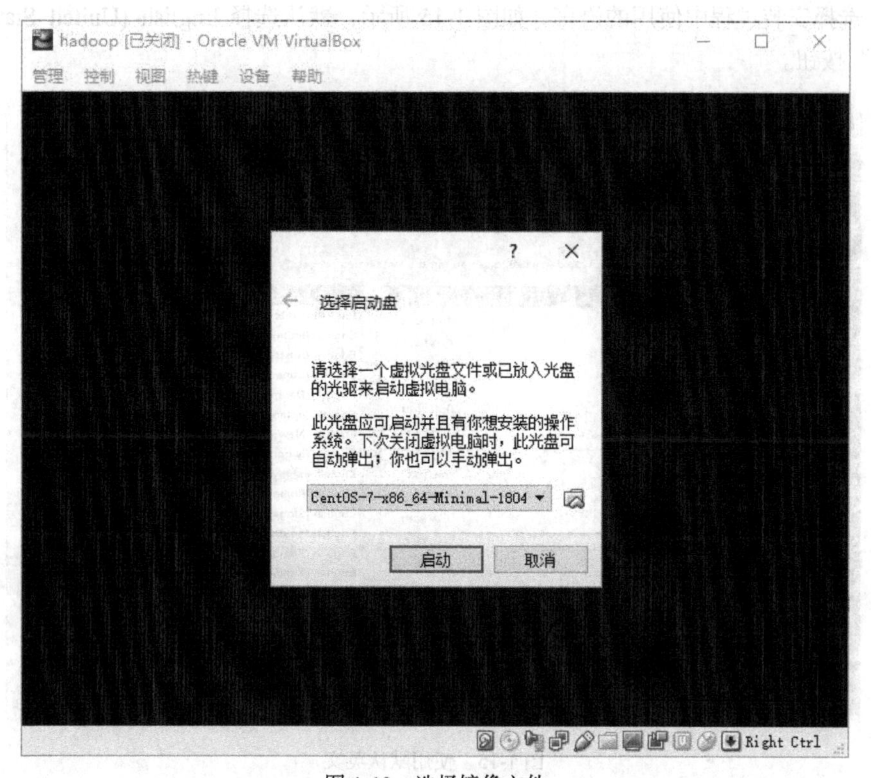

图 1-13　选择镜像文件

（10）如图 1-14 所示，用上、下方向键来选择 Install CentOS 7，然后按 Enter 键。

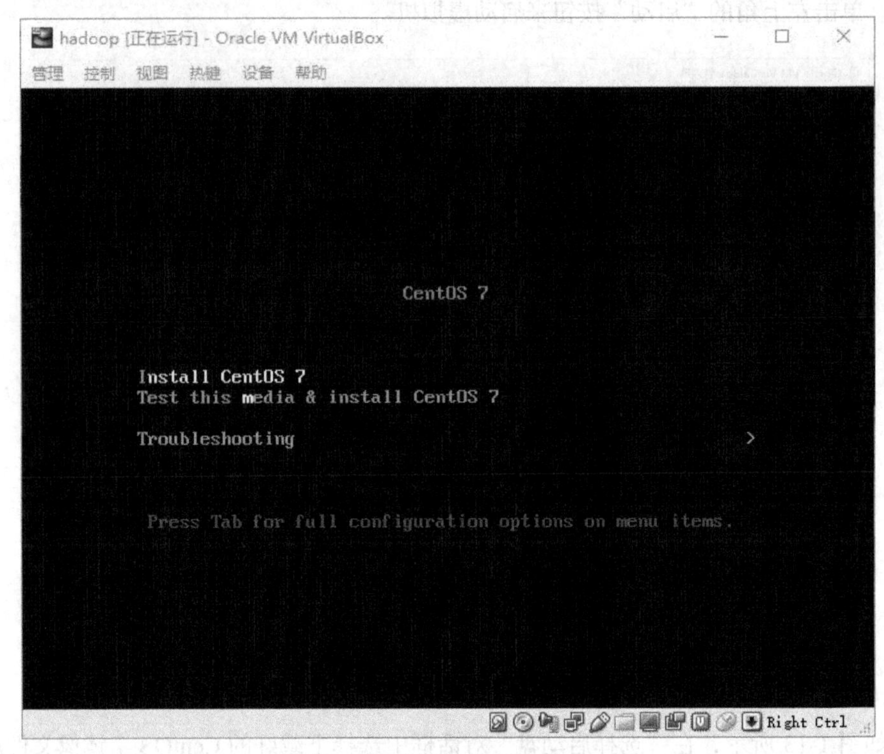

图 1-14　开始安装 CentOS 7

（11）选择安装过程中使用的语言，如图 1-15 所示，默认选择 English (United States)。单击“Continue”按钮。

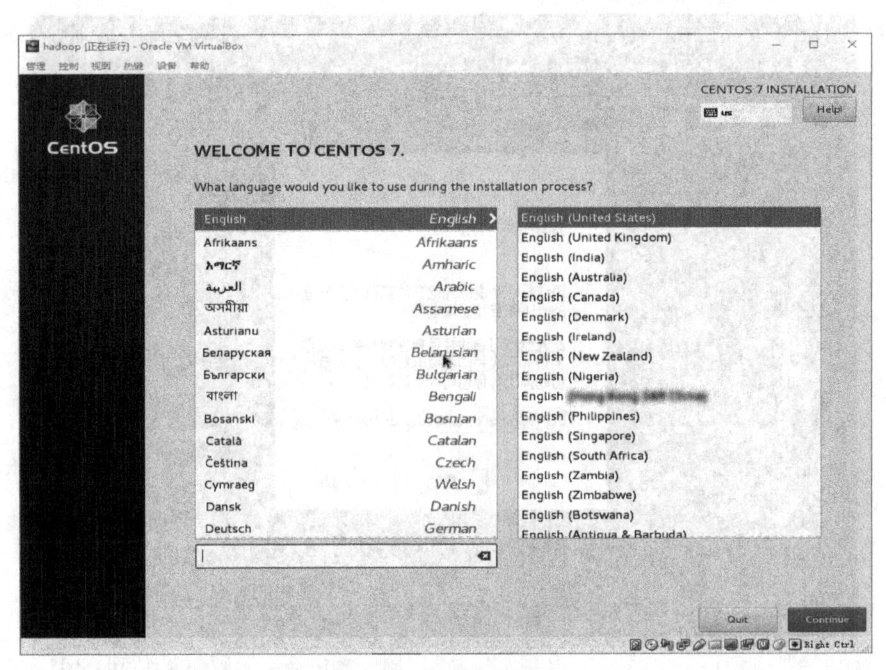

图 1-15　使用默认英文

（12）如图 1-16 所示，配置基本信息，单击“Begin Installation”按钮。

此处由于使用了 CentOS Minimal 镜像文件，默认最小化安装，只安装系统基本的安装包，不包含 X Windows 和桌面管理器。后续如果需要使用相关的安装包，都可以通过 YUM、RPM 等工具进行安装。

单击 INSTALLATION DESTINATION 配置项，进入系统的分区设置。

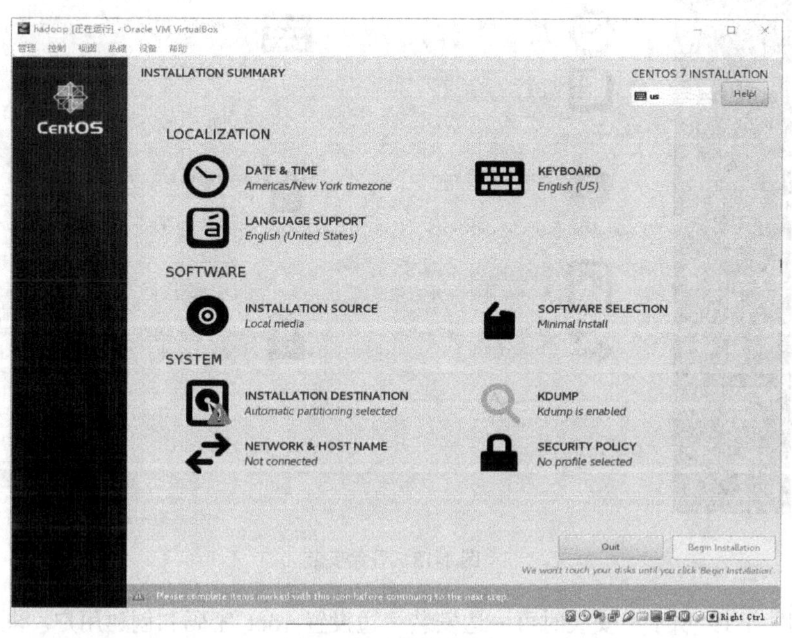

图 1-16　配置基本信息

单击左上角的“Done”按钮，使用默认的自动分区设置，如图 1-17 所示。

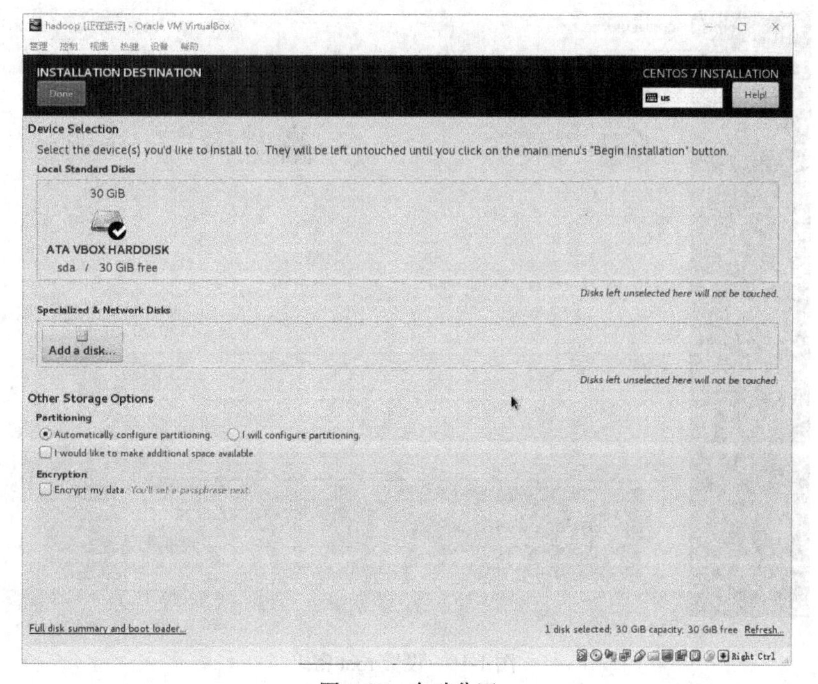

图 1-17　自动分区

返回安装界面，单击“Begin Installation”按钮开始安装，如图 1-18 所示。

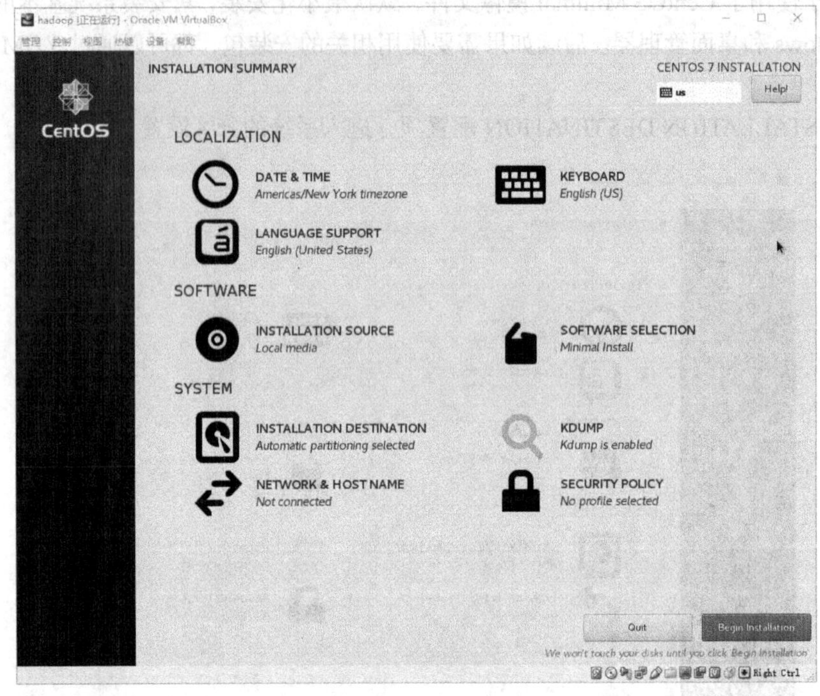

图 1-18　开始安装

（13）如图 1-19 所示，安装时可一边安装一边设置 root 密码和创建用户。弱密码需要双击“Done”按钮，界面下方有提示信息。

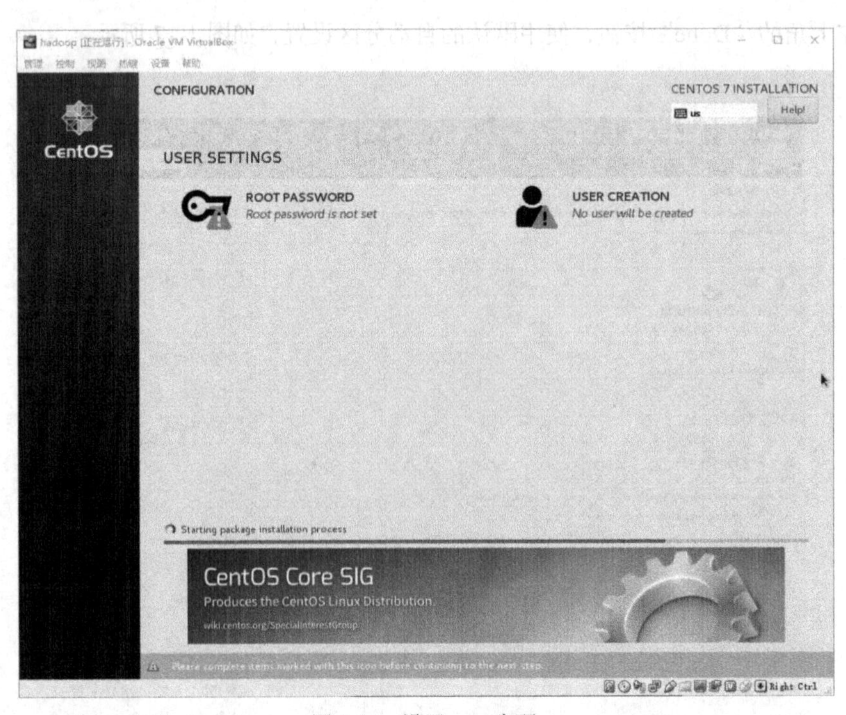

图 1-19　设置 root 密码

（14）安装完成后，重启进入系统，输入设置好的 root 密码登录。输入 cat /etc/redhat-release 命令查看系统版本为 CentOS 7.5，如图 1-20 所示。

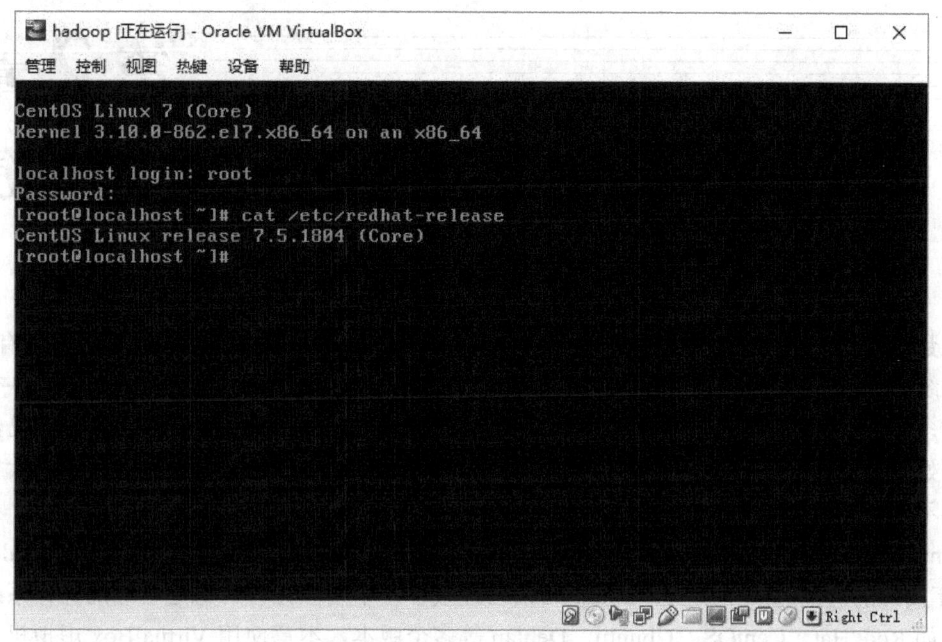

图 1-20　登录系统

至此，完成了基于 VirtualBox 的 Linux 虚拟机的安装任务。

习　题

1. 简述大数据的 4V 特性。
2. 简述大数据处理流程并思考每个步骤所解决的问题。
3. 简述 Hadoop 核心组件及其主要功能。
4. 使用 VirtualBox 创建 CentOS 虚拟机。

第2章 Linux 基础与集群搭建

大数据处理所需的大数据计算硬件平台、文件系统、数据库系统等，通常由多台装有 Linux 操作系统的计算机节点组成。Linux 是一个自由、免费、开源的类 UNIX 操作系统，是一个支持抢占式多任务、多线程、虚拟内存、换页、动态链接和 TCP/IP 的现代操作系统，目前已经成为大数据处理系统的软件基础。因而熟悉 Linux 安装、配置和基本操作成了学习、掌握大数据技术的必不可缺的环节。

Linux 操作系统核心最初是由芬兰赫尔辛基大学学生林纳斯·托瓦兹（Linus Torvalds）在 1990 年设计的。经过多年发展，Linux 目前已经成了一个功能完善、稳定可靠的操作系统，并演化出如 Red Hat、CentOS、Ubuntu、Debian 等多个版本。本章使用 VirtualBox 虚拟机来介绍 CentOS 7。

2.1 Linux 常用命令

本节以 CentOS 7 为例，介绍一些在大数据环境中可能会用到的 Linux Shell 命令，例如用户管理、文件管理、磁盘管理、网络配置和安全设置等命令，相关命令的详细参数和高级用法，请读者查阅相关资料。安装 CentOS 操作系统的过程，已在 1.4 节中详细讲解。读者也可以先复习 1.4 节 CentOS 操作系统安装的有关内容，再学习本节内容并上机练习。以下命令在 CentOS 提供的默认命令行界面执行。若需要对文件进行编辑，可以使用 Vim 或 Gedit（Gedit 需要图形化界面支持）。

以下对 Linux 常用命令的操作全部在 1.4 节搭建的虚拟机上完成。

2.1.1 用户和用户组

1. 创建用户

为 Linux 操作系统创建用户的基本命令为 useradd 和 passwd，分别用于创建用户和设置用户密码。在 Linux 命令行界面，root 用户提示符用“#”表示，普通用户提示符用“$”表示。本书部分注释用“%”开头。使用 root 用户创建 hadoop 用户的示例命令如下：

```
# useradd hadoop
# passwd hadoop
```

创建用户操作如图 2-1 所示。

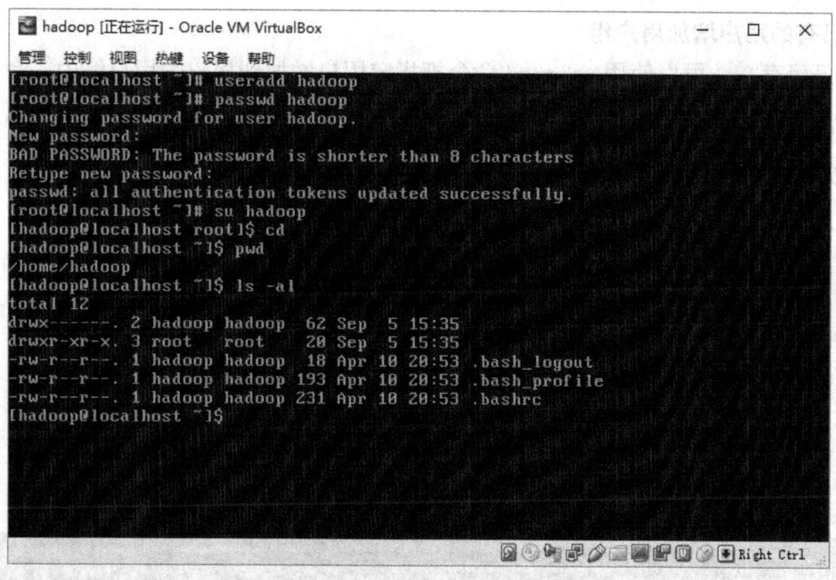

图 2-1　创建用户操作

2. 创建用户组

Linux 文件系统的安全管理权限有用户组管理权限，可以通过 groupadd 命令创建用户组，以方便用户管理。命令示例如下：

```
# groupadd testgroup        %创建 test 用户组
```

3. 创建用户的同时增加用户组

在创建用户 test 时将其增加到用户组 testgroup 中的命令如下：

```
# useradd -g testgroup test    %创建 test 用户并将其增加到 testgroup 用户组中
```

创建用户组并在创建用户的同时增加用户组的操作示例如图 2-2 所示。

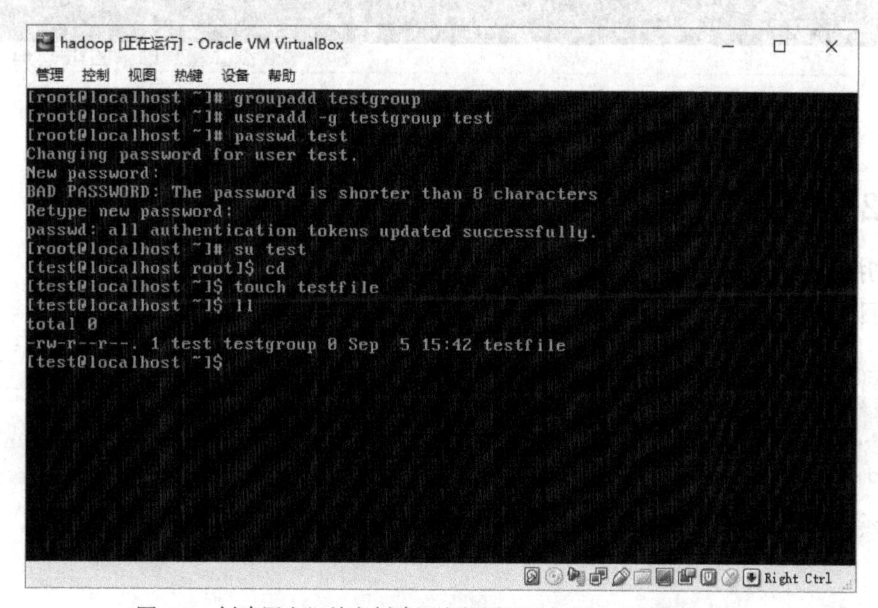

图 2-2　创建用户组并在创建用户的同时增加用户组的操作示例

4. 给已有的用户增加用户组

若用户已经存在，可以使用 usermod 命令把指定用户增加到相应的用户组中，命令如下：

```
# usermod -G groupname username        %参数说明:-G 后面的组名必须是现有用户组中存在的组名
```

5. 永久删除用户和用户组

可以使用 userdel 和 groupdel 删除用户和用户组，命令示例如下：

```
# userdel test
# userdel -r test                      %删除用户的主目录和邮件池
# groupdel testgroup
```

永久删除用户和用户组操作示例如图 2-3 所示。

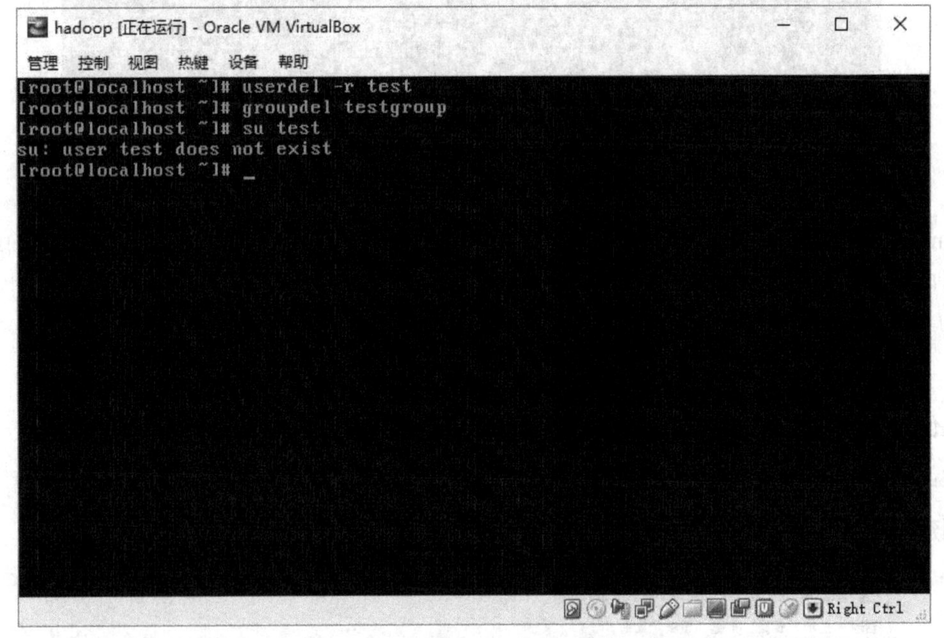

图 2-3 永久删除用户和用户组操作示例

2.1.2 文件与目录

1. 切换目录

切换目录用 cd 命令，命令的用法如下：

```
# cd /home              %进入 home 目录
# cd ..                 %返回上级目录
# cd ../..              %返回上两级目录
# pwd                   %显示当前工作目录名称
```

cd 命令操作示例如图 2-4 所示。

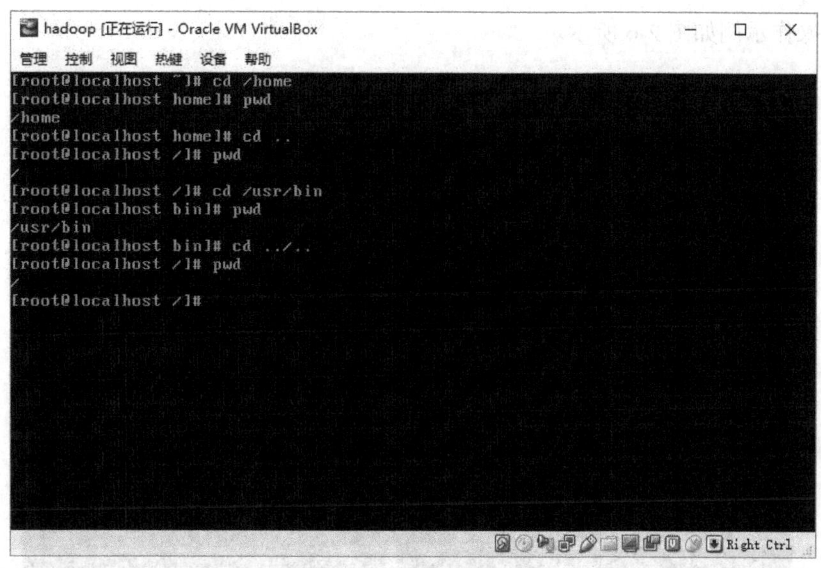

图 2-4　cd 命令操作示例

2. 查看目录中的文件信息

查看目录中的文件信息用 ls 命令，用法示例如下：

```
# ls -a        %查看当前目录下的所有文件（包括隐藏文件）
# ls -al       %显示文件的详细信息
# ls -alrt     %按时间显示文件（l 表示详细列表，r 表示反向排序，t 表示按时间排序）
```

ls 命令操作示例如图 2-5 所示。

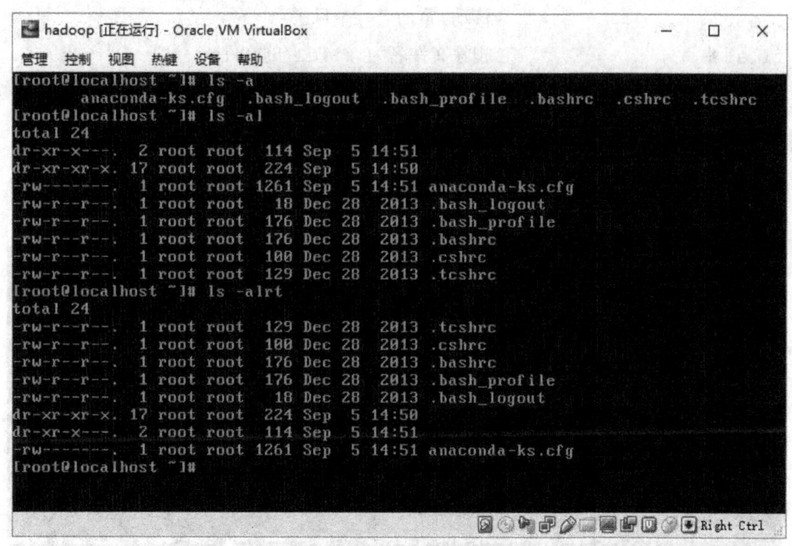

图 2-5　ls 命令操作示例

3. 文件或目录复制

可以用 cp 命令进行文件或目录复制，命令示例如下：

```
# cp file1 file2         % 将 file1 复制为 file2
# cp -a dir1 dir2        % 复制一个目录（包括目录下所有子目录和文件）
# cp -a /tmp/dir1 .      % 复制一个目录到当前工作目录（.代表当前目录）
```

cp 命令操作示例如图 2-6 所示。

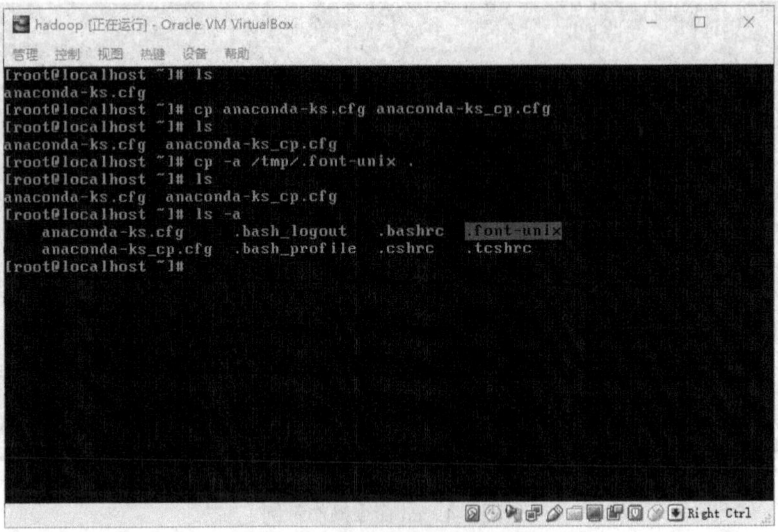

图 2-6　cp 命令操作示例

4. 文件或目录的创建、移动、删除

创建、移动、删除文件或目录涉及 mkdir、mv 和 rm 这 3 个命令。用法示例如下：

```
# mkdir dir1                        %创建 dir1 目录
# mkdir dir1 dir2                   %同时创建两个目录
# mkdir -p /tmp/dir1/dir2           %创建一个目录树
# mv dir1 dir2                      %移动/重命名一个目录
# rm -f file1                       %删除文件名为 file1 的文件
# rm -rf dir2                       %删除 dir2 目录及其子目录的内容
```

文件或目录的创建、移动、删除操作示例如图 2-7 所示。

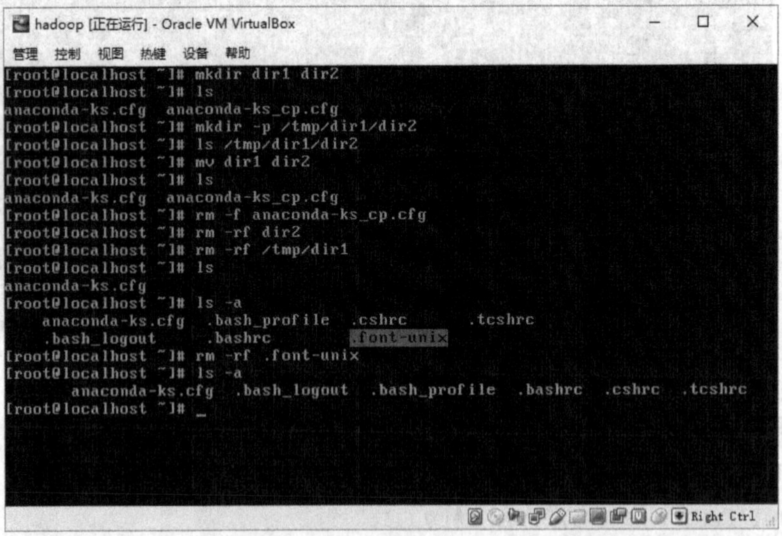

图 2-7　文件或目录的创建、移动、删除操作示例

5. 查看文件内容

可以使用 cat、tac 和 more 命令查看文件内容。cat 命令按照文本文件的行顺序依次显示文件内容；tac 命令是“cat”反向拼写，表示从最后一行开始倒序依次显示文本文件的内容。more 命令可以分页显示文本文件内容。用法示例如下：

```
# cat file1              %从第一行开始正向查看文件的内容
# tac file1              %从最后一行开始反向查看一个文件的内容
# more file1             %查看一个长文件的内容
```

查看/etc/hosts 文件内容操作示例如图 2-8 所示。

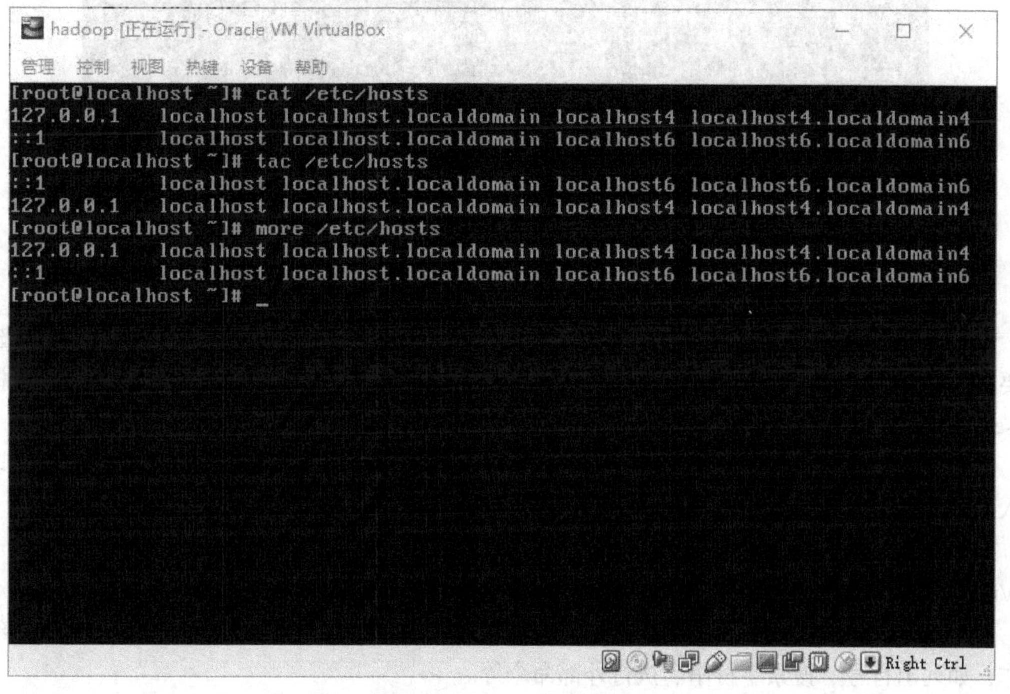

图 2-8　查看/etc/hosts 文件内容操作示例

6. 文本内容处理

在 Linux 下经常需要从文本文件中查找相关字符串，或比较文件的差异。常用命令为 grep 和 diff 命令。用法示例如下：

```
# grep str /tmp/test         %在文件 /tmp/test 中查找 "str"
# grep ^str /tmp/test        %在文件 /tmp/test 中查找以"str"开始的行
# grep [0-9] /tmp/test       %查找 /tmp/test 文件中所有包含数字的行
# grep str -r /tmp/*         %在目录 /tmp 及其子目录中查找"str"
# diff file1 file2           %找出两个文件的不同处
# sdiff file1 file2          %以对比的方式显示两个文件的不同处
```

文本内容处理操作示例如图 2-9 所示。

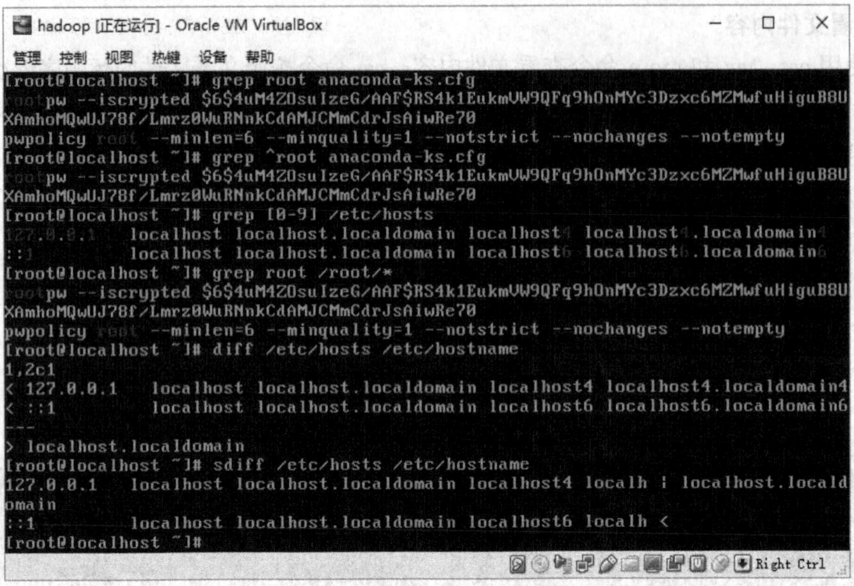

```
hadoop [正在运行] - Oracle VM VirtualBox
管理 控制 视图 热键 设备 帮助
[root@localhost ~]# grep root anaconda-ks.cfg
rootpw --iscrypted $6$4uM4ZOsuIzeG/AAF$RS4k1EukmVW9QFq9hOnMYc3Dzxc6MZMwfuHiguB8U
XAmhoMQwUJ78f/Lmrz0WuRNnkCdAMJCMmCdrJsAiwRe70
pwpolicy root --minlen=6 --minquality=1 --notstrict --nochanges --notempty
[root@localhost ~]# grep ^root anaconda-ks.cfg
rootpw --iscrypted $6$4uM4ZOsuIzeG/AAF$RS4k1EukmVW9QFq9hOnMYc3Dzxc6MZMwfuHiguB8U
XAmhoMQwUJ78f/Lmrz0WuRNnkCdAMJCMmCdrJsAiwRe70
[root@localhost ~]# grep [0-9] /etc/hosts
127.0.0.1   localhost localhost.localdomain localhost4 localhost4.localdomain4
::1         localhost localhost.localdomain localhost6 localhost6.localdomain6
[root@localhost ~]# grep root /root/*
rootpw --iscrypted $6$4uM4ZOsuIzeG/AAF$RS4k1EukmVW9QFq9hOnMYc3Dzxc6MZMwfuHiguB8U
XAmhoMQwUJ78f/Lmrz0WuRNnkCdAMJCMmCdrJsAiwRe70
pwpolicy root --minlen=6 --minquality=1 --notstrict --nochanges --notempty
[root@localhost ~]# diff /etc/hosts /etc/hostname
1,2c1
< 127.0.0.1   localhost localhost.localdomain localhost4 localhost4.localdomain4
< ::1         localhost localhost.localdomain localhost6 localhost6.localdomain6
---
> localhost.localdomain
[root@localhost ~]# sdiff /etc/hosts /etc/hostname
127.0.0.1   localhost localhost.localdomain localhost4 localh | localhost.locald
omain
::1         localhost localhost.localdomain localhost6 localh <
[root@localhost ~]#
Right Ctrl
```

图 2-9　文本内容处理操作示例

7. Vim 文件编辑操作

Vim 是 Linux 操作系统常用的文本编辑器（Linux 操作系统自带 Vi 编辑器，Vim 为其升级版，需要通过 yum install vim 命令安装后才能使用）。Vim 有命令模式（Command Mode）、插入模式（Insert Mode）和底行模式（Last Line Mode）3 种工作模式。

命令模式：在此模式下只能控制光标的移动，进行文本的删除、复制等文字编辑工作，以及进入插入模式，或者回到底行模式。

插入模式：只有在插入模式下，才可以输入文字。在此模式下按 Esc 键可回到命令模式。打开 Vim 编辑器时 Vim 处于命令模式，需要按 i 键进入插入模式。

底行模式：在此模式下可以保存文件或退出 Vim，同时也可以设置编辑环境和进行一些编译工作，如列出行号、搜索字符串、执行外部命令等。

例如，打开一个文件：

```
# vim test.txt
```

进入 Vim 界面后，按 i 键，Vim 进入插入模式，用户就可以像使用其他编辑器一样编辑文件。编辑完毕后按 Esc 键退出插入模式。并按 : 键进入底行模式。常用操作命令如下。

（1）:w 用于保存当前修改。

（2）:q! 用于不保存并强制退出 Vim。

（3）:wq 用于保存当前修改并退出 Vim。

8. 查询

可以通过 find 命令查找相关的文件或文件目录。命令示例如下：

```
# find / -name file1        %从/开始进入根文件系统查找文件和目录
# find / -user user1        %查找属于用户 user1 的文件和目录
# find /home/user1 -name *.bin  %在目录/home/user1 中查找以.bin 为扩展名的文件
```

find 命令操作示例如图 2-10 所示。

```
hadoop [正在运行] - Oracle VM VirtualBox
管理 控制 视图 热键 设备 帮助
[root@localhost ~]# find /etc -name hosts
/etc/hosts
[root@localhost ~]# find /root -user root
/root
/root/.bash_logout
/root/.bash_profile
/root/.bashrc
/root/.cshrc
/root/.tcshrc
/root/anaconda-ks.cfg
[root@localhost ~]# find /root -name *.cfg
/root/anaconda-ks.cfg
[root@localhost ~]#
Right Ctrl
```

图 2-10　find 命令操作示例

9. 压缩、解压

可以利用 tar 命令对文件进行压缩、解压。tar 命令可以解压.tar、.tar.gz、.tar.bz2 文件，其参数 z 和 j 分别代表.tar.gz 和.bz2 文件，示例如下：

```
# tar -cvf archive.tar file1   %把 file1 压缩成 archive.tar（-c：建立压缩档案。-v：显示所有过程。-f：使用档案名字，它是必须的，也是最后一个参数）
# tar -tf archive.tar          %显示一个包中的内容
# tar -xvf archive.tar         %解压一个包（-x：解压）
# tar -xzvf archive.tar.gz     %解压一个.tar.gz 压缩包(-x 解压 gz:表示解压或压缩的为.tar.gz 文件)
# tar -xjvf archive.tar.bz2  -C /tmp    %把 archive.tar.bz2 压缩包解压到 /tmp 目录下
```

压缩、解压操作示例如图 2-11 所示。

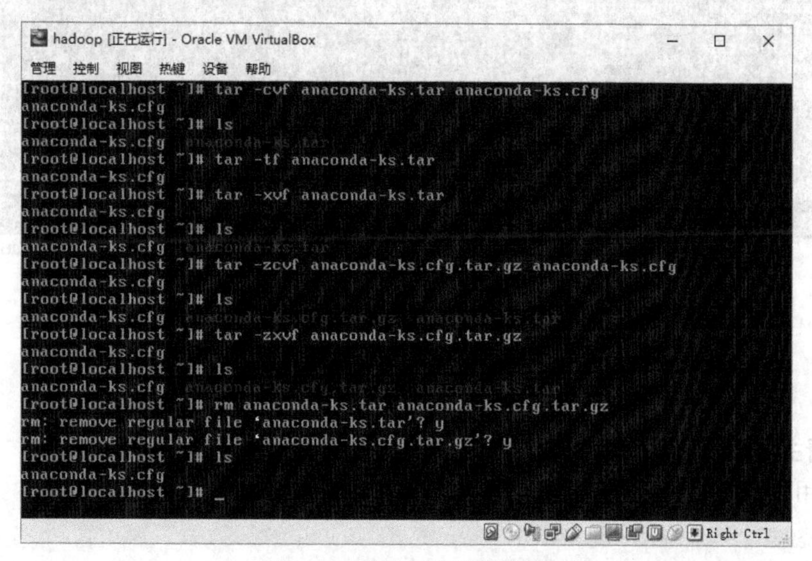

图 2-11　压缩、解压操作示例

10. 修改文件或目录权限

Linux 文件被创建时，文件所有者自动拥有对该文件的读、写和可执行权限，以便于对文件进行读、写和执行操作。

Linux 操作系统根据对文件安全的设置将用户分成 3 种不同的类型：文件所有者、同组用户、其他用户。文件所有者一般是文件的创建者，不仅可以允许同组用户访问文件，还能将文件的访问权限赋予 Linux 操作系统中的其他用户。

每一个文件或目录的访问权限都有 3 组，每组访问权限都使用 3 位数据表示，分别为文件所有者的读、写和执行权限，文件所有者同组用户的读、写和执行权限，以及其他用户的读、写和执行权限。当用 ls-l 命令显示文件或目录的详细信息时，最左边的一列即文件的访问权限。可以使用 chmod、chgrp、chown 分别修改所有用户的权限、修改同组用户的读、写、执行的权限，以及修改文件的所有者。命令示例如下：

```
# chmod 777 test  %把 test 文件修改为所有用户可读、写、执行，是 chmod 的数字设定法
# chmod a+rwx test    %同上，是 chmod 的文字设定法（a：所有用户。g：同组用户。o:其他用户。r：读。w：写。x：执行）
# chgrp student  /opt/book     %把/opt/book 的用户组修改为 student（需要 root 权限）
# chown zhangsan /opt/book     %把/opt/book 的文件所有者修改为张三（需要 root 权限）
```

修改文件或目录权限操作示例如图 2-12 所示。

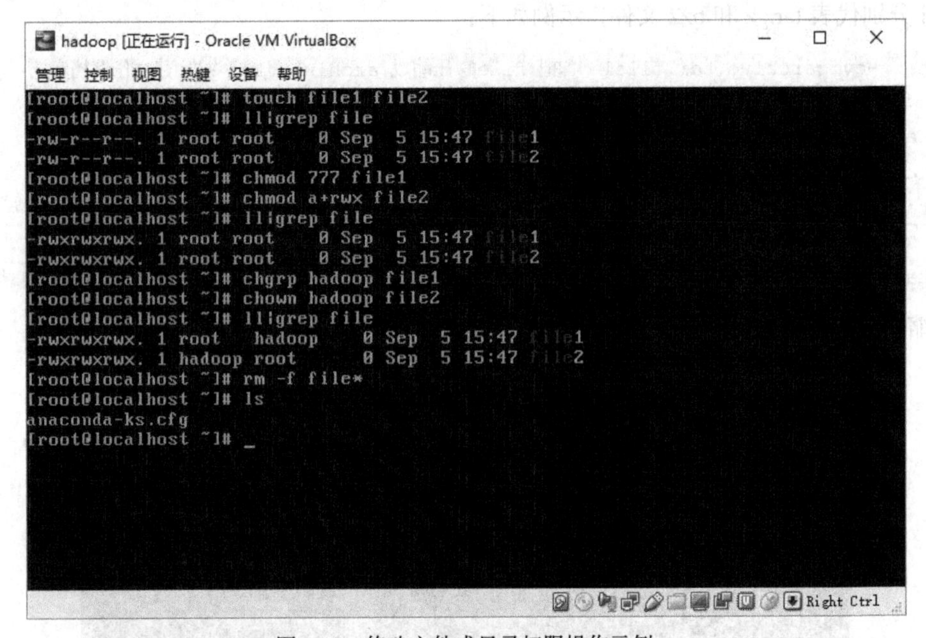

图 2-12 修改文件或目录权限操作示例

2.1.3 主机名

1. 查看主机名

可以使用 hostname 查看当前主机名，命令如下：

```
# hostname
```

2. 永久修改主机名

可以使用 hostnamectl 永久修改主机名，修改后的主机名存储在/etc/hostname 文件中，命令如下：

```
# hostnamectl set-hostname hadoop          %修改主机名为 hadoop
# cat /etc/hostname                        %用 cat 命令在命令行界面显示文件内容 hadoop
```

永久修改主机名操作示例如图 2-13 所示。

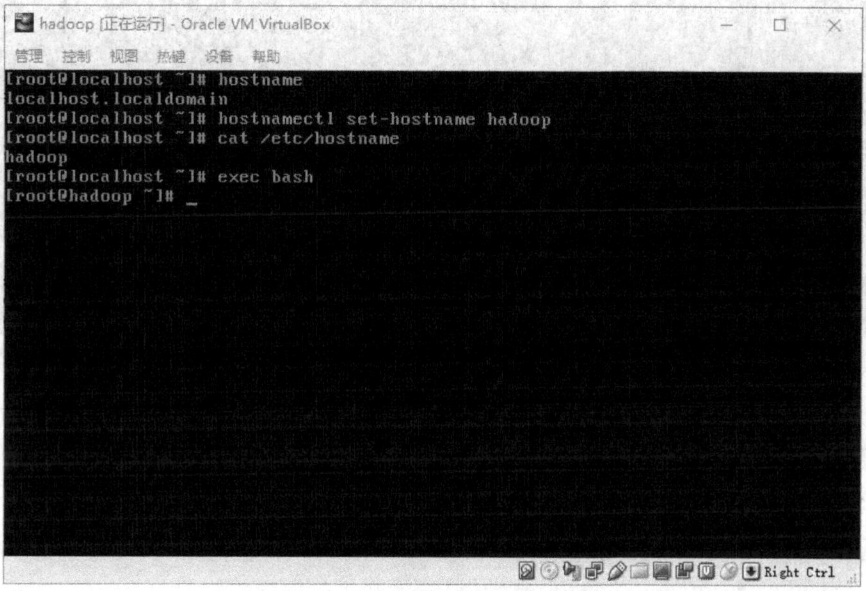

```
[root@localhost ~]# hostname
localhost.localdomain
[root@localhost ~]# hostnamectl set-hostname hadoop
[root@localhost ~]# cat /etc/hostname
hadoop
[root@localhost ~]# exec bash
[root@hadoop ~]# _
```

图 2-13　永久修改主机名操作示例

也可以通过直接修改/etc/hosts 文件中的主机名来修改主机 IP 地址对应的域名。还可以使用 Vim 等编辑工具打开该文件，修改对应 IP 地址后的主机名。

```
# vim /etc/hosts              %注意：在打开文件并修改主机名后，需保存文件
# cat /etc/hosts
```

执行命令后，内容如下：

```
127.0.0.1   localhost localhost.localdomain localhost4 localhost4.localdomain4
::1         localhost localhost.localdomain localhost6 localhost6.localdomain6
192.168.142.106 controller      %修改对应的主机名为 controller
192.168.142.107 compute
```

注意：下划线标注部分为对应的修改内容。

2.1.4　分区管理

Linux 用户可以使用 df、fdisk、mount 等命令查看、分区、挂载硬盘。

1. 查看硬盘的使用状况

使用 df 命令查看当前硬盘的使用状况，命令如下：

```
# df -h  %-h：将容量显示为易读的格式（如 1K、234M、2G 等）
```

df 命令操作示例如图 2-14 所示。

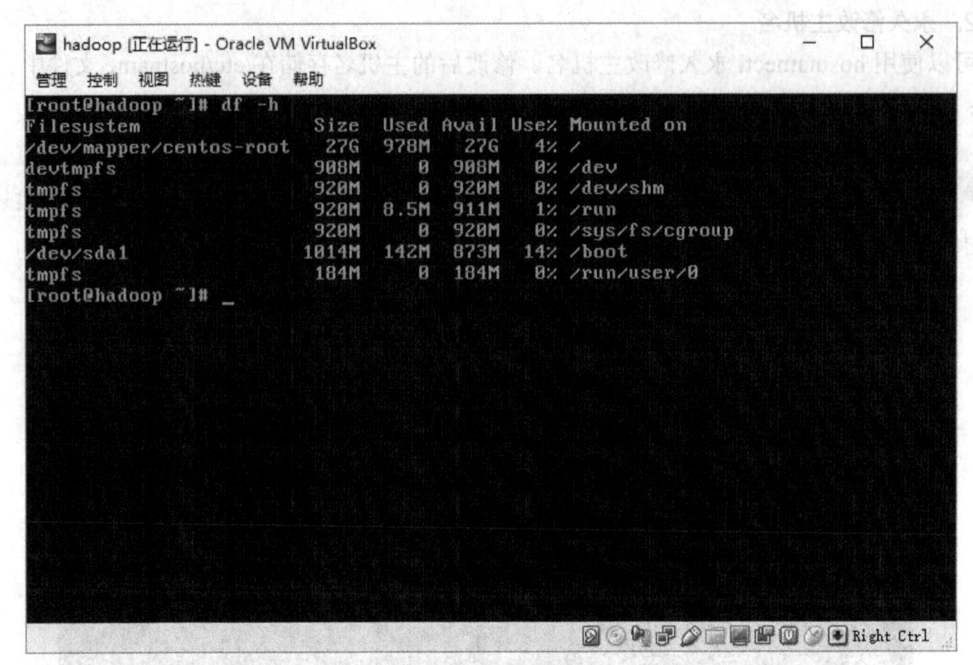

图 2-14 df 命令操作示例

2. 硬盘分区

使用 fdisk 命令可以对硬盘进行分区，命令如下：

```
# fdisk -l   %查看所有分区
```

fdisk 命令操作示例如图 2-15 所示。

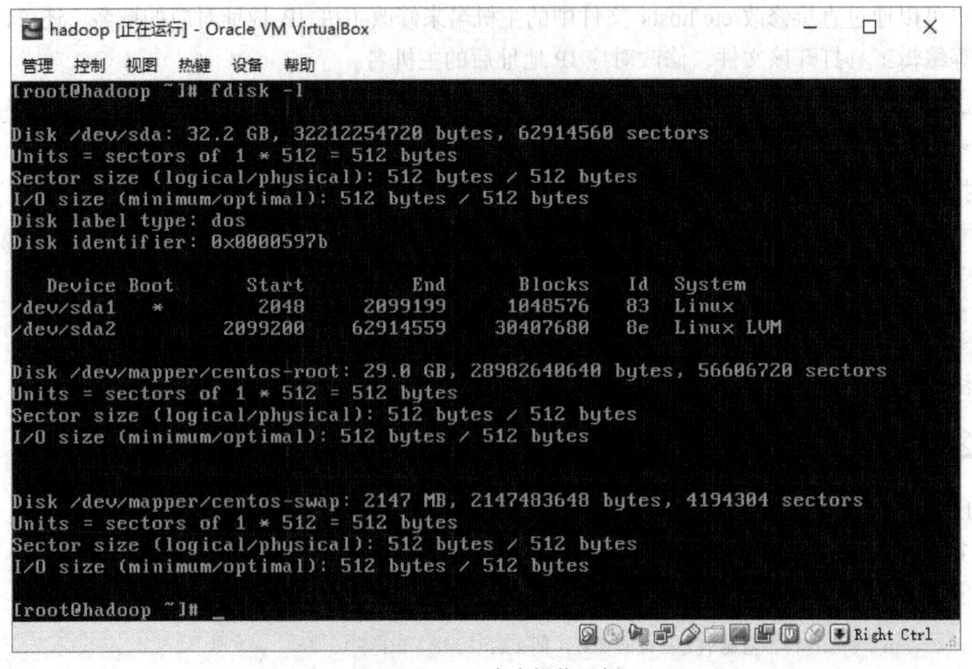

图 2-15 fdisk 命令操作示例

列出所有分区后，可以通过以下命令来实现对指定分区的操作：

```
# fdisk /dev/sda3                    %使用 fdisk 管理/dev/sda3
```

此时会进入 fdisk 的命令行界面，输入 m 后按 Enter 键可以看到帮助信息，使用 w 或者 q 命令可以保存或放弃 fdisk 的分区操作。分区操作完成后通常要使用 mkfs 命令进行文件系统的格式化（请读者自行查阅相关资料）。

3. 挂载硬盘

把/media/rhel-server-5.3-x86_64-dvd.iso 挂载到/mnt/vcdrom 的命令如下：

```
# mkdir /mnt/vcdrom
# mount -o loop -t iso9660 /media/rhel-server-5.3-x86_64-dvd.iso /mnt/vcdrom
```

注意：-o loop 允许用户以一个普通磁盘文件虚拟一个块设备，-t iso9660 指定挂载格式为 iso 镜像

挂载网络文件系统（Network File System，NFS）分区（需要配置好 NFS 服务器），命令如下：

```
# mount -t nfs 192.168.123.2:/tmp /mnt
```

注意：192.168.123.2 为 NFS 服务器地址

挂载文件分配表（File Allocation Table，FAT）32 分区（假设 U 盘已经插入，且通过 fdisk 命令得知 U 盘的标识为/dev/sdb1），命令如下：

```
# mkdir /mnt/usb
# mount -t vfat /dev/sdb1 /mnt/usb
```

2.2 网络配置

2.2.1 基本网络配置

1. 关闭防火墙

（1）查看防火墙

命令如下：

```
# firewall-cmd --list-all
```

（2）关闭防火墙/禁止防火墙开机自启动

命令如下：

```
# systemctl stop firewalld
# systemctl disable firewalld
```

2. 关闭 SELinux

安全增强式 Linux（Security-Enhanced Linux，SELinux）是美国国家安全局（National Security Agency，NSA）对强制访问控制的一种实现，是 Linux 历史上最杰出的安全子系统之一。SELinux 默认安装在 CentOS、Fedora 和 Red Hat Enterprise Linux 上。然而，SELinux 会阻碍 Hadoop 组件的安装与配置，因而需要掌握关闭 SELinux 的相关方法。

（1）查看状态

在命令行中输入 getenfore 就可以查看 SELinux 的状态。

```
# getenforce
Enforcing
```

（2）关闭 SELinux

可以临时或永久关闭 SELinux。使用 setenforce 0 命令可以临时关闭 SELinux；永久关闭 SELinux 可以通过编辑/etc/selinux/config 文件，将 SELINUX=enforcing 修改为 SELINUX=disabled 实现。修改后需要保存 config 文件并重启虚拟机才能生效。

```
# setenforce 0                    % 临时关闭
# vim /etc/selinux/config         % 打开 config 文件，参考上面的内容修改设置，永久关闭 SELinux
```

修改 config 文件关闭 SELinux 操作示例如图 2-16 所示。

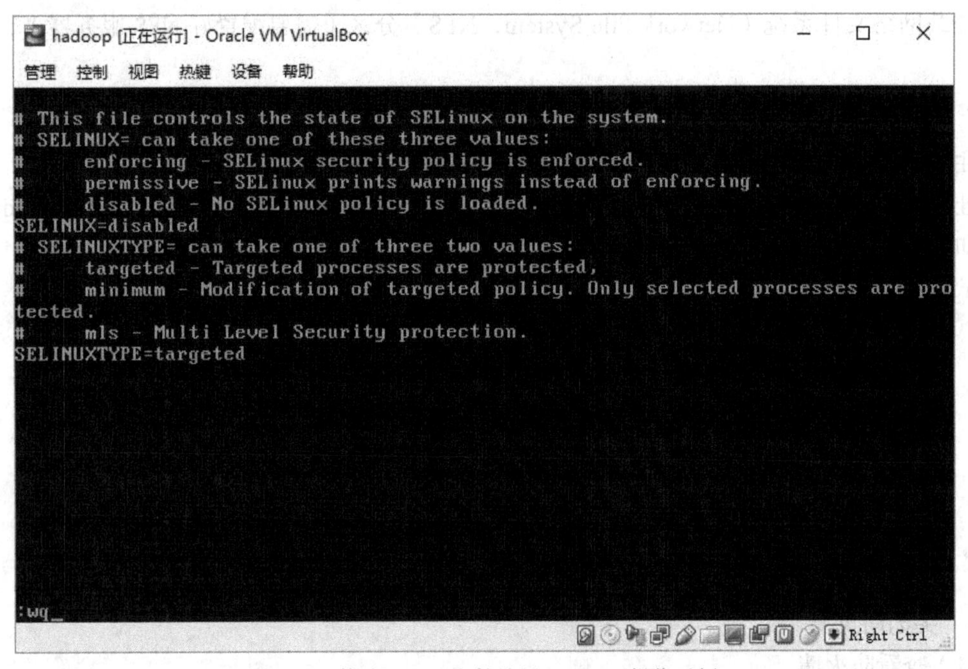

图 2-16　修改 config 文件关闭 SELinux 操作示例

可以看到，在/etc/selinux/config 文件中设置 SELINUX=disabled，并重启虚拟机后可以永久关闭 SELinux。需要说明的是，为了降低环境配置难度和学习门槛，本书中默认关闭防火墙和 SELinux。但这种做法存在严重的安全隐患，在生产环境中绝不允许这样做。

2.2.2　集群网络配置

1. VirtualBox 的桥接网卡模式

在桥接网卡模式下，虚拟机通过主机网卡架设一条网桥，直接连入主机所在网络。虚拟机能获取网络中的一个独立 IP 地址，其所处的网络环境和在网络中的物理机完全一样。由于虚拟机在真实网络中有独立 IP 地址，因此虚拟机与主机之间可以相互访问，虚拟机与虚拟机之间也可以相互访问。

对配置大数据学习环境而言，选择桥接网卡模式后，不同物理机上的虚拟机间可以组成一个 Linux 集群，不同用户可以一起搭建由多个虚拟机构建的 Linux 集群，而不需要在一台物理机上创建所有的虚拟机。这种模式的缺点是，当网络环境变化时，需要重新配置所有虚拟机的 IP 地址。

VirtualBox 默认是使用网络地址转换（Network Address Translation，NAT）模式的，如图 2-17 所示，也可以通过设置来修改为桥接网卡模式。

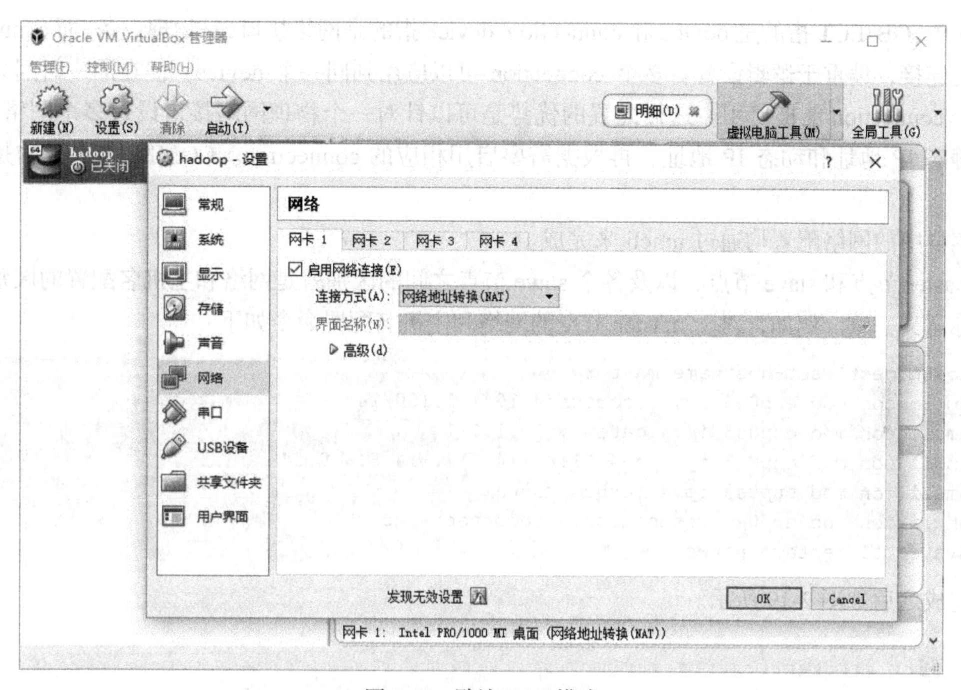

图 2-17　默认 NAT 模式

修改完成后如图 2-18 所示。

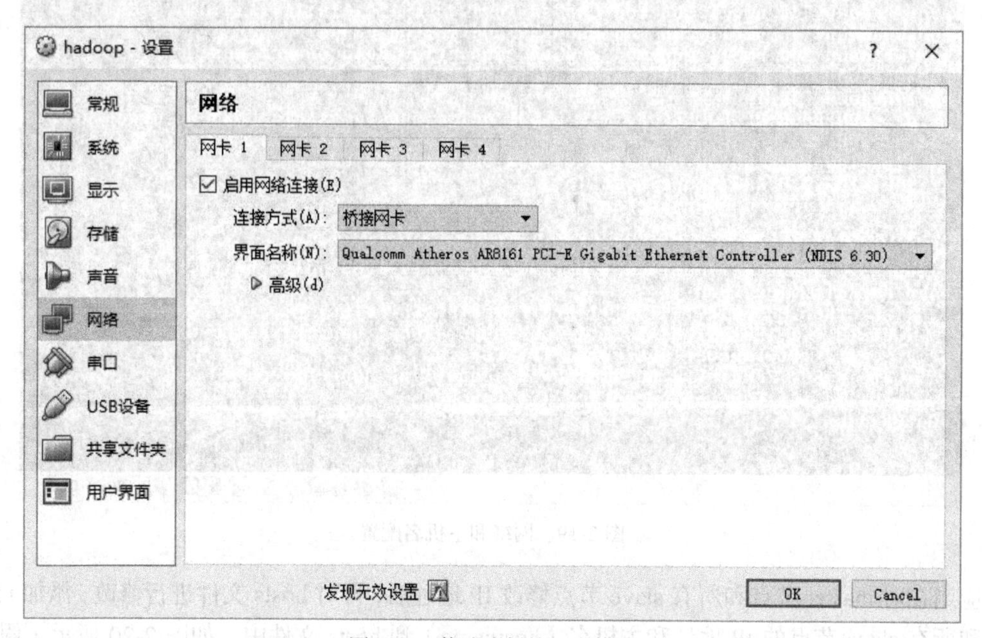

图 2-18　修改为桥接网卡模式

桥接的物理机网卡界面名称会根据计算机硬件的不同而有所区别。

2. 网络管理命令行工具

CentOS 中的 nmcli 网络管理命令行工具（Network Manager Command Tools），比传统网络管理命令 ifconfig 的功能要更加强大、复杂。其命令语法如下：

```
nmcli [OPTIONS] OBJECT { COMMAND | help }
```

其中，OBJECT 指的是 device 和 connection。device 指的是网络接口，是物理设备；而 connection 指的是连接，偏重于逻辑设置。多个 connection 可以应用到同一个 device，但在同一时间只有其中一个 connection 能被启用。这样配置的优势是可以针对一个物理网络接口设置多个网络连接，比如静态 IP 地址和动态 IP 地址，再根据需要启用相应的 connection。COMMAND 指的是具体命令。

集群中的网络配置均通过 nmcli 来完成。

master 节点和 slave 节点，以及各个 slave 节点之间的区别就是网络和主机名配置的区别，与其名称意义一致。例如，对于 master 节点的网络和主机名配置命令如下：

```
hostnamectl set-hostname master
nmcli con mod enp0s3 ipv4.addresses 10.1.1.150/24
nmcli con mod enp0s3 ipv4.gateway 10.1.1.1
nmcli con mod enp0s3 ipv4.dns "114.114.114.114 8.8.8.8"
nmcli con mod enp0s3 ipv4.method manual
nmcli con mod enp0s3 connection.autoconnect yes
systemctl restart network
```

完成之后如图 2-19 所示。

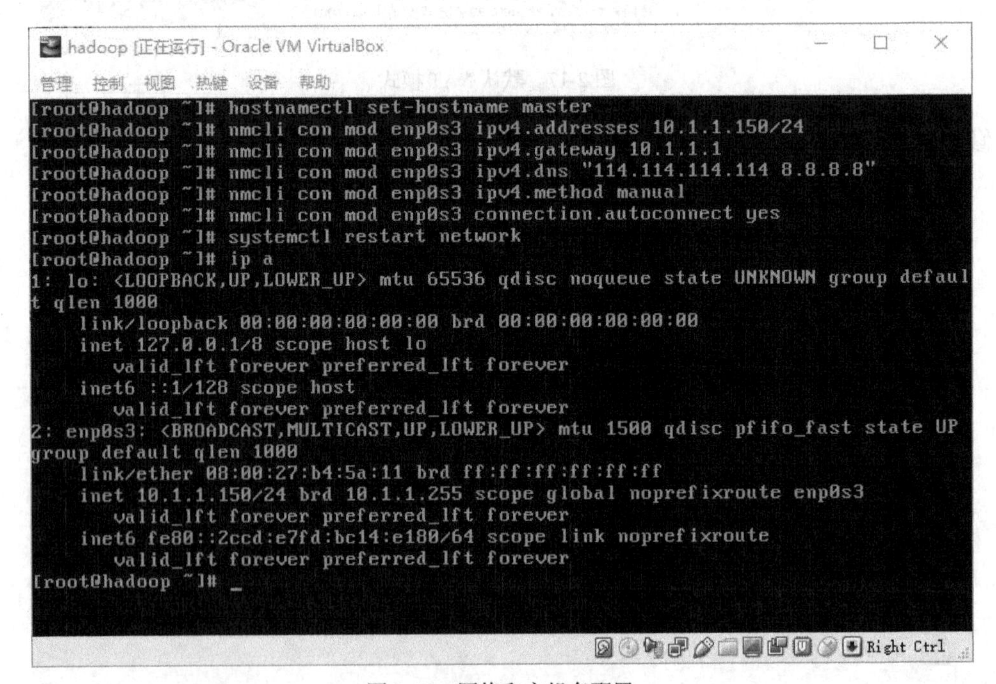

图 2-19 网络和主机名配置

应当在对 master 节点和所有 slave 节点修改 IP 地址后，再对 hosts 文件进行修改，添加 master 节点和所有 slave 节点的 IP 地址和主机名（hostname）到 hosts 文件中，如图 2-20 所示（假设有两个 slave 节点）。

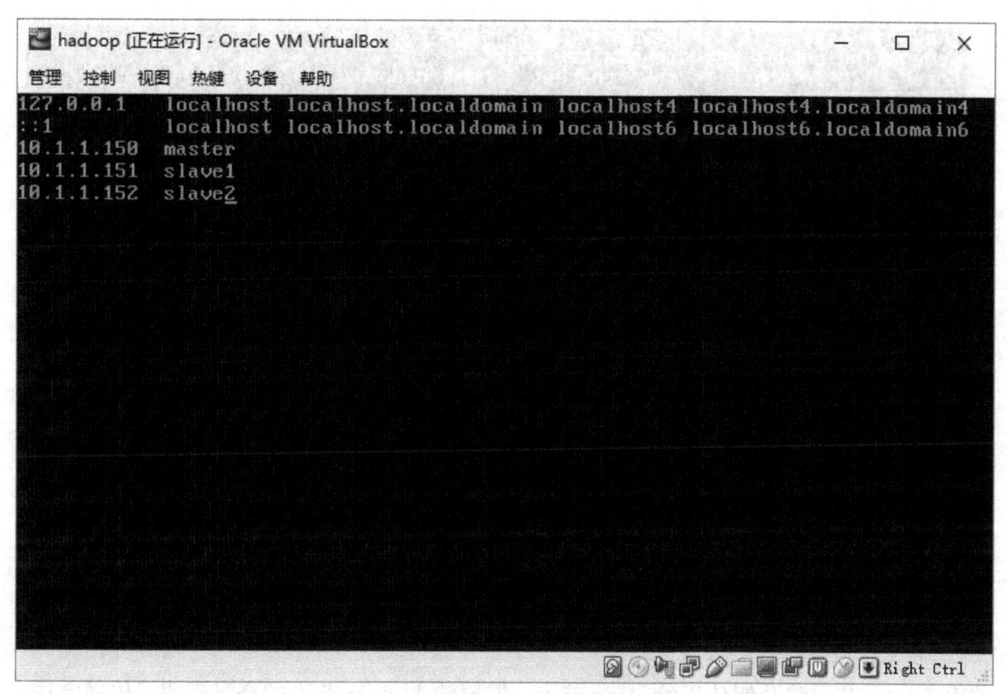

图 2-20　查看 hosts 文件

2.3　Linux 集群配置

Hadoop 的运行对 Linux 集群的环境要求较为复杂，以下是对环境的分步配置。若需要一个快速的配置方案，请看 2.4 节。该方案能够在机房环境中，通过虚拟机快照备份和相应脚本进行集群的快速配置。

以下步骤均在 1.4 节安装的虚拟机的基础上进行配置，master、slave1、slave2 节点只存在 IP 地址和主机名的不同。

2.3.1　SSH 免密码登录

安全外壳（Secure Shell，SSH）密钥是一个可靠、安全的 SSH 验证方式。SSH 密钥实际上是两个加密的安全密码，由公钥和私钥组成，可用于 SSH 服务器验证客户端。私钥由客户端保留，公钥被上传到希望能够使用 SSH 登录的远程服务器，并放置在用户账户目录下的.ssh/authorized_keys 文件中。

当客户端尝试使用 SSH 密钥进行身份验证时，服务器可以通过客户端提供的私钥验证，生成一个终端会话或执行请求的命令。

在 Linux 集群间配置免密码登录，是 Hadoop 集群运维的基础。以下操作在 master 节点进行，实现从 master 节点免密码登录 slave1、slave2 节点。生成 SSH 密钥的命令如下：

```
[hadoop@master~]$ ssh-keygen        % master 节点生成 SHH 密钥
```

生成 SSH 密钥如图 2-21 所示。

```
[hadoop@master ~]$ ssh-keygen
Generating public/private rsa key pair.
Enter file in which to save the key (/home/hadoop/.ssh/id_rsa):
/home/hadoop/.ssh/id_rsa already exists.
Overwrite (y/n)? y
Enter passphrase (empty for no passphrase):
Enter same passphrase again:
Your identification has been saved in /home/hadoop/.ssh/id_rsa.
Your public key has been saved in /home/hadoop/.ssh/id_rsa.pub.
The key fingerprint is:
85:48:01:05:42:bc:16:76:43:fc:c6:fb:b9:68:a1:7c hadoop@master
The key's randomart image is:
+--[ RSA 2048]----+
| ooo+++.         |
|  +.+. . .       |
| . + +. . .      |
|  o   +  .       |
| .     . .S      |
|        o        |
|     . . o .     |
|      o E.o      |
|       o. ..     |
+-----------------+
```

图 2-21　生成 SSH 密钥

将生成的公钥传送到 slave1 节点的命令如下：

```
[hadoop@master~]$ ssh-copy-id root@slave1          %将公钥传送到 slave1 节点
```

首次 master 节点将公钥传送给 slave 节点，需要输入 slave 节点的登录密码，传送完毕可实现免密码登录，如图 2-22 所示。

```
[hadoop@master ~]$ ssh-copy-id hadoop@slave1
The authenticity of host 'slave1 (192.168.2.57)' can't be established.
ECDSA key fingerprint is fc:85:de:4b:ea:f8:27:21:3b:fa:99:96:dc:92:69:a5.
Are you sure you want to continue connecting (yes/no)? yes
/bin/ssh-copy-id: INFO: attempting to log in with the new key(s), to filter out
any that are already installed
/bin/ssh-copy-id: INFO: 1 key(s) remain to be installed -- if you are prompted n
ow it is to install the new keys
hadoop@slave1's password:

Number of key(s) added: 1

Now try logging into the machine, with:   "ssh 'hadoop@slave1'"
and check to make sure that only the key(s) you wanted were added.

[hadoop@master ~]$ ssh-copy-id hadoop@slave2
The authenticity of host 'slave2 (192.168.2.58)' can't be established.
ECDSA key fingerprint is e3:ba:68:43:f8:87:ff:6e:03:f5:ed:47:ff:b3:ee:81.
Are you sure you want to continue connecting (yes/no)? yes
/bin/ssh-copy-id: INFO: attempting to log in with the new key(s), to filter out
any that are already installed
/bin/ssh-copy-id: INFO: 1 key(s) remain to be installed -- if you are prompted n
ow it is to install the new keys
hadoop@slave2's password:

Number of key(s) added: 1

Now try logging into the machine, with:   "ssh 'hadoop@slave2'"
and check to make sure that only the key(s) you wanted were added.
```

图 2-22　传送公钥

登录 slave1 节点的命令如下：

```
[hadoop@master~]$ ssh root@slave1        %测试免密码登录
```

成功登录 slave1 节点，如图 2-23 所示。

```
[hadoop@master ~]$ ssh hadoop@slave1
Last login: Wed Mar 28 10:02:23 2018
[hadoop@slave1 ~]$
```

图 2-23　免密码登录 slave1 节点

slave2 节点实现免密码登录的操作方式与上述一致。

2.3.2　Java 环境安装

因为 Hadoop 的环境需要依赖于 Java 开发工具包（Java Development Kit，JDK），所以要确保虚拟机中已经正确安装了 JDK，并配置了环境变量。本小节以 master 节点为例，查看 Java 版本的命令如下：

```
[root@master~]# java -version        %查看 Java 版本
```

执行结果如图 2-24 所示。

```
[root@master ~]# java -version
java version "1.8.0_161"
Java(TM) SE Runtime Environment (build 1.8.0_161-b12)
Java HotSpot(TM) 64-Bit Server VM (build 25.161-b12, mixed mode)
[root@master ~]#
```

图 2-24　查看 Java 版本

出现以上情况，表示 JDK 已经安装完毕。若没有安装，请先安装 JDK，并配置 Java 环境变量。以 master 节点为例，JDK 的安装步骤如下。

（1）查询系统自带的 JDK，命令如下：

```
[root@master~]# rpm -qa | grep java
```

执行结果如图 2-25 所示。

```
[root@slave1 ~]# rpm -qa | grep java
python-javapackages-3.4.1-11.el7.noarch
java-1.8.0-openjdk-1.8.0.121-0.b13.el7_3.x86_64
tzdata-java-2017b-1.el7.noarch
java-1.8.0-openjdk-headless-1.8.0.121-0.b13.el7_3.x86_64
javapackages-tools-3.4.1-11.el7.noarch
[root@slave1 ~]#
```

图 2-25　查询系统自带的 JDK

移除自带的 JDK，命令如下：

```
[root@master~]# yum remove java-1.*
```

（2）在 Oracle 官网下载 JDK。

选择需要的 JDK 版本下载，安装命令如下：

```
[root@master~]# mkdir /usr/java
[root@master~]# mv jdk-8u161-linux-x64.tar.gz /usr/java
[root@master~]# cd /usr/java
[root@master~]# tar -zxvf jdk-8u161-linux-x64.tar.gz       %解压缩
```

（3）配置 Java 环境变量到/etc/profile 文件中，命令如下：

```
[root@master~] #vim~/etc/profile
```

在文件尾部添加以下内容：

```
export JAVA_HOME=/usr/java/jdk1.8.0_161
export CLASSPATH=.:$JAVA_HOME/jre/lib/rt.jar:$JAVA_HOME/lib/dt.jar:$JAVA_HOME/lib/tools.jar
export PATH=$PATH:$JAVA_HOME/bin
```

（4）使环境变量生效，命令如下：

```
[root@master~]# source /etc/profile     %使环境变量生效
[root@master~]# java -version           %查看 Java 版本
```

2.3.3 MySQL 服务

1. 安装 MySQL

首先在 MySQL 官网中下载所需版本的 YUM 源，本次安装的版本为 MySQL 5.7.21 社区版，且选择安装在 master 节点上。

下载 MySQL 及安装命令如下：

```
[root@master~]# wget http://dev.mysql.com/get/mysql57-community-release-el7-8.noarch.rpm
[root@master~]# yum localinstall mysql57-community-release-el7-8.noarch.rpm   %安装 MySQL 源
```

命令执行完毕后，执行以下命令检查 YUM 源是否安装成功，执行结果如图 2-26 所示。

```
[root@master~]# yum repolist enabled | grep "mysql.*-community.*"
```

```
[root@master ~]# yum repolist enabled | grep "mysql.*-community.*"
mysql-connectors-community/x86_64        MySQL Connectors Community        45
mysql-tools-community/x86_64             MySQL Tools Community             59
mysql57-community/x86_64                 MySQL 5.7 Community Server       247
[root@master ~]#
```

图 2-26　MySQL 源安装成功

然后，安装、启动、设置 MySQL，命令如下：

```
[root@master~]# yum install mysql-community-server        %安装 MySQL
[root@master~]# systemctl start mysqld                    %启动 MySQL
[root@master~]# systemctl enable mysqld                   %设置开机启动
```

安装完成之后，MySQL 会默认在/var/log/mysqld.log 文件中给 root 用户生成一个临时密码。可以通过如下命令找到临时密码，然后登录 MySQL 进行修改，查看 MySQL 临时密码如图 2-27 所示。

```
[root@master~]# grep 'temporary password' /var/log/mysqld.log
```

```
[root@master ~]# grep 'temporary password' /var/log/mysqld.log
2018-03-28T01:45:05.314906Z 1 [Note] A temporary password is generated for root@
localhost: X_u40rX0bs7#
[root@master ~]#
```

图 2-27　查看 MySQL 临时密码

运行以下命令登录 MySQL 并设置密码：

```
[root@master~]# mysql -u root -p                         %登录 MySQL 数据库
[root@master~]# SET PASSWORD = PASSWORD('Hadoop@123');   %设置新密码
```

2. 创建 hadoop 用户并设置权限

在 MySQL 中创建 hadoop 用户并设置所有权限，具体步骤如下。

（1）创建用户的语法格式如下：

```
CREATE USER 'username'@'host' IDENTIFIED BY 'password';
```

其中的参数说明如下。

- username：用户名。
- host：指定在哪个主机上可以登录，本机可用 localhost，%通配所有远程主机。
- password：用户登录密码。

下面创建用户名并设置密码：

```
mysql> CREATE USER hadoop@'master' IDENTIFIED BY 'Hadoop@123';
```

（2）设置权限的语法格式如下：

```
GRANT ALL PRIVILEGES ON *.* TO 'username'@'%' IDENTIFIED BY 'password';
```

语法格式说明：GRANT 权限 ON 数据库名.表名 TO 用户名@用户 IP 地址或主机名 IDENTIFIED BY 用户密码。

其中，*.*代表所有数据库表的权限，@后面是访问 MySQL 的用户 IP 地址（或是主机名）；%匹配任意的用户，如果将%修改为 localhost 则表示本地访问，那么此用户就没有远程访问该 MySQL 数据库的权限。

```
mysql>grant all on *.* to hadoop@'%' identified by 'Hadoop@123';
mysql>grant all on *.* to hadoop@'localhost' identified by 'Hadoop@123';
mysql>grant all on *.* to hadoop@'master' identified by 'Hadoop@123';
```

（3）刷新权限的语法格式如下：

```
FLUSH PRIVILEGES;
```

例如：

```
mysql>flush privileges;
```

完成如上 MySQL 的安装与配置后，用户就可以对 2.3 节中 SQL 的内容进行上机操作。

2.3.4 配置时钟同步

Linux 集群中节点的时钟同步对分布式组间协同工作来说意义重大。例如，在 HBase 分布式部署中，一定要求节点的时钟同步。后文介绍时钟同步工具 NTP 的安装与配置。

NTP 是网络时间协议（Network Time Protocol）的缩写，目的是保证 master 端（服务器）与 slave 端（客户）的时钟同步。

1. master 端配置

命令如下：

```
[hadoop@master~]$ yum install ntp              %安装 NTP 软件
[hadoop@master~]$ chkconfigntpd on             %启动 NTP
[hadoop@master~]$ vim /etc/ntp.conf            %配置服务器
```

由于 NTP 服务的设置需要上游服务器的支持，所以注释掉原先的服务器设置，添加 master 端配置，如图 2-28 所示。

```
[hadoop@master ~]$ sudo vi /etc/ntp.conf
[sudo] password for hadoop:
# For more information about this file, see the man pages
# ntp.conf(5), ntp_acc(5), ntp_auth(5), ntp_clock(5), ntp_misc(5), ntp_mon(5).

driftfile /var/lib/ntp/drift

# Permit time synchronization with our time source, but do not
# permit the source to query or modify the service on this system.
restrict default nomodify notrap nopeer noquery

# Permit all access over the loopback interface.  This could
# be tightened as well, but to do so would effect some of
# the administrative functions.
restrict 127.0.0.1
restrict ::1

# Hosts on local network are less restricted.
#restrict 192.168.1.0 mask 255.255.255.0 nomodify notrap

# Use public servers from the pool.ntp.org project.
# Please consider joining the pool (http://www.pool.ntp.org/join.html).
server cn.pool.ntp.org iburst
#server 1.centos.pool.ntp.org iburst
#server 2.centos.pool.ntp.org iburst
#server 3.centos.pool.ntp.org iburst

#broadcast 192.168.1.255 autokey        # broadcast server
#broadcastclient                        # broadcast client
#broadcast 224.0.1.1 autokey            # multicast server
#multicastclient 224.0.1.1              # multicast client
#manycastserver 239.255.254.254         # manycast server
#manycastclient 239.255.254.254 autokey # manycast client

# Enable public key cryptography.
```

图 2-28　master 端配置

重启 NTP 服务，并查看其是否启动，如图 2-29 所示。

```
[hadoop@master~]$ service ntpd restart                %重启 NTP 服务
[hadoop@master~]$ systemctl status ntpd.service       %查看 NTP 服务
```

```
[hadoop@master ~]$ sudo service ntpd restart
Redirecting to /bin/systemctl restart  ntpd.service
[hadoop@master ~]$ sudo systemctl status ntpd.service
● ntpd.service - Network Time Service
   Loaded: loaded (/usr/lib/systemd/system/ntpd.service; enabled; vendor preset: disabled)
   Active: active (running) since 三 2018-03-28 11:41:42 CST; 45s ago
  Process: 20749 ExecStart=/usr/sbin/ntpd -u ntp:ntp $OPTIONS (code=exited, status=0/SUCCESS)
 Main PID: 20750 (ntpd)
   CGroup: /system.slice/ntpd.service
           └─20750 /usr/sbin/ntpd -u ntp:ntp -g

3月 28 11:41:43 master ntpd[20750]: Listen normally on 2 lo 127.0.0.1 UDP 123
3月 28 11:41:43 master ntpd[20750]: Listen normally on 3 eth0 10.1.250.22 UDP 123
3月 28 11:41:43 master ntpd[20750]: Listen normally on 4 virbr0 192.168.122.1 UDP 123
3月 28 11:41:43 master ntpd[20750]: Listen normally on 5 lo ::1 UDP 123
3月 28 11:41:43 master ntpd[20750]: Listen normally on 6 eth0 fe80::f816:3eff:fe90:d609 UDP 123
3月 28 11:41:43 master ntpd[20750]: Listening on routing socket on fd #23 for interface updates
3月 28 11:41:43 master ntpd[20750]: 0.0.0.0 c016 06 restart
3月 28 11:41:43 master ntpd[20750]: 0.0.0.0 c012 02 freq_set kernel 0.000 PPM
3月 28 11:41:43 master ntpd[20750]: 0.0.0.0 c011 01 freq_not_set
3月 28 11:41:53 master ntpd[20750]: 0.0.0.0 c614 04 freq_mode
```

图 2-29　重启并查看 NTP 服务

2. slave 端配置

分别修改各个主机的/etc/chrony.conf 文件，设置其时钟同步服务器为 master 端，命令如下：

```
[root@slave1]# vim /etc/chrony.conf
```

编辑后保存结果如下：

```
# Use public servers from the pool.ntp.org project.
# Please consider joining the pool (http://www.pool.ntp.org/join.html).
server master iburst
```

3. 重启并验证

命令如下：

```
[root@slave1]# chronyc sources
```

返回结果如下：

```
210 Number of sources = 1
MS Name/IP address         Stratum Poll Reach LastRx Last sample
===============================================================================
^* master                     10   6    17    42  -1741ns[  -72us] +/-  332us
```

2.4 快速配置 Linux 集群

在机房等环境中，一步一步进行环境配置显然是不符合时间和学习效率的要求的，对此，本书将提供 Linux 集群的快速配置方案。

若虚拟机中创建了 hadoop 用户，在使用 root 用户首次运行配置脚本后，即可切换为 hadoop 用户进行实践。

2.4.1 导入虚拟机

下载本书所提供的虚拟机备份文件 master.ova 和 slave.ova，它们分别对应集群的主节点（master 节点）和从节点（slave 节点）（下载方法：登录 www.ryjiaoyu.com，然后搜索本书即可看到相关资源）。master 节点和 slave 节点的虚拟机备份文件是两个独立的、以.ova 为扩展名的镜像文件，如图 2-30 所示。

名称	修改日期	类型	大小
master.ova	2018/9/3 20:38	Open Virtualizati...	1,788,123
slave.ova	2018/9/3 20:37	Open Virtualizati...	1,550,229

图 2-30 镜像文件

如图 2-31 所示，打开 VirtualBox，单击“管理”菜单下的“导入虚拟电脑”选项。

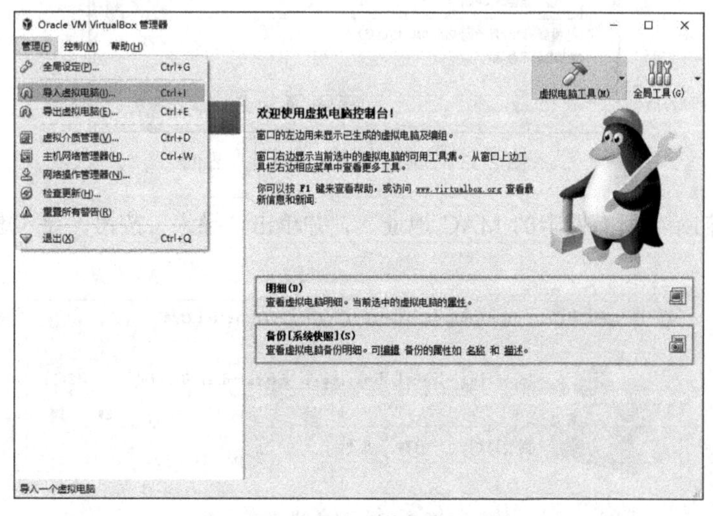

图 2-31 “导入虚拟电脑”选项

以导入 master 节点为例，选择要导入的镜像文件，如图 2-32 所示。

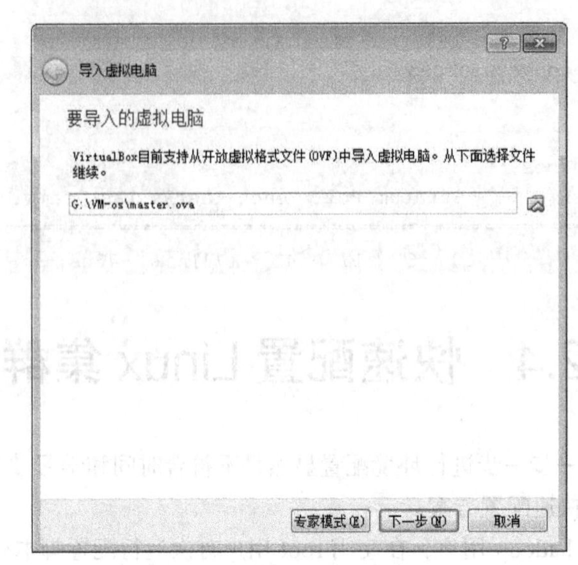

图 2-32　选择要导入的镜像文件

单击“下一步”按钮，进入“虚拟电脑导入设置”界面，如图 2-33 所示。

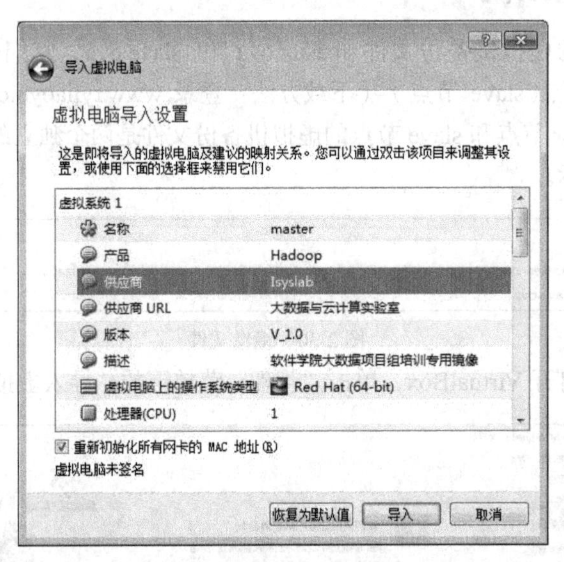

图 2-33　“虚拟电脑导入设置”界面

勾选“重新初始化所有网卡的 MAC 地址”，并单击“导入”按钮，导入虚拟机，导入进度如图 2-34 所示。

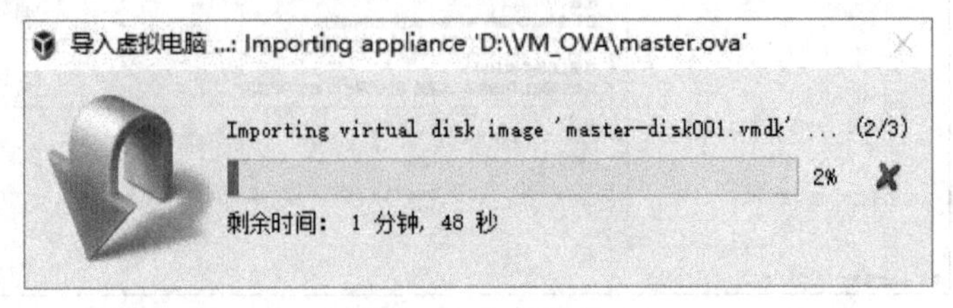

图 2-34　导入进度

导入完成后的虚拟机显示在 VirtualBox 主界面中，如图 2-35 所示。

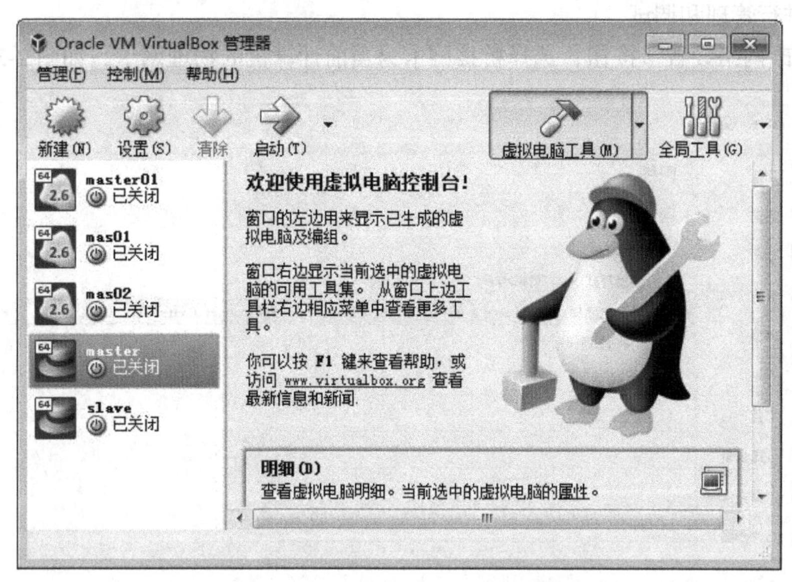

图 2-35　导入成功的 master 节点虚拟机

slave 节点虚拟机的导入和 master 节点虚拟机一致，不赘述。

2.4.2　快速配置

1. master 节点

第一次启动虚拟机会提示更改物理网卡，如图 2-36 所示。

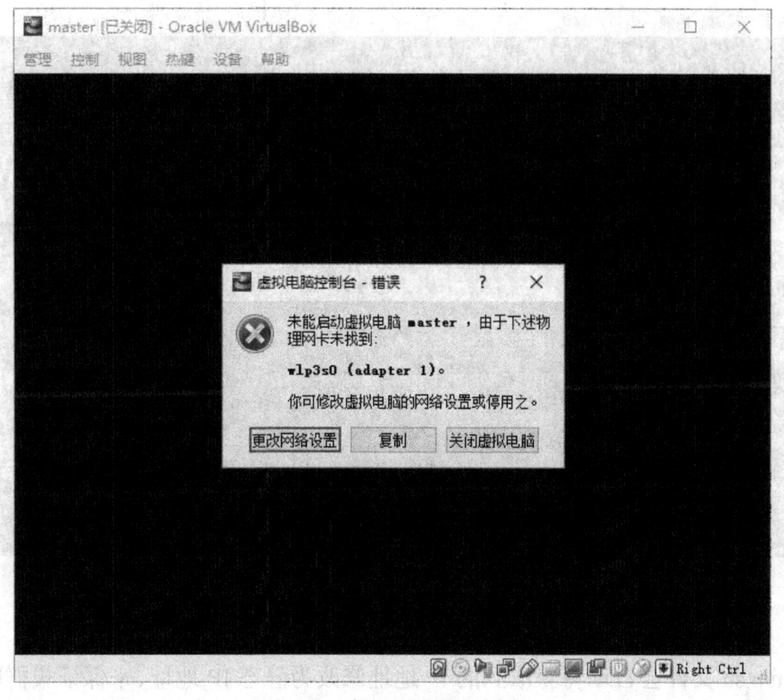

图 2-36　提示更改物理网卡

本书使用桥接网卡模式来配置集群的网络，将使虚拟机和物理机处于同一局域网，方便对Hadoop集群进行管理和调试。

单击“更改网络设置”按钮，选择桥接网卡为当前计算机的物理网卡，如图2-37所示。

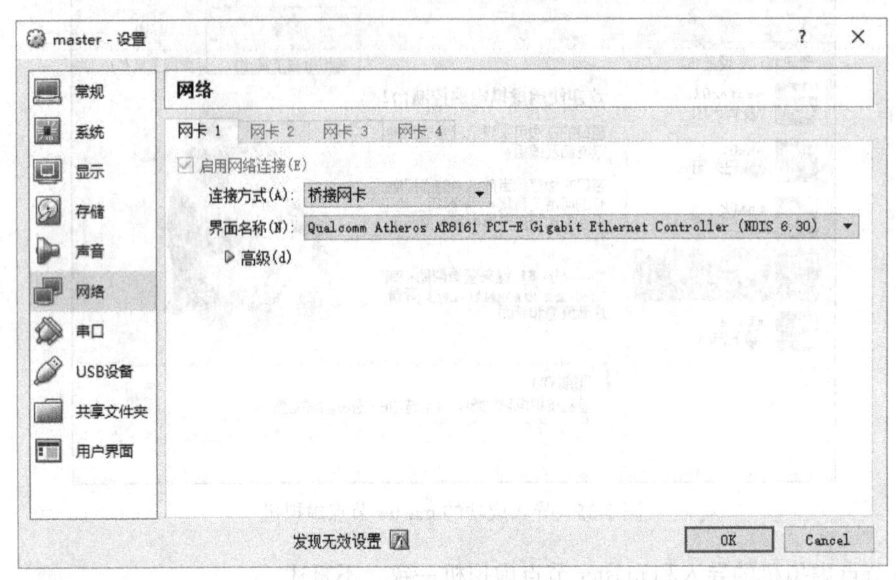

图2-37 选择桥接网卡

更改完成后单击“OK”按钮，再次启动虚拟机。

使用root用户登录master节点后，执行master_setup.sh文件，如图2-38所示。

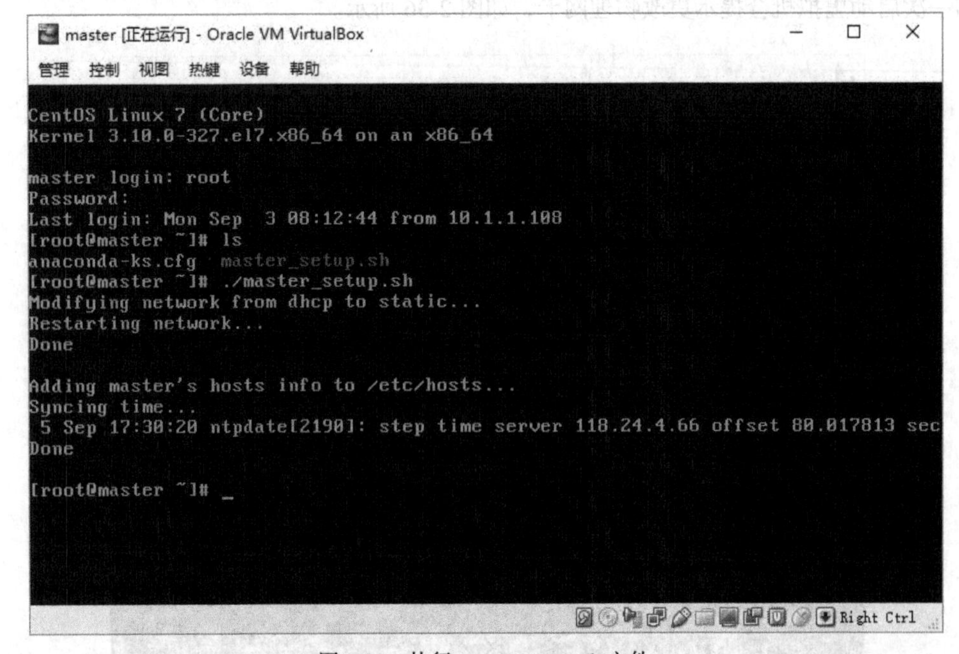

图2-38 执行master_setup.sh文件

master_setup.sh文件会自动将获取到的IP地址修改为静态IP地址，来保证集群的稳定性和持久化。

文件执行完成后查看 master 节点的 IP 地址为 10.1.1.209，如图 2-39 所示。

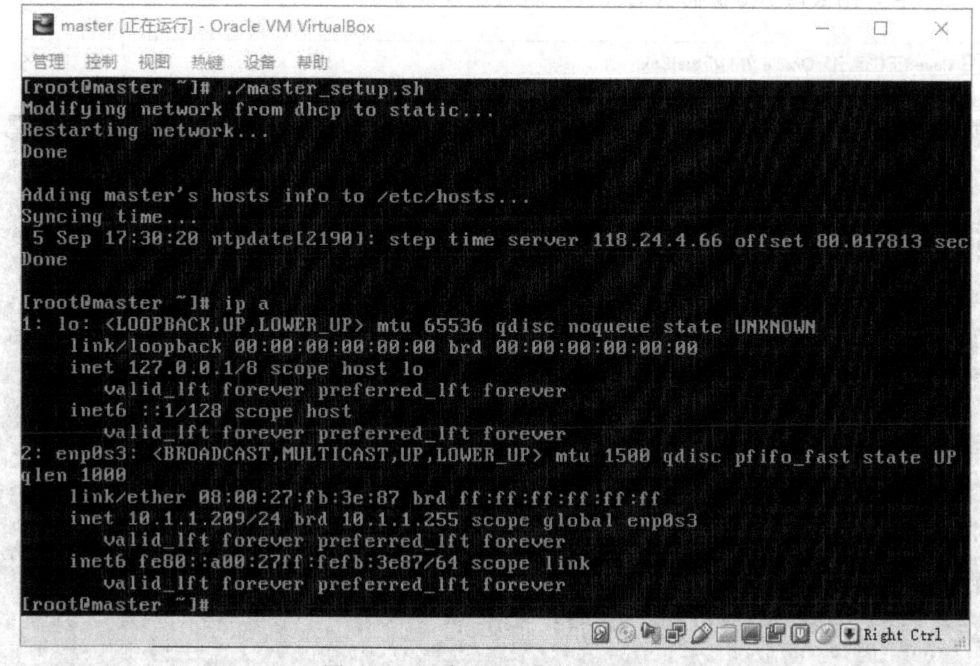

图 2-39　查看 master 节点的 IP 地址

2. slave 节点

同 master 节点一样，启动虚拟机后，以 root 用户登录 slave 节点，执行 slave_setup.sh 文件，并输入 master 节点的 IP 地址，如图 2-40 所示。

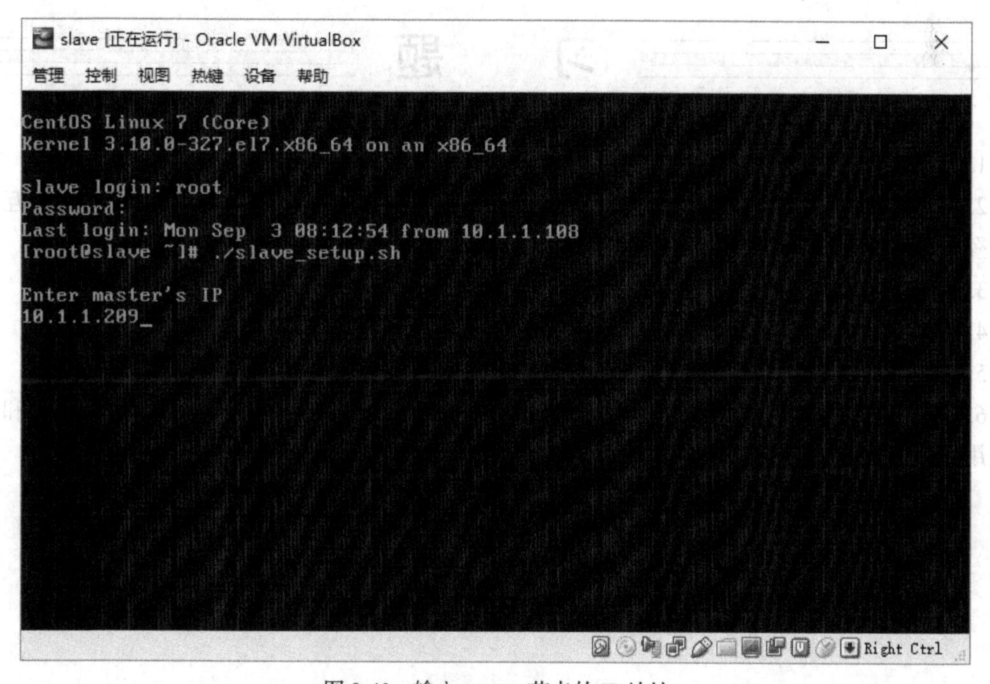

图 2-40　输入 master 节点的 IP 地址

按 Enter 键后，会提示输入 slave 节点的主机名，并且能够自动检测是否已经有重名的 slave 节点。如果有，则会提示重新输入主机名，如图 2-41 所示。

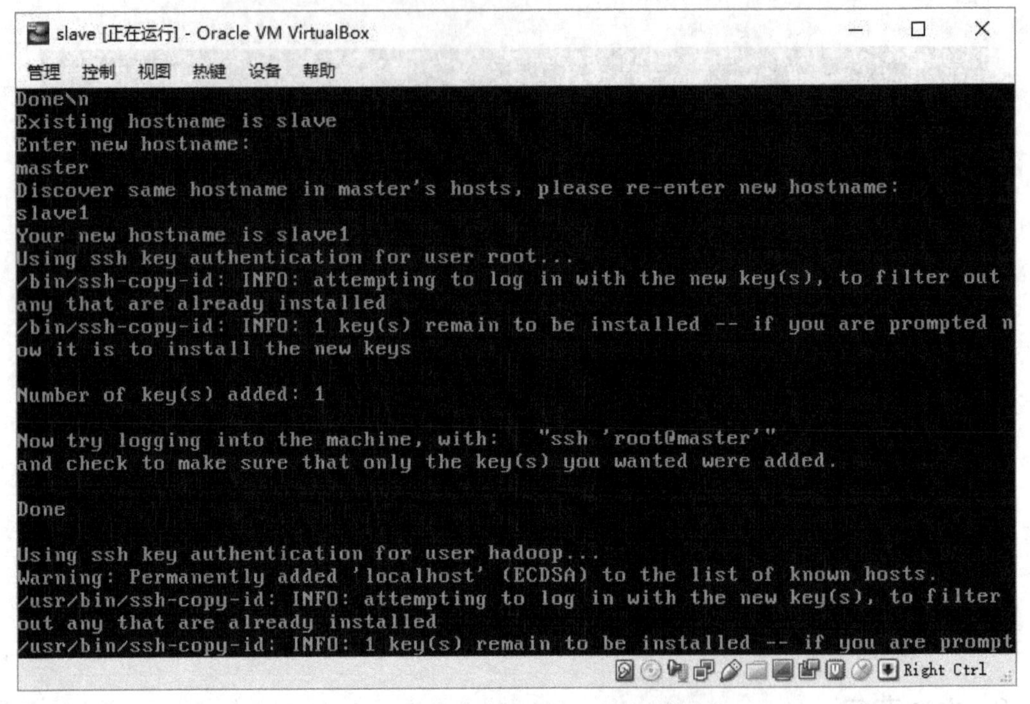

图 2-41 执行 slave_setup.sh 文件检测主机名

至此，整个环境配置完成，对于一个 master 节点，可以配置多个 slave 节点。

习 题

1. 在 CentOS 7 上熟悉 Linux 基本命令。
2. 在 VirtualBox 上安装 Linux 集群，配置一个 master 节点和一个 slave 节点，并实现互相免密码登录。
3. 在 master 节点上安装 MySQL，并练习 2.3 节的操作。
4. 尝试 3 个人一组，在 3 台以上的物理机上采用 VirtualBox 搭建 Linux 虚拟机集群。
5. 分析在搭建集群时配置时钟同步的作用。
6. 分析在配置网络时，桥接网卡模式和 NAT 模式、内部网络模式和主机模式的区别和各自的使用场景。

第 3 章 Hadoop 集群配置

Hadoop 是 Apache 软件基金会旗下的一个开源分布式计算平台，以 HDFS 分布式文件系统和 MapReduce 分布式计算引擎为核心，为用户提供了系统底层细节透明的分布式基础架构。HDFS 的高容错性、高伸缩性等优点允许用户将 Hadoop 分布式系统部署在低廉的硬件上。第 1 章介绍了 Hadoop 发展历史和版本信息，本章将介绍 Hadoop 集群的安装和配置。

3.1 Hadoop 集群安装

为了顺利安装 Hadoop 集群，需要做好充分的软件、硬件准备。本书实践环境采用的是 VirtualBox 虚拟机，Linux 操作系统是 CentOS 7 的 64 位版本。系统准备好之后，需要修改主机名、修改/etc/hosts 文件、在主机之间配置互信、配置 Java 环境等。为方便操作，本实践环境的 slave 节点只有一个。主机名、IP 地址、JDK 和 Hadoop 安装路径等信息如表 3-1 所示。

表 3-1 master 节点、slave 节点信息

主机名	IP 地址	JDK 安装路径	Hadoop 安装路径
master	192.168.2.176	/home/hadoop/java/	/home/hadoop/software/
slave	192.168.2.177		

3.1.1 基础环境准备

操作下面的步骤之前，请确保在自己的 VirtualBox 上安装了两个 CentOS 7 操作系统虚拟机，分别作为 master 节点和 slave 节点。修改主机名的步骤如下。

（1）查看主机的网卡名，并记住第一块网卡，命令如下：

```
[root@localhost ~]#cd /etc/sysconfig/network-scripts/
[root@controller network-scripts]# ls
```

在/etc/sysconfig/network-scripts/目录中可以查看当前环境的所有网卡，其中第一块网卡为 ifcfg-eno16777984，如图 3-1 所示。

```
[root@localhost ~]# cd /etc/sysconfig/network-scripts/
[root@localhost network-scripts]# ls
ifcfg-eno16777984  ifdown-ipv6      ifup-aliases  ifup-ppp
ifcfg-eno33557248  ifdown-isdn      ifup-bnep     ifup-routes
ifcfg-eno50336512  ifdown-post      ifup-eth      ifup-sit
ifcfg-eno67115776  ifdown-ppp       ifup-ib       ifup-Team
ifcfg-lo           ifdown-routes    ifup-ippp     ifup-TeamPort
ifdown             ifdown-sit       ifup-ipv6     ifup-tunnel
ifdown-bnep        ifdown-Team      ifup-isdn     ifup-wireless
ifdown-eth         ifdown-TeamPort  ifup-plip     init.ipv6-global
ifdown-ib          ifdown-tunnel    ifup-plusb    network-functions
ifdown-ippp        ifup             ifup-post     network-functions-ipv6
```

图 3-1　查看网卡

（2）配置第一块网卡的 IP 地址、子网掩码、网关、DNS 等信息，命令如下：

```
[root@localhost network-scripts]# vi ifcfg-eno16777984
```

使用 Vi 编辑器配置第一块网卡，需要把 BOOTPROTO 设置为 static，把 ONBOOT 设置为 yes，另外需要添加 IPADDR、NETMASK、GATEWAY、DNS 等内容，本实验环境网关地址为 192.168.2.1。master 节点的网卡信息如图 3-2 所示，slave 节点的网卡信息与之相似，只需更换 IP 地址即可。

```
TYPE=Ethernet
BOOTPROTO=static
DEFROUTE=yes
PEERDNS=yes
PEERROUTES=yes
IPV4_FAILURE_FATAL=no
IPV6INIT=no
NAME=eno16777984
UUID=60e31674-873d-4d0f-bcc3-c77d0e36aa18
DEVICE=eno16777984
ONBOOT=yes
IPADDR=192.168.2.176
NETMASK=255.255.255.0
GATEWAY=192.168.2.1
DNS=114.114.114.114
```

图 3-2　master 节点的网卡信息

注意：Vi 是一款编辑器，输入 vi ifcfg-eno16777984 命令后按 i 键可以进入插入模式。对文本进行编辑之后，按 Esc 键可以退出插入模式，输入:wq 便可以保存文本并退出。若不想保存，可以输入:q!强制退出。

（3）重启网络，能 ping 通外部网络说明配置有效，命令如下：

```
[root@localhost network-scripts]# service network restart
[root@localhost network-scripts]# ping www.baidu.com
```

两个节点重启网络之后，能相互 ping 通，并能 ping 通外部网络。如果不能 ping 通外部网络，可能是因为 IP 地址设置有问题。本环境能够正常 ping 通外部网络，如图 3-3 所示。

```
[root@localhost network-scripts]# service network restart
Restarting network (via systemctl):                        [  OK  ]
[root@localhost network-scripts]# ping www.baidu.com
PING www.a.shifen.com (180.97.33.108) 56(84) bytes of data.
64 bytes from 180.97.33.108: icmp_seq=1 ttl=55 time=18.4 ms
64 bytes from 180.97.33.108: icmp_seq=2 ttl=55 time=18.2 ms
64 bytes from 180.97.33.108: icmp_seq=3 ttl=55 time=18.1 ms
```

图 3-3　重启网络并 ping 通外部网络

（4）修改主机名。在 master 节点中将/etc/hostname 文件内容改为 master，在 slave 节点中则修改为 slave。修改完成，重启虚拟机后主机才会显示修改后的主机名，命令如下：

```
[root@localhost ~]# vi /etc/hostname
```

图 3-4 显示了在 master 节点中修改的内容，修改之后并不会立即生效，需要重启 CentOS。重启命令为 reboot。

```
master
~
~
~
~
~
~
~
~
:wq
```

图 3-4　修改主机名

（5）分别在两个节点上修改/etc/hosts 文件，修改成功后，可以通过主机名 ping 通对方主机。

```
[root@master ~]# vi /etc/hosts
[root@master ~]# ping slave
```

master 节点和 slave 节点的/etc/hosts 文件均修改为图 3-5 所示的内容。

```
127.0.0.1   localhost localhost.localdomain localhost4 localhost4.localdomain4
::1         localhost localhost.localdomain localhost6 localhost6.localdomain6
192.168.2.176 master
192.168.2.177 slave
~
~
```

图 3-5　修改/etc/hosts 文件

两个节点修改完毕之后，可以互相 ping 通。图 3-6 所示为 master 节点通过主机名 ping 通 slave 节点。

```
[root@master ~]# ping slave
PING slave (192.168.2.177) 56(84) bytes of data.
64 bytes from slave (192.168.2.177): icmp_seq=1 ttl=64 time=0.342 ms
64 bytes from slave (192.168.2.177): icmp_seq=2 ttl=64 time=0.350 ms
64 bytes from slave (192.168.2.177): icmp_seq=3 ttl=64 time=0.315 ms
64 bytes from slave (192.168.2.177): icmp_seq=4 ttl=64 time=0.368 ms
64 bytes from slave (192.168.2.177): icmp_seq=5 ttl=64 time=0.356 ms
^C
--- slave ping statistics ---
5 packets transmitted, 5 received, 0% packet loss, time 4000ms
rtt min/avg/max/mdev = 0.315/0.346/0.368/0.021 ms
```

图 3-6　master 节点通过主机名 ping 通 slave 节点

（6）关闭防火墙，两个节点的防火墙都要关闭，命令如下：

```
[root@master ~]# systemctl disable firewalld
[root@master ~]# systemctl stop firewalld
[root@master ~]# systemctl status firewalld
```

第 1 条命令表示禁止启动防火墙，第 2 条命令表示关闭防火墙，第 3 条命令表示显示当前防火墙状态。从图 3-7 中可以看出，当前环境中防火墙状态是“inactive”，即已经关闭了防火墙。

```
[root@master ~]# systemctl disable firewalld
[root@master ~]# systemctl stop firewalld
[root@master ~]# systemctl status firewalld
● firewalld.service - firewalld - dynamic firewall daemon
   Loaded: loaded (/usr/lib/systemd/system/firewalld.service; disabled; vendor preset: enabled)
   Active: inactive (dead)

Jun 06 14:55:03 master systemd[1]: Stopped firewalld - dynamic firewall daemon.
```

图 3-7　关闭防火墙

（7）关闭 SELinux。编辑/etc/sysconfig/selinux 文件，将其中的 SELINUX 选项内容改为“disabled”，两个节点都要关闭 SELinux。修改完后需要重启虚拟机才能生效。命令如下：

```
[root@master ~]# vi /etc/sysconfig/selinux
```

防火墙和 SELinux 均是系统安全的一部分。在实践环境中，若不关闭这两项，则会影响后续安装 Hadoop。/etc/sysconfig/selinux 修改后的内容如图 3-8 所示。

```
# This file controls the state of SELinux on the system.
# SELINUX= can take one of these three values:
#     enforcing - SELinux security policy is enforced.
#     permissive - SELinux prints warnings instead of enforcing.
#     disabled - No SELinux policy is loaded.
SELINUX=disabled
# SELINUXTYPE= can take one of three two values:
#     targeted - Targeted processes are protected,
#     minimum - Modification of targeted policy. Only selected processes are protected.
#     mls - Multi Level Security protection.
SELINUXTYPE=targeted
```

图 3-8　关闭 SELinux

（8）新建 hadoop 用户并设置密码，两个节点均需要进行此操作，命令如下：

```
[root@master ~]# useradd hadoop
[root@master ~]# passwd hadoop
[root@master ~]# su - hadoop
```

第 1 条命令表示新建 hadoop 用户，第 2 条命令表示为 hadoop 用户设置密码，第 3 条命令表示切换到 hadoop 用户，具体操作如图 3-9 所示。

```
[root@master ~]# useradd hadoop
[root@master ~]# passwd hadoop
Changing password for user hadoop.
New password:
BAD PASSWORD: The password is shorter than 8 characters
Retype new password:
passwd: all authentication tokens updated successfully.
[root@master ~]# su - hadoop
[hadoop@master ~]$
```

图 3-9　新建 hadoop 用户并设置密码

（9）在 master 节点和 slave 节点之间配置 SSH 互信，两个节点均需要操作，命令如下：

```
[hadoop@master ~]$ ssh-keygen -t rsa
```

切换到 hadoop 用户之后，使用命令生成公钥和私钥，如图 3-10 所示。

```
[hadoop@master ~]$ ssh-keygen -t rsa
Generating public/private rsa key pair.
Enter file in which to save the key (/home/hadoop/.ssh/id_rsa):
Created directory '/home/hadoop/.ssh'.              提示：此处直接按 Enter 键
Enter passphrase (empty for no passphrase):
Enter same passphrase again:
Your identification has been saved in /home/hadoop/.ssh/id_rsa.
Your public key has been saved in /home/hadoop/.ssh/id_rsa.pub.
The key fingerprint is:
32:41:4f:6b:3a:f1:28:f1:84:0f:94:7d:dc:69:c4:14 hadoop@master
The key's randomart image is:
+--[ RSA 2048]----+
|     .o...=Eo    |
|    ..o.oo.=     |
|     + +.+.      |
|      * B        |
|     . o S       |
|      . +        |
|                 |
|                 |
|                 |
+-----------------+
```

图 3-10　生成公钥和私钥

将生成的公钥分发给 slave 节点和 master 节点。在分发的过程中需要根据提示信息输入信任主机的密码，实现 master 节点到 master 节点、master 节点到 slave 节点之间的互信，命令如下：

```
[hadoop@master ~]$ ssh-copy-id slave
[hadoop@master ~]$ ssh-copy-id master
```

在 master 节点中将其生成的公钥分发给 master 节点和 slave 节点，在 slave 节点中将其生成的公钥分发给 master 节点，做到 master 节点到 master 节点、master 节点到 slave 节点、slave 节点到 master 节点之间的互信。从 master 节点将公钥分发给 slave 节点的操作如图 3-11 所示。

```
[hadoop@master ~]$ ssh-copy-id slave
The authenticity of host 'slave (192.168.2.177)' can't be established.
ECDSA key fingerprint is 28:0a:0d:77:9d:a3:a6:06:44:c0:7b:c3:74:f3:0a:c3.
Are you sure you want to continue connecting (yes/no)? yes  输入yes
/bin/ssh-copy-id: INFO: attempting to log in with the new key(s), to filte
r out any that are already installed
/bin/ssh-copy-id: INFO: 1 key(s) remain to be installed -- if you are prom
pted now it is to install the new keys
hadoop@slave's password:        输入slave节点的密码

Number of key(s) added: 1

Now try logging into the machine, with:   "ssh 'slave'"
and check to make sure that only the key(s) you wanted were added.
```

图 3-11　从 master 节点将公钥分发给 slave 节点的操作

验证是否能够免密码登录slave节点，命令如下：

```
[hadoop@master ~]$ ssh slave
```

通过ssh可以实现从master节点免密码登录到slave节点，从图3-12可以看出当前使用的主机为slave节点。

```
[hadoop@master root]$ ssh slave
Last login: Fri May 18 09:43:52 2018 from master
[hadoop@slave ~]$
```

图3-12　验证免密码登录

在使用ssh切换到slave节点的过程中，如果不需要输入密码就能切换，则表明可以从master节点到slave节点进行免密码登录。从slave节点免密码登录master节点的方法与上述相同。

（10）配置时钟同步，保证master节点和slave节点的时间保持一致，命令如下：

```
[root@master ~]# yum install chrony ntp -y
[root@master ~]# systemctl start chrony
[root@master ~]# vi /etc/chrony.conf
```

第1条命令使用yum安装chrony和NTP软件。第2条命令启动chrony。第3条命令编辑/etc/chrony.conf文件，让master节点与公共时钟同步服务器保持时间一致，同时允许slave节点同步自己的时间。master节点的chrony.conf文件内容如图3-13所示。

```
# Please consider joining the pool (http://www.pool.ntp.org/join.html).
server times.aliyun.com
server cn.pool.ntp.org
#server 2.centos.pool.ntp.org iburst
#server 3.centos.pool.ntp.org iburst

# Record the rate at which the system clock gains/losses time.
driftfile /var/lib/chrony/drift

# Allow the system clock to be stepped in the first three updates
# if its offset is larger than 1 second.
makestep 1.0 3

# Enable kernel synchronization of the real-time clock (RTC).
rtcsync

# Enable hardware timestamping on all interfaces that support it.
#hwtimestamp *

# Increase the minimum number of selectable sources required to adjust
# the system clock.
#minsources 2

# Allow NTP client access from local network.
allow 192.168.2.0/24

# Serve time even if not synchronized to a time source.
local stratum 10

# Specify file containing keys for NTP authentication.
#keyfile /etc/chrony.keys
-- INSERT --
```

图3-13　master节点的chrony.conf文件内容

在slave节点安装好chrony软件，启动chrony服务，配置/etc/chrony.conf，最后重启chrony，命令如下：

```
[root@slave ~]# yum install chrony -y
[root@slave ~]# systemctl start chrony
[root@slave ~]# vi /etc/chrony.conf
[root@slave ~]# systemctl restart chrony
```

slave 节点的配置文件比较简单，只需要将上游时间服务器的 IP 地址改为 master 节点的 IP 地址即可，master 节点的 chrony.conf 文件内容如图 3-14 所示。

```
# Use public servers from the pool.ntp.org project.
# Please consider joining the pool (http://www.pool.ntp.org/join.html).
server 192.168.2.176 iburst
#server 1.centos.pool.ntp.org iburst
#server 2.centos.pool.ntp.org iburst
#server 3.centos.pool.ntp.org iburst

# Record the rate at which the system clock gains/losses time.
driftfile /var/lib/chrony/drift

# Allow the system clock to be stepped in the first three updates
# if its offset is larger than 1 second.
makestep 1.0 3
```

图 3-14 master 节点的 chrony.conf 文件内容

3.1.2 配置 Java 环境

因为 Hadoop 的环境依赖于 JDK，所以需要确保虚拟机中已经正确安装了 JDK，并配置了环境变量，配置步骤如下。

（1）下载 JDK 安装包。

本实践采用的是 jdk_1.8.161，软件包的安装路径为/home/hadoop/java。使用 FileZilla 工具将下载好的安装包从本地传到服务器，如图 3-15 所示。

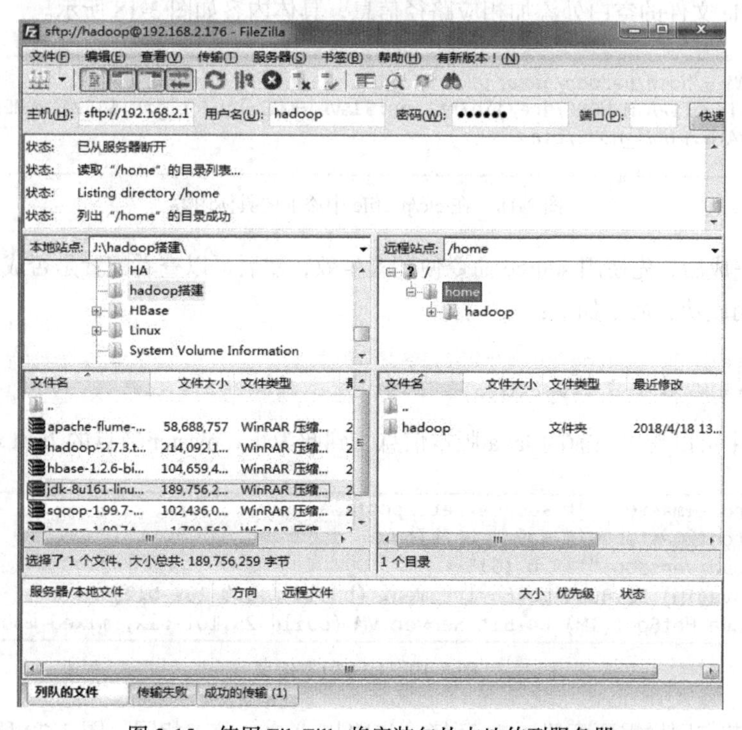

图 3-15 使用 FileZilla 将安装包从本地传到服务器

（2）解压 JDK 安装包。

进入 JDK 安装包所在目录并使用 tar 命令解压，解压后可以看到在对应目录下多了一个文件夹，如图 3-16 所示。操作命令如下：

```
[hadoop@master java] tar -zxvf jdk-8u161-linux-x64.tar.gz
```

```
[hadoop@master java]$ ls
jdk-8u161-linux-x64.tar.gz
[hadoop@master java]$ tar -zxf jdk-8u161-linux-x64.tar.gz
[hadoop@master java]$ ls          解压后的文件夹
jdk1.8.0_161  jdk-8u161-linux-x64.tar.gz
```

图 3-16 解压 JDK 安装包

（3）复制完整的 JDK 路径信息/home/hadoop/java/jdk1.8.0_161，如图 3-17 所示。

```
[hadoop@master java]$ ls
jdk1.8.0_161  jdk-8u161-linux-x64.tar.gz
[hadoop@master java]$ cd jdk1.8.0_161/
[hadoop@master jdk1.8.0_161]$ pwd
/home/hadoop/java/jdk1.8.0_161 ——→ 复制此路径信息
[hadoop@master jdk1.8.0_161]$
```

图 3-17 复制 JDK 路径信息

（4）切换到 root 用户下，对/etc/profile 文件进行配置，在文件末尾添加环境变量信息，命令如下：

```
[hadoop@master jdk1.8.0_161]$ su
[root@master jdk1.8.0_161]# vi /etc/profile
```

在/etc/profile 文件的空白处添加相应路径信息，具体内容如图 3-18 所示。

```
export JAVA_HOME=/home/hadoop/java/jdk1.8.0_161
export CLASSPATH=.:$PATH_HOME/jre/lib/rt.jar:$JAVA_HOME/lib/dt.jar:$JAVA_HOME/lib/tools.jar
export PATH=$PATH:$JAVA_HOME/bin
```

图 3-18 在/etc/profile 中添加的具体内容

（5）配置完成后，先使用 source 命令使配置生效，然后可以查看配置是否成功。若显示 Java 版本信息则表明成功，命令如下：

```
[root@master ~]# source /etc/profile
[root@master ~]# java -version
```

从图 3-19 中可以看出当前的 Java 版本信息。到此为止，master 节点的 Java 环境配置完毕。

```
[root@master ~]# source /etc/profile
[root@master ~]# java -version
java version "1.8.0_161"
Java(TM) SE Runtime Environment (build 1.8.0_161-b12)
Java HotSpot(TM) 64-Bit Server VM (build 25.161-b12, mixed mode)
```

图 3-19 Java 版本信息

（6）slave 节点同样需要配置 Java 环境，步骤与 master 节点相同。图 3-20 所示为在 slave 节点配置好 Java 环境后进行的检查。

```
[root@slave ~]# java -version
java version "1.8.0_161"
Java(TM) SE Runtime Environment (build 1.8.0_161-b12)
Java HotSpot(TM) 64-Bit Server VM (build 25.161-b12, mixed mode)
```

图 3-20　在 slave 节点中检查 Java 环境

3.1.3　安装 Hadoop

经过前文的准备，基础环境已经搭建完毕，下面开始安装 Hadoop。本实践环境采用的版本为 Hadoop 2.9.0，Hadoop 安装包可通过官网下载到/home/hadoop/software/路径下。具体安装步骤如下。

（1）切换到安装包所在路径/home/hadoop/software/，并进行解压。

```
[hadoop@master ~]$ cd ./software/
[hadoop@master software]$ tar -zxvf hadoop-2.9.0.tar.gz
```

从图 3-21 中可以看到经过解压后生成的新文件夹。

```
[hadoop@master ~]$ cd ./software/
[hadoop@master software]$ ls
hadoop-2.9.0.tar.gz
[hadoop@master software]$ tar -zxf hadoop-2.9.0.tar.gz
[hadoop@master software]$ ls
hadoop-2.9.0  hadoop-2.9.0.tar.gz    解压后生成新的文件夹
```

图 3-21　解压 Hadoop 安装包

（2）修改文件 hadoop-2.9.0/etc/hadoop/slaves，输入 slave 节点的主机名。如果有多个 slave 节点，则每一行输入一个 slave 节点主机名，命令如下：

```
[hadoop@master software]$ cd ./hadoop-2.9.0/etc/hadoop
[hadoop@master hadoop]$ ls
[hadoop@master hadoop]$ vi slaves
```

本环境中只有一个 slave 节点，因此只需要添加一行即可，如图 3-22 所示。

```
[hadoop@master software]$ cd ./hadoop-2.9.0/etc/hadoop
[hadoop@master hadoop]$ ls
capacity-scheduler.xml    hadoop-metrics2.properties    httpfs-signature.secret    log4j.properti
configuration.xsl         hadoop-metrics.properties     httpfs-site.xml            mapred-env.cmd
container-executor.cfg    hadoop-policy.xml             kms-acls.xml               mapred-env.sh
core-site.xml             hdfs-site.xml                 kms-env.sh                 mapred-queues.
hadoop-env.cmd            httpfs-env.sh                 kms-log4j.properties       mapred-site.xm
hadoop-env.sh             httpfs-log4j.properties       kms-site.xml               slaves
[hadoop@master hadoop]$ vi slaves
```

```
slave
~
~
~
~
~
~
~
```

删掉文件中的内容，添加 slave 节点主机名。若有多个 slave 节点，则另起一行，添加其他 slave 节点主机名

图 3-22　修改 slaves 文件

（3）修改 hadoop-env.sh 文件，方法如下：用 ls 命令找到该文件，如图 3-23 上部所示，然后用如下 vi 命令修改该文件。

```
[hadoop@master hadoop]$ vi hadoop-env.sh
```

```
[hadoop@master hadoop]$ ls
capacity-scheduler.xml        httpfs-env.sh               mapred-env.sh
configuration.xsl             httpfs-log4j.properties     mapred-queues.xml.template
container-executor.cfg        httpfs-signature.secret     mapred-site.xml.template
core-site.xml                 httpfs-site.xml             slaves
hadoop-env.cmd                kms-acls.xml                ssl-client.xml.example
hadoop-env.sh                 kms-env.sh                  ssl-server.xml.example
hadoop-metrics2.properties    kms-log4j.properties        yarn-env.cmd
hadoop-metrics.properties     kms-site.xml                yarn-env.sh
hadoop-policy.xml             log4j.properties            yarn-site.xml
hdfs-site.xml                 mapred-env.cmd
[hadoop@master hadoop]$ vi hadoop-env.sh
```

图 3-23 修改 hadoop-env.sh 文件

找到图 3-24 所示的内容，将里面的 export JAVA_HOME=${JAVA_HOME}修改为 export JAVA_HOME=/home/hadoop/java/jdk1.8.0_161。

```
# The only required environment variable is JAVA_HOME.  All others are
# optional.  When running a distributed configuration it is best to
# set JAVA_HOME in this file, so that it is correctly defined on
# remote nodes.

# The java implementation to use.
export JAVA_HOME=${JAVA_HOME}        修改前

# The jsvc implementation to use. Jsvc is required to run secure datanodes
# that bind to privileged ports to provide authentication of data transfer
# protocol.  Jsvc is not required if SASL is configured for authentication of
# data transfer protocol using non-privileged ports.
#export JSVC_HOME=${JSVC_HOME}
```

图 3-24 修改前的内容

修改后的内容如图 3-25 所示。需要注意的是，导入的是前文配置 Java 环境时的 JDK 路径。

```
# The only required environment variable is JAVA_HOME.  All others are
# optional.  When running a distributed configuration it is best to
# set JAVA_HOME in this file, so that it is correctly defined on
# remote nodes.

# The java implementation to use.
export JAVA_HOME=/home/hadoop/java/jdk1.8.0_161        导入 JDK 路径

# The jsvc implementation to use. Jsvc is required to run secure datanodes
# that bind to privileged ports to provide authentication of data transfer
# protocol.  Jsvc is not required if SASL is configured for authentication of
# data transfer protocol using non-privileged ports.
#export JSVC_HOME=${JSVC_HOME}
```

图 3-25 修改后的内容

（4）修改 yarn-env.sh 文件，去掉环境变量行首的“#”之后再加上 JDK 路径，方法如下：用 ls 命令找到该文件，如图 3-26 上部所示，然后用如下 vi 命令修改该文件，修改后的 yarn-env.sh 文件如图 3-26 下部所示。

```
[hadoop@master hadoop]$ vi yarn-env.sh
```

```
[hadoop@master hadoop]$ ls
capacity-scheduler.xml      httpfs-env.sh              mapred-env.sh
configuration.xsl           httpfs-log4j.properties    mapred-queues.xml.template
container-executor.cfg      httpfs-signature.secret    mapred-site.xml.template
core-site.xml               httpfs-site.xml            slaves
hadoop-env.cmd              kms-acls.xml               ssl-client.xml.example
hadoop-env.sh               kms-env.sh                 ssl-server.xml.example
hadoop-metrics2.properties  kms-log4j.properties       yarn-env.cmd
hadoop-metrics.properties   kms-site.xml               yarn-env.sh
hadoop-policy.xml           log4j.properties           yarn-site.xml
hdfs-site.xml               mapred-env.cmd
[hadoop@master hadoop]$ vi yarn-env.sh
```

```
# User for YARN daemons
export HADOOP_YARN_USER=${HADOOP_YARN_USER:-yarn}

# resolve links - $0 may be a softlink
export YARN_CONF_DIR="${YARN_CONF_DIR:-$HADOOP_YARN_HOME/conf}"

# some Java parameters
export JAVA_HOME=/home/hadoop/java/jdk1.8.0_161    加入 JDK 路径
if [ "$JAVA_HOME" != "" ]; then
  #echo "run java in $JAVA_HOME"
  JAVA_HOME=$JAVA_HOME
fi

if [ "$JAVA_HOME" = "" ]; then
-- INSERT --
```

图 3-26　修改后的 yarn-env.sh 文件

（5）修改配置文件 core-site.xml，方法如下：用 ls 命令找到该文件，如图 3-27 上部所示，然后用如下 vi 命令修改该文件。

```
[hadoop@master hadoop]$ vi core-site.xml
```

```
[hadoop@master hadoop]$ ls
capacity-scheduler.xml      httpfs-env.sh              mapred-env.sh
configuration.xsl           httpfs-log4j.properties    mapred-queues.xml.template
container-executor.cfg      httpfs-signature.secret    mapred-site.xml.template
core-site.xml               httpfs-site.xml            slaves
hadoop-env.cmd              kms-acls.xml               ssl-client.xml.example
hadoop-env.sh               kms-env.sh                 ssl-server.xml.example
hadoop-metrics2.properties  kms-log4j.properties       yarn-env.cmd
hadoop-metrics.properties   kms-site.xml               yarn-env.sh
hadoop-policy.xml           log4j.properties           yarn-site.xml
hdfs-site.xml               mapred-env.cmd
[hadoop@master hadoop]$ vi core-site.xml
```

图 3-27　修改 core-site.xml 文件

在文件相应位置添加如下内容，修改后的 core-site.xml 文件如图 3-28 所示。

```
<property>
 <name>fs.defaultFS</name>
 <value>hdfs: //master: 9000</value>
</property>
```

```
<property>
  <name>io.file.buffer.size</name>
  <value>131072</value>
</property>
<property>
  <name>hadoop.tmp.dir</name>
  <value>/home/hadoop/hadoop/tmp</value>
  <description>Abasefor other temporary directories.</description>
</property>
```

```
<configuration>
<property>
  <name>fs.defaultFS</name>
  <value>hdfs://master:9000</value>
 </property>
 <property>
   <name>io.file.buffer.size</name>
   <value>131072</value>
 </property>
 <property>
   <name>hadoop.tmp.dir</name>
   <value>/home/hadoop/hadoop/tmp</value>
   <description>Abasefor other temporary directories.</description>
 </property>

</configuration>
```

图 3-28　修改后的 core-site.xml 文件

（6）修改配置文件 hdfs-site.xml，命令如下：

```
[hadoop@master hadoop]$ vi hdfs-site.xml
```

添加下列内容：

```
<property>
   <name>dfs.namenode.secondary.http-address</name>
   <value>master: 9001</value>
 </property>
 <property>
   <name>dfs.namenode.name.dir</name>
   <value>/home/hadoop/hadoop/dfs/name</value>
 </property>
 <property>
   <name>dfs.datanode.data.dir</name>
   <value>/home/hadoop/hadoop/dfs/data</value>
 </property>
 <property>
   <name>dfs.replication</name>
   <value>2</value>
 </property>
```

修改后的 hdfs-site.xml 文件如图 3-29 所示。

```
<!-- Put site-specific property overrides in this file. -->

<configuration>
<property>
   <name>dfs.namenode.secondary.http-address</name>
   <value>master:9001</value>
 </property>
 <property>
   <name>dfs.namenode.name.dir</name>
   <value>/home/hadoop/hadoop/dfs/name</value>
 </property>
 <property>
   <name>dfs.datanode.data.dir</name>
   <value>/home/hadoop/hadoop/dfs/data</value>
 </property>
 <property>
   <name>dfs.replication</name>
   <value>2</value>
 </property>

</configuration>
-- INSERT --
```

图 3-29　修改后的 hdfs-site.xml 文件

（7）修改配置文件 mapred-site.xml，因为默认没有这个文件，所以需要复制一份，如图 3-30 所示。命令如下：

```
[hadoop@master hadoop]$ cp mapred-site.xml.template mapred-site.xml
```

```
[hadoop@master hadoop]$ cp mapred-site.xml.template mapred-site.xml
[hadoop@master hadoop]$ ls
capacity-scheduler.xml  hadoop-metrics2.properties  httpfs-signature.secret  log4j.properties
configuration.xsl       hadoop-metrics.properties   httpfs-site.xml          mapred-env.cmd
container-executor.cfg  hadoop-policy.xml           kms-acls.xml             mapred-env.sh
core-site.xml           hdfs-site.xml               kms-env.sh               mapred-queues.xml.template
hadoop-env.cmd          httpfs-env.sh               kms-log4j.properties     mapred-site.xml
hadoop-env.sh           httpfs-log4j.properties     kms-site.xml             mapred-site.xml.template
```

图 3-30　复制 mapred-site.xml 文件

复制完 mapred-site.xml 文件之后需要对它进行配置，命令如下：

```
[hadoop@master hadoop]$ vi mapred-site.xml
```

添加下列内容：

```
<property>
   <name>mapreduce.framework.name</name>
   <value>yarn</value>
 </property>
 <property>
   <name>mapreduce.jobhistory.address</name>
```

```
    <value>master: 10020</value>
  </property>
  <property>
    <name>mapreduce.jobhistory.webapp.address</name>
    <value>master: 19888</value>
  </property>
```

修改后的 mapred-site.xml 文件如图 3-31 所示。

```
<!-- Put site-specific property overrides in this file. -->

<configuration>
<property>
   <name>mapreduce.framework.name</name>
   <value>yarn</value>
 </property>
 <property>
   <name>mapreduce.jobhistory.address</name>
   <value>master:10020</value>
 </property>
 <property>
   <name>mapreduce.jobhistory.webapp.address</name>
   <value>master:19888</value>
 </property>

</configuration>
~
```

图 3-31　修改后的 mapred-site.xml 文件

（8）修改配置文件 yarn-site.xml，命令如下：

```
[hadoop@master hadoop]$ vi yarn-site.xml
```

添加下列内容：

```
<property>
    <name>yarn.nodemanager.aux-services</name>
    <value>mapreduce_shuffle</value>
  </property>
  <property>
    <name>yarn.nodemanager.aux-services.mapreduce.shuffle.class</name>
    <value>org.apache.hadoop.mapred.ShuffleHandler</value>
  </property>
  <property>
    <name>yarn.resourcemanager.address</name>
    <value>master: 8032</value>
  </property>
  <property>
    <name>yarn.resourcemanager.scheduler.address</name>
    <value>master: 8030</value>
  </property>
  <property>
    <name>yarn.resourcemanager.resource-tracker.address</name>
```

```
    <value>master: 8035</value>
  </property>
  <property>
    <name>yarn.resourcemanager.admin.address</name>
    <value>master: 8033</value>
  </property>
  <property>
    <name>yarn.resourcemanager.webapp.address</name>
    <value>master: 8088</value>
  </property>
```

修改后的 yarn-site.xml 文件如图 3-32 所示。

```
<configuration>
<!-- Site specific YARN configuration properties -->
<property>
   <name>yarn.nodemanager.aux-services</name>
   <value>mapreduce_shuffle</value>
 </property>
 <property>
   <name>yarn.nodemanager.aux-services.mapreduce.shuffle.class</name>
   <value>org.apache.hadoop.mapred.ShuffleHandler</value>
 </property>
 <property>
   <name>yarn.resourcemanager.address</name>
   <value>master:8032</value>
 </property>
 <property>
   <name>yarn.resourcemanager.scheduler.address</name>
   <value>master:8030</value>
 </property>
 <property>
   <name>yarn.resourcemanager.resource-tracker.address</name>
   <value>master:8035</value>
 </property>
 <property>
   <name>yarn.resourcemanager.admin.address</name>
   <value>master:8033</value>
 </property>
 <property>
   <name>yarn.resourcemanager.webapp.address</name>
   <value>master:8088</value>
 </property>
</configuration>
```

图 3-32　修改后的 yarn-site.xml 文件

（9）修改完 core-site.xml、hdfs-site.xml、mapred-site.xml、yarn-site.xml 文件之后，将配置好的 Hadoop 文件从 master 节点远程复制到 slave 节点，如图 3-33 所示，命令如下：

```
[hadoop@master ~]$scp -r ~/software/hadoop-2.9.0 hadoop@slave: /home/hadoop/software/
```

```
[hadoop@master ~]$ scp -r ~/software/hadoop-2.9.0 hadoop@slave:/home/hadoop/software/
```

图 3-33　将配置好的 Hadoop 文件远程复制到 slave 节点

此时可以发现，在 slave 节点的/home/hadoop/software 路径下多了一个 hadoop-2.9.0 文件夹，如图 3-34 所示。

```
[hadoop@slave software]$ pwd
/home/hadoop/software
[hadoop@slave software]$ ls
hadoop-2.9.0
```

图 3-34　slave 节点的 hadoop-2.9.0 文件夹

（10）配置 Hadoop 启动的系统环境变量，需要同时在 master 节点和 slave 节点上操作，命令如下：

```
[hadoop@master ~]$ vi ~/.bash_profile
```

在.bash-profile 文件末尾添加 Hadoop 的路径：

```
export HADOOP_HOME=/home/hadoop/software/hadoop-2.9.0
export PATH=$HADOOP_HOME/bin: $HADOOP_HOME/sbin: $PATH
```

修改后的.bash-profile 文件如图 3-35 所示。

```
# .bash_profile

# Get the aliases and functions
if [ -f ~/.bashrc ]; then
        . ~/.bashrc
fi

# User specific environment and startup programs

PATH=$PATH:$HOME/.local/bin:$HOME/bin

export PATH



export HADOOP_HOME=/home/hadoop/software/hadoop-2.9.0
export PATH=$HADOOP_HOME/bin:$HADOOP_HOME/sbin:$PATH
~
~
~
```

图 3-35　修改后的.bash_ profile 文件

3.1.4　启动 Hadoop

经过 3.1.3 小节的操作，Hadoop 环境配置完毕。本小节开始启动 Hadoop 集群，步骤如下。

（1）执行命令使环境变量生效，命令如下：

```
[hadoop@master ~]$ source ~/.bash_profile
```

（2）格式化 HDFS，如图 3-36 所示。命令如下：

```
[hadoop@master ~]$ hdfsnamenode -format
```

```
[hadoop@master ~]$ hdfs namenode -format
18/04/19 14:52:36 INFO namenode.NameNode: STARTUP_MSG:
/************************************************************
STARTUP_MSG: Starting NameNode
STARTUP_MSG:   host = master/192.168.2.176
STARTUP_MSG:   args = [-format]
STARTUP_MSG:   version = 2.9.0
STARTUP_MSG:   classpath = /home/hadoop/software/hadoop-2.9.0/et
imbus-jose-jwt-3.9.jar:/home/hadoop/software/hadoop-2.9.0/share/
p-2.9.0/share/hadoop/common/lib/commons-configuration-1.6.jar:/h
1.2.jar:/home/hadoop/software/hadoop-2.9.0/share/hadoop/common/l
p/common/lib/jersey-core-1.9.jar:/home/hadoop/software/hadoop-2.
doop-2.9.0/share/hadoop/common/lib/gson-2.2.4.jar:/home/hadoop/s
jar:/home/hadoop/software/hadoop-2.9.0/share/hadoop/common/lib/l
n/lib/woodstox-core-5.0.3.jar:/home/hadoop/software/hadoop-2.9.0
hadoop-2.9.0/share/hadoop/common/lib/jettison-1.1.jar:/home/hado
.jar:/home/hadoop/software/hadoop-2.9.0/share/hadoop/common/lib/
ommon/lib/slf4j-log4j12-1.7.25.jar:/home/hadoop/software/hadoop-
-2.9.0/share/hadoop/common/lib/jackson-jaxrs-1.9.13.jar:/home/ha
jar:/home/hadoop/software/hadoop-2.9.0/share/hadoop/common/lib/a
adoop/common/lib/stax-api-1.0-2.jar:/home/hadoop/software/hadoop
/hadoop-2.9.0/share/hadoop/common/lib/api-asn1-api-1.0.0-M20.jar
```

图 3-36　格式化 HDFS

（3）切换到/home/hadoop/software/hadoop-2.9.0，启动 Hadoop 集群，命令如下：

```
[hadoop@master ~]$ cd /home/hadoop/software/hadoop-2.9.0
[hadoop@master hadoop-2.9.0]$ sbin/start-all.sh
```

启动集群的过程如图 3-37 所示。

```
[hadoop@master ~]$ cd /home/hadoop/software/hadoop-2.9.0
[hadoop@master hadoop-2.9.0]$ sbin/start-all.sh
This script is Deprecated. Instead use start-dfs.sh and start-ya
rn.sh
Starting namenodes on [master]
hadoop@master's password:
master: starting namenode, logging to /home/hadoop/software/hado
op-2.9.0/logs/hadoop-hadoop-namenode-master.out
slave: starting datanode, logging to /home/hadoop/software/hadoo
p-2.9.0/logs/hadoop-hadoop-datanode-slave.out
Starting secondary namenodes [master]
hadoop@master's password:
master: starting secondarynamenode, logging to /home/hadoop/soft
ware/hadoop-2.9.0/logs/hadoop-hadoop-secondarynamenode-master.ou
t
starting yarn daemons
starting resourcemanager, logging to /home/hadoop/software/hadoo
p-2.9.0/logs/yarn-hadoop-resourcemanager-master.out
slave: starting nodemanager, logging to /home/hadoop/software/ha
doop-2.9.0/logs/yarn-hadoop-nodemanager-slave.out
```

图 3-37　启动集群的过程

（4）分别在 master 节点和 slave 节点使用 jps 命令查看进程是否启动成功。若 master 节点显示 4 个进程，slave 节点显示 3 个进程，则表明进程启动成功，命令如下：

```
[hadoop@master hadoop-2.9.0]$ jps
```

图 3-38 显示查看 master 节点进程，当前启动了 Jps、ResourceManager、SecondaryNameNode、NameNode 这 4 个进程。

```
[hadoop@master hadoop-2.9.0]$ jps
11474 Jps
11206 ResourceManager
11053 SecondaryNameNode
10862 NameNode
```

图 3-38 查看 master 节点进程

图 3-39 显示查看 slave 节点进程，当前启动了 DataNode、Jps、NodeManager 这 3 个进程。

```
[hadoop@slave ~]$ jps
12400 DataNode
12625 Jps
12509 NodeManager
```

图 3-39 查看 slave 节点进程

（5）使用命令查看集群信息，命令如下：

```
[hadoop@master hadoop-2.9.0]$ bin/hadoop dfsadmin -report
```

从图 3-40 中可以看出，当前集群状态正常。

```
[hadoop@master hadoop-2.9.0]$ bin/hadoop dfsadmin -report
DEPRECATED: Use of this script to execute hdfs command is depreca
ted.
Instead use the hdfs command for it.

Configured Capacity: 205665792000 (191.54 GB)
Present Capacity: 203203633152 (189.25 GB)
DFS Remaining: 203203629056 (189.25 GB)
DFS Used: 4096 (4 KB)
DFS Used%: 0.00%
Under replicated blocks: 0
Blocks with corrupt replicas: 0
Missing blocks: 0
Missing blocks (with replication factor 1): 0
Pending deletion blocks: 0

-------------------------------------------------
Live datanodes (1):

Name: 192.168.2.177:50010 (slave)
Hostname: slave
Decommission Status : Normal
```

图 3-40 集群状态

（6）在浏览器中输入 192.168.2.176:50070，可以查看集群基本信息；输入 192.168.2.176:8088，可以检查 YARN 是否正常，如图 3-41、图 3-42 所示。

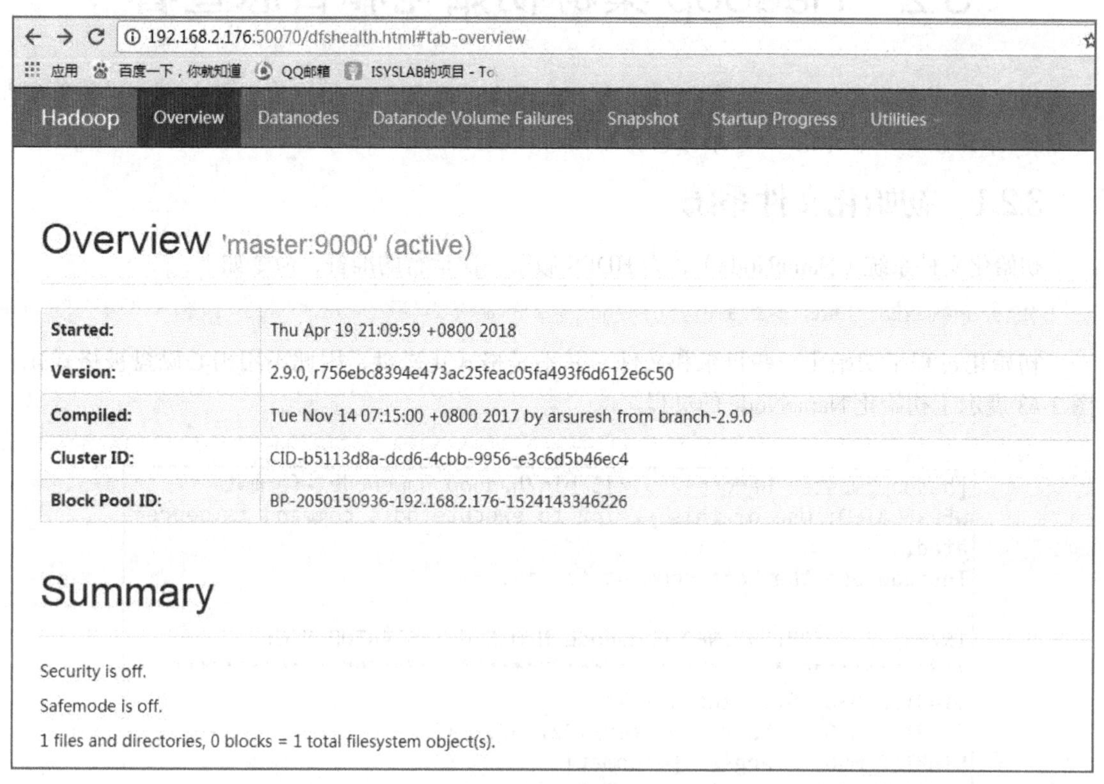

图 3-41　查看集群基本信息

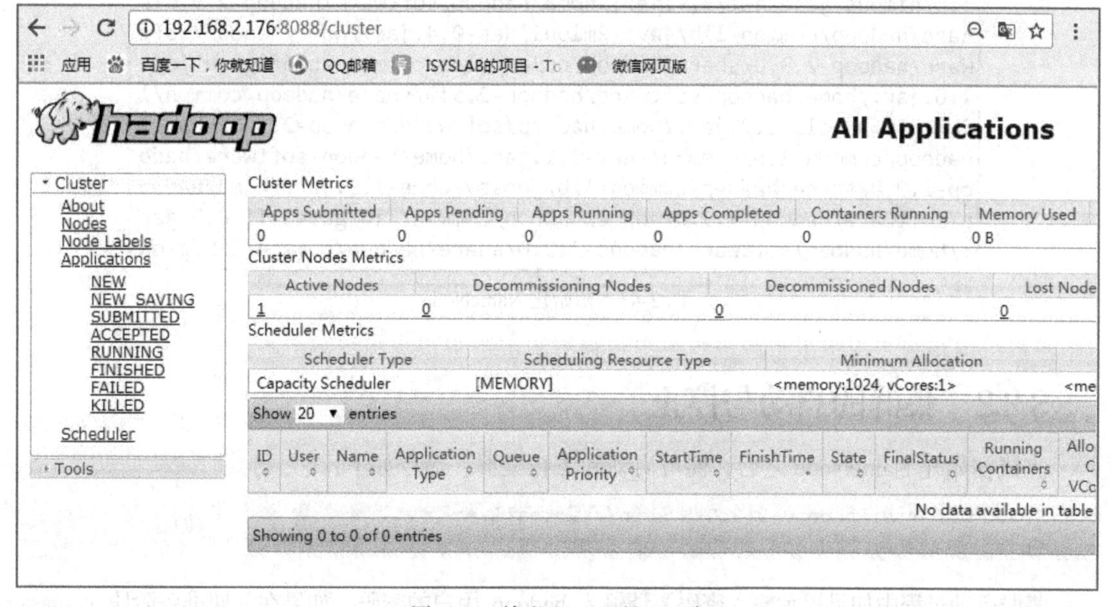

图 3-42　检查 YARN 是否正常

3.2 Hadoop 集群初始化和日志查看

经过 3.1 节的实践，我们配置了 Hadoop 基础环境，完成了 Hadoop 集群的安装，并成功启动了 Hadoop 集群。本节介绍 Hadoop 集群初始化和日志查看。

3.2.1 初始化文件系统

初始化文件系统（NameNode），为 HDFS 做第一次运行的准备，命令如下：

```
[hadoop@master hadoop-2.9.0]$ bin/hadoop namenode -format
```

初始化过程指初始化一些目录和文件，并不是格式化磁盘，因此不用担心磁盘被格式化。图 3-43 展示了初始化 NameNode 的过程。

```
[hadoop@master hadoop-2.9.0]$ bin/hadoop namenode -format
DEPRECATED: Use of this script to execute hdfs command is deprec
ated.
Instead use the hdfs command for it.

18/04/20 13:20:59 INFO namenode.NameNode: STARTUP_MSG:
/************************************************************
STARTUP_MSG: Starting NameNode
STARTUP_MSG:   host = master/192.168.2.176
STARTUP_MSG:   args = [-format]
STARTUP_MSG:   version = 2.9.0
STARTUP_MSG:   classpath = /home/hadoop/software/hadoop-2.9.0/et
c/hadoop:/home/hadoop/software/hadoop-2.9.0/share/hadoop/common/
lib/nimbus-jose-jwt-3.9.jar:/home/hadoop/software/hadoop-2.9.0/s
hare/hadoop/common/lib/java-xmlbuilder-0.4.jar:/home/hadoop/soft
ware/hadoop-2.9.0/share/hadoop/common/lib/commons-configuration-
1.6.jar:/home/hadoop/software/hadoop-2.9.0/share/hadoop/common/l
ib/commons-cli-1.2.jar:/home/hadoop/software/hadoop-2.9.0/share/
hadoop/common/lib/commons-net-3.1.jar:/home/hadoop/software/hado
op-2.9.0/share/hadoop/common/lib/jersey-core-1.9.jar:/home/hadoo
p/software/hadoop-2.9.0/share/hadoop/common/lib/guava-11.0.2.jar
:/home/hadoop/software/hadoop-2.9.0/share/hadoop/common/lib/gson
```

图 3-43　初始化 NameNode

3.2.2 集群的启动与停止

完成初始化 NameNode 之后，便可启动和停止集群，命令如下：

```
[hadoop@master hadoop-2.9.0]$ sbin/start-all.sh
[hadoop@master hadoop-2.9.0]$ sbin/stop-all.sh
```

集群启动过程中如果提示输入密码，请输入 hadoop 用户的密码。如果在前期准备阶段对 master 节点进行了自身互信，那么不需要输入密码，启动集群和停止集群如图 3-44、图 3-45 所示。

```
[hadoop@master hadoop-2.9.0]$ sbin/start-all.sh
This script is Deprecated. Instead use start-dfs.sh and start-ya
rn.sh
Starting namenodes on [master]
hadoop@master's password:
master: starting namenode, logging to /home/hadoop/software/hado
op-2.9.0/logs/hadoop-hadoop-namenode-master.out
slave: starting datanode, logging to /home/hadoop/software/hadoo
p-2.9.0/logs/hadoop-hadoop-datanode-slave.out
Starting secondary namenodes [master]
hadoop@master's password:
master: starting secondarynamenode, logging to /home/hadoop/soft
ware/hadoop-2.9.0/logs/hadoop-hadoop-secondarynamenode-master.ou
t
starting yarn daemons
starting resourcemanager, logging to /home/hadoop/software/hadoo
p-2.9.0/logs/yarn-hadoop-resourcemanager-master.out
slave: starting nodemanager, logging to /home/hadoop/software/ha
doop-2.9.0/logs/yarn-hadoop-nodemanager-slave.out
```

图 3-44　启动集群

```
[hadoop@master hadoop-2.9.0]$ sbin/stop-all.sh
This script is Deprecated. Instead use stop-dfs.sh and stop-yarn
.sh
Stopping namenodes on [master]
hadoop@master's password:
master: stopping namenode
slave: stopping datanode
Stopping secondary namenodes [master]
hadoop@master's password:
master: stopping secondarynamenode
stopping yarn daemons
stopping resourcemanager
slave: stopping nodemanager
slave: nodemanager did not stop gracefully after 5 seconds: kill
ing with kill -9
no proxyserver to stop
```

图 3-45　停止集群

3.2.3　查看日志

在 Hadoop 安装目录的 logs 子目录下有 Hadoop 运行期间产生的各种日志，可以通过命令查看日志信息，命令如下：

```
[hadoop@master hadoop-2.9.0]$ ll logs/
[hadoop@master hadoop-2.9.0]$ cat logs/hadoop-hadoop-namenode-master.log
```

图 3-46 列出了 Hadoop 在运行期间的各种日志，图 3-47 显示了 NameNode 日志信息。当集群出现问题时，这些日志信息可以为解决问题提供巨大帮助。

```
[hadoop@master hadoop-2.9.0]$ ll logs/
total 132
-rw-rw-r-- 1 hadoop hadoop 38823 Apr 19 21:11 hadoop-hadoop-namenode-master.log
-rw-rw-r-- 1 hadoop hadoop   717 Apr 19 21:09 hadoop-hadoop-namenode-master.out
-rw-rw-r-- 1 hadoop hadoop 32719 Apr 19 21:11 hadoop-hadoop-secondarynamenode-master.log
-rw-rw-r-- 1 hadoop hadoop   717 Apr 19 21:10 hadoop-hadoop-secondarynamenode-master.out
-rw-rw-r-- 1 hadoop hadoop     0 Apr 19 21:09 SecurityAuth-hadoop.audit
-rw-rw-r-- 1 hadoop hadoop 47772 Apr 19 21:20 yarn-hadoop-resourcemanager-master.log
-rw-rw-r-- 1 hadoop hadoop  1524 Apr 19 21:10 yarn-hadoop-resourcemanager-master.out
```

图 3-46 Hadoop 在运行期间的各种日志

```
2018-04-20 13:35:21,257 INFO org.apache.hadoop.hdfs.server.namenode.NameNode:
 SHUTDOWN_MSG:
/************************************************************
SHUTDOWN_MSG: Shutting down NameNode at master/192.168.2.176
************************************************************/
2018-04-20 13:36:14,160 INFO org.apache.hadoop.hdfs.server.namenode.NameNode:
 STARTUP_MSG:
/************************************************************
STARTUP_MSG: Starting NameNode
STARTUP_MSG:   host = master/192.168.2.176
STARTUP_MSG:   args = []
STARTUP_MSG:   version = 2.9.0
STARTUP_MSG:   classpath = /home/hadoop/software/hadoop-2.9.0/etc/hadoop:/hom
e/hadoop/software/hadoop-2.9.0/share/hadoop/common/lib/nimbus-jose-jwt-3.9.ja
r:/home/hadoop/software/hadoop-2.9.0/share/hadoop/common/lib/java-xmlbuilder-
0.4.jar:/home/hadoop/software/hadoop-2.9.0/share/hadoop/common/lib/commons-co
nfiguration-1.6.jar:/home/hadoop/software/hadoop-2.9.0/share/hadoop/common/li
b/commons-cli-1.2.jar:/home/hadoop/software/hadoop-2.9.0/share/hadoop/common/
lib/commons-net-3.1.jar:/home/hadoop/software/hadoop-2.9.0/share/hadoop/commo
n/lib/jersey-core-1.9.jar:/home/hadoop/software/hadoop-2.9.0/share/hadoop/com
mon/lib/guava-11.0.2.jar:/home/hadoop/software/hadoop-2.9.0/share/hadoop/comm
on/lib/gson-2.2.4.jar:/home/hadoop/software/hadoop-2.9.0/share/hadoop/common/
```

图 3-47 NameNode 日志信息

本章讲述了 Hadoop 集群的安装过程。其实对 Hadoop 来说，有 3 种安装模式，分别为单机模式、伪分布式模式和真正分布式模式。单机模式没有 HDFS，通常用来开发、调试 MapReduce 程序的应用逻辑。伪分布式模式是指在一台机器上安装完整的 Hadoop 服务，模拟分布式环境。真正分布式模式需要两台及两台以上机器安装 Hadoop 集群，其中一台用来做 NameNode，其余用来做 DataNode。为了简化操作，实践时可选用一台机器做 NameNode，一台机器做 DataNode。在大规模生产环境中，DataNode 数量要根据业务需求进行调整。

习　题

1. 简单梳理 Hadoop 安装流程。
2. 下列哪项通常是集群的最主要的性能瓶颈？（　　）

   A. CPU　　B. 网络　　C. 磁盘　　D. 内存
3. Hadoop 项目孵化时主要包括哪两个项目？
4. Hadoop 有几种安装模式？
5. 请完成 Hadoop 集群搭建，其中包括一个 NameNode、两个 DataNode。

第 4 章 HDFS

Hadoop 之所以能够分布式处理海量数据，最重要的原因之一在于其拥有强大的分布式存储能力，即有 HDFS 的支持。本章首先介绍 HDFS 的基本概念，接着介绍 HDFS 文件的读取和写入过程以及数据备份，然后侧重介绍 HDFS 的基本命令、数据平衡，以及 HDFS API 的使用。

4.1 HDFS 简介

4.1.1 HDFS 的基本概念

HDFS 以流式数据访问模式来存储超大文件，运行于商业硬件集群上。HDFS 集群拥有一个管理文件系统元数据的 NameNode，拥有多个存储实际数据的 DataNode。存储在 HDFS 上的文件首先被划分为不同的数据块，也可简称块（Block），然后以数据块的形式存储在不同 DataNode 上，客户端访问文件系统时需要同 NameNode 和 DataNode 进行交互。下面简单介绍 HDFS 中的基本概念。

1. 数据块

在用磁盘进行数据存储时，每个磁盘都有默认的数据块大小，数据块是磁盘进行数据读/写操作的最小单位。HDFS 中默认的数据块大小为 64MB，可根据实际情况调整数据块大小。存储在 HDFS 上的文件被划分为块大小的多个分块并作为独立的存储单元，HDFS 中小于一个块大小的文件不会占据整个块的空间。HDFS 的块比磁盘的块大，目的是最小化寻址开销，即如果块设置得足够大，从磁盘传输数据的时间会明显大于定位块开始位置所需的时间。对 HDFS 中的块进行抽象会带来两个好处：一个好处是文件的大小可以大于网络中任意一个磁盘的容量，文件的所有块并不需要存储在同一个磁盘上，它们可以利用集群上的任意磁盘进行存储；另一个好处是大大简化了存储子系统的设计。

2. NameNode 和 DataNode

在 HDFS 集群中有两类节点。一类是 NameNode，扮演“管理者”角色，管理文件系统元数据；另一类是 DataNode，扮演“工作者”角色，存储实际数据。NameNode 维护文件系统树及整棵树内所有的文件和目录，并管理文件系统的命名空间。这些信息以命名空间镜像文件和编辑日志文件的形式永久保存在本地磁盘上。NameNode 记录每个文件中各个块所在的 DataNode 信息，但并不永久保存块的位置信息，因为这些位置信息会在系统启动时由 DataNode 重建。DataNode 是文件系统的工作节点，根据需要存储并检索数据块，并且定期向 NameNode 发送它们所存储的块的列表。

如果没有 NameNode，文件系统将无法使用。如果运行 NameNode 服务的机器毁坏，文件系统上所有的文件将会丢失，因此 Hadoop 提供了多种机制实现 NameNode 容错。

3. 联邦 HDFS

NameNode 在内存中保存文件系统中每个文件和每个数据块的引用关系。这意味着对一个拥有大量文件的超大集群来说，内存将成为限制系统横向扩展的瓶颈。引入联邦 HDFS 允许系统通过添加 NameNode 实现扩展，其中每个 NameNode 管理文件系统命名空间中的一部分。例如，一个 NameNode 可能管理/user 目录下的所有文件，另一个 NameNode 可能管理/share 目录下的所有文件。

4.1.2 HDFS 文件的读取

当数据块存储到 HDFS 中以后，文件的读取流程如图 4-1 所示。首先，客户端通过调用 FileSystem 对象中的 open( )函数来打开希望读取的文件。对 HDFS 来说，FileSystem 是分布式文件系统的一个实例，对应着图中的第 1 步。然后进行第 2 步，DistributedFileSystem 通过远程过程调用（Remote Procedure Call，RPC）调用 NameNode，以确定文件起始块的位置。对于每一个块，NameNode 返回存有该块副本的 DataNode 的地址。随后，这些返回的 DataNode 会按照 Hadoop 定义的集群网络拓扑结构计算自己与客户端的距离并进行排序。如果客户端本身就是一个 DataNode，当保存有相应数据块的一个副本时，该节点就会直接从本地读取数据。

接着，DistributedFileSystem 会向客户端返回一个支持文件定位的输入流对象 FSDataInputStream，用于给客户端读取数据。FSDataInputStream 类转而封装 DFSInputStream 对象，该对象管理着 NameNode 和 DataNode 之间的 I/O。

当获取到数据块的位置后进行第 3 步，客户端会在这个输入流之上调用 read( )函数，存储着文件起始块 DataNode 的地址的 DFSInputStream 对象随即连接距离最近的 DataNode。

第 4 步，连接完成后，在数据流中反复调用 read( )函数，将数据从 DataNode 传输到客户端，直到这个块全部读取完毕。第 5 步，当最后一个数据块读取完毕时，DFSInputStream 会关闭与该 DataNode 的连接，然后寻找下一个数据块距离客户端最近的 DataNode。

客户端从流中读取数据时，块是按照打开 DFSInputStream 与 DataNode 新建连接的顺序读取的。它会根据需要询问 NameNode 来检索下一批数据块的 DataNode 的位置。第 6 步，一旦客户端完成读取，就会对 FSDataInputStream 调用 close( )函数。

在读取数据的时候，如果 DFSInputStream 与 DataNode 通信时遇到错误，会尝试从这个块的另外一个最近邻 DataNode 读取数据，同时也会记住那个发生故障的 DataNode，以保证以后不会去读取该节点上后续的块。收到数据块以后，DFSInputStream 也会校验和确认从 DataNode 发来的数据是否完整。如果有损坏的块，DFSInputStream 就试图从其他 DataNode 读取该副本，并向 NameNode 报告该信息。

对于文件的读取，HDFS 是通过 NameNode 引导获取合适的 DataNode，然后直接连接 DataNode 去读取数据的。这样设计的好处是，可以使 HDFS 扩展到更大规模的客户端并行处理，因为数据的流动是在所有 DataNode 之间分散进行的。同时 NameNode 的压力也减小了。NameNode 只需提供请求数据块所在的位置信息，而不需要提供数据，避免了 NameNode 随着客户端数量的增长而成为系统的瓶颈。

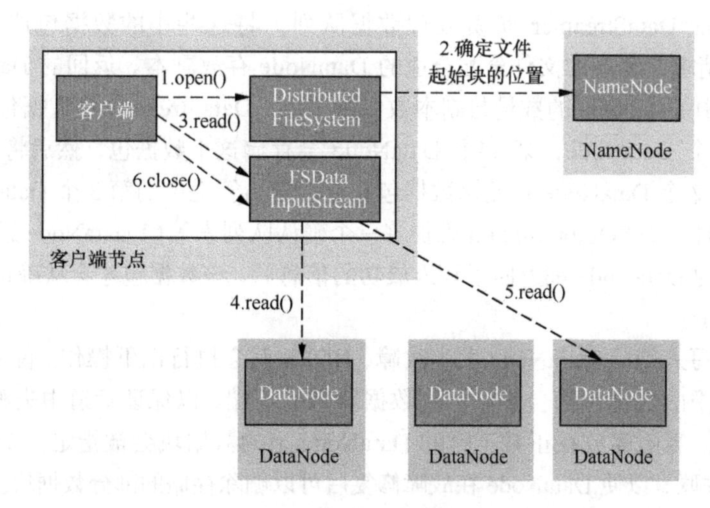

图 4-1 文件的读取流程

4.1.3 HDFS 文件的写入

图 4-2 展示的是客户端在 HDFS 中写入一个新文件的过程。第 1 步，客户端通过对 DistributedFileSystem 对象调用 create( )函数创建一个文件。然后 DistributedFileSystem 对 NameNode 创建一个 RPC 调用，在文件系统的命名空间中创建一个文件，此时该文件没有相应的数据块，也没有相关的 DataNode 与之关联，此过程对应着图中的第 2 步。为确保这个新文件不存在于文件系统中，NameNode 会执行各种不同的检查。当所有检查通过时，NameNode 会创建一个新文件的记录；否则，文件创建失败并向客户端抛出一个 IOException 异常。如果创建成功，则 DistributedFileSystem 向客户端返回一个 FSDataOutputStream 对象，客户端开始借助这个对象向 HDFS 写入数据，同样，FSDataOutputStream 封装着一个 DFSOutputStream 数据流对象，负责处理 NameNode 和 DataNode 之间的通信。

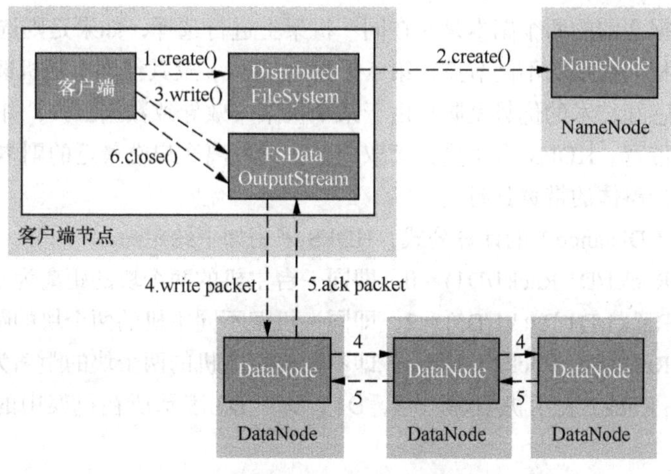

图 4-2 文件的写入流程

第 3 步，当客户端写入数据时，DFSOutputStream 会将文件按指定大小分割成数据包，并写入一个数据队列。DataStreamer 负责处理数据队列，将这些小的数据包放入数据流，并对 NameNode 发出请求，为新的文件分配合适的 DataNode 存放副本。返回的 DataNode 列表形成一个管道，管道中 DataNode 的数量与副本数保持一致，DataStreamer 将数据包以流的方式传送给队列中的第 1 个 DataNode，第 1 个 DataNode 会存储这个数据包，然后将它传送到第 2 个 DataNode 中，第 2 个 DataNode 存储该数据包并且传送给管道中的第 3 个 DataNode，对应着图中的第 4 步。同时，DFSOutputStream 维护着一个确认队列来等待 DataNode 返回确认信息。只有当管道中所有的 DataNode 都返回了写入成功的信息后，该数据包才会从确认队列中删除，这对应图中的第 5 步。

如果在数据写入期间 DataNode 发生故障，HDFS 就会执行以下操作。首先管道会被关闭，任何在确认队列中的数据包都会被添加到数据队列的前端，以保证管道中失败的 DataNode 的数据包不会丢失。当前存放在正常工作的 DataNode 上的数据块会被指定一个新的标识，并和 NameNode 进行关联，以便 DataNode 在故障修复后可以删除存储的部分数据块。然后，管道会把失败的 DataNode 删除，文件被继续写到另外两个 DataNode 中。最后，NameNode 注意到现在的数据块副本没有达到配置属性要求，会在另外的 DataNode 上重新安排创建一个副本，后续的数据块继续正常接收处理。

客户端成功完成数据写入操作以后，就到了第 6 步，对数据流调用 close( )函数。该操作将剩余的所有数据包写入 DataNode 管道，并连接 NameNode 等待通知确认信息。

4.1.4 HDFS 数据备份

由于许多公司的业务场景受限制，在众多场景下的小集群运行环境中，没有不间断电源（Uninterruptible Power Supply，UPS）等市电中断下的续航设备，导致集群意外停止，在市电恢复后可能出现系统无法启动、系统数据目录损坏等情况，此时可能由于集群 HDFS 数据损坏导致集群无法正常启动。因此，针对此情况，就必须对 HDFS 的数据进行一定的备份操作。

HDFS 采用机架感知技术来增强数据的可靠性、可用性和提高网络带宽的利用率。通过机架感知技术，NameNode 可确定每个 DataNode 所属的机架（Rack）ID，HDFS 会把副本放在不同的机架上。如图 4-3 所示，第 1 个副本（B1）在本地机架上，第 2 个副本（B2）在远端机架上，第 3 个副本（B3）根据之前的两个副本是否在同一机架上进行选择，如果是则选择其他机架，否则选择和第 1 个副本相同机架的不同节点，第 4 个副本（B4）及其他副本随机选择存放位置。

HDFS 的机架感知技术的优势是防止由于某个机架失效导致数据丢失，并允许读取数据时充分利用多个机架的带宽。HDFS 会尽量让读取任务去读取离客户端最近的副本数据，以减少整体带宽消耗，从而减少整体的带宽延时。

对于副本距离（Distance）的计算公式，HDFS 采用如下约定。

（1）Distance (Rack1/D1 Rack1/D1) = 0，即同一台主机的两个块的距离为 0。

（2）Distance (Rack1/D1 Rack1/D3) = 2，即同一机架不同主机的两个块的距离为 2。

（3）Distance (Rack1/D1 Rack2/D1) = 4，即不同机架主机的两个块的距离为 4。

其中，Rack1、Rack2 表示机架标识号，D1、D2、D3 表示所在机架中的 DataNode 主机的编号。

通过机架感知技术，处于工作状态的 HDFS 总是设法确保数据块的 3 个副本（或更多副本）中至少有两个在同一机架上，至少有一个在不同机架上。

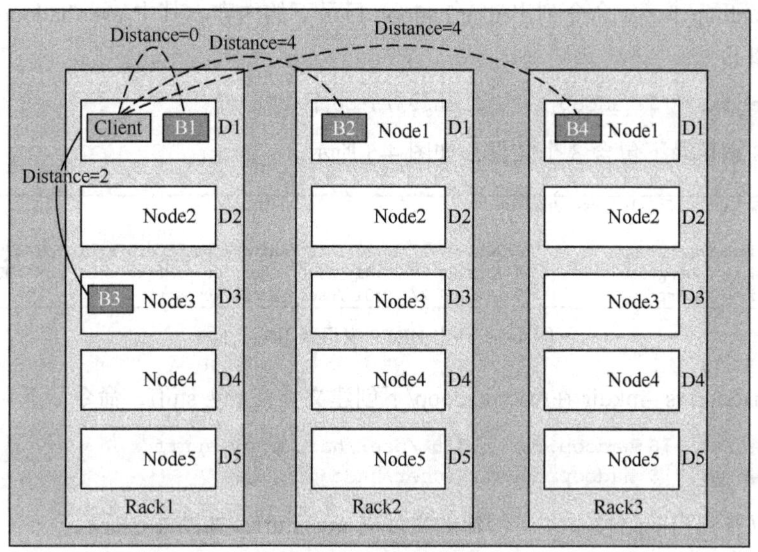

图 4-3 机架感知技术

4.2 HDFS 基本命令

本节介绍 HDFS 的基本命令，包括常见的用于增、删、改、查的命令，同时还会介绍修改文件权限的命令。限于篇幅，无法一一介绍每条命令的详细使用方法，读者可以利用系统自带的帮助（命令为 hadoop fs –help）学习本书没有提到的命令。

（1）使用 hadoop fs 查看 HDFS 中存在的命令，命令如下：

```
[hadoop@master ~]$ hadoop fs
```

从图 4-4 中可以看到 HDFS 的常用命令及其使用方法，熟悉 Linux 操作的读者可以看出 HDFS 命令与 Linux 命令有很多相似之处。

```
[hadoop@master ~]$ hadoop fs
Usage: hadoop fs [generic options]
 [-appendToFile <localsrc> ... <dst>]
 [-cat [-ignoreCrc] <src> ...]
 [-checksum <src> ...]
 [-chgrp [-R] GROUP PATH...]
 [-chmod [-R] <MODE[,MODE]... | OCTALMODE> PATH...]
 [-chown [-R] [OWNER][:[GROUP]] PATH...]
 [-copyFromLocal [-f] [-p] [-l] [-d] <localsrc> ... <dst>]
 [-copyToLocal [-f] [-p] [-ignoreCrc] [-crc] <src> ... <localdst>]
 [-count [-q] [-h] [-v] [-t [<storage type>]] [-u] [-x] <path> ...]
 [-cp [-f] [-p | -p[topax]] [-d] <src> ... <dst>]
 [-createSnapshot <snapshotDir> [<snapshotName>]]
 [-deleteSnapshot <snapshotDir> <snapshotName>]
 [-df [-h] [<path> ...]]
 [-du [-s] [-h] [-x] <path> ...]
 [-expunge]
 [-find <path> ... <expression> ...]
 [-get [-f] [-p] [-ignoreCrc] [-crc] <src> ... <localdst>]
 [-getfacl [-R] <path>]
 [-getfattr [-R] {-n name | -d} [-e en] <path>]
 [-getmerge [-nl] [-skip-empty-file] <src> <localdst>]
```

图 4-4 使用 hadoop fs 命令

（2）使用 hadoop fs –ls 命令列出/user/hadoop 目录下的内容。其中/user/hadoop 是 HDFS 下的家目录，命令如下：

```
[hadoop@master ~]$ hadoop fs -ls /user/hadoop
```

当前 HDFS 家目录下包含 3 个文件，如图 4-5 所示。

```
[hadoop@master ~]$ hadoop fs -ls /user/hadoop
Found 3 items
drwxr-xr-x   - hadoop supergroup          0 2018-04-23 15:09 /user/hadoop/QuasiMonteCarlo_1524467367134_1218466016
drwxr-xr-x   - hadoop supergroup          0 2018-04-23 15:41 /user/hadoop/QuasiMonteCarlo_1524469260966_1396029560
drwxr-xr-x   - hadoop supergroup          0 2018-04-22 09:31 /user/hadoop/terasort
```

图 4-5　列出 HDFS 家目录下的文件

（3）使用 hadoop fs –mkdir 在/user/hadoop/下创建文件夹 my_stuff，命令如下：

```
[hadoop@master ~]$ hadoop fs -mkdir /user/hadoop/my_stuff
[hadoop@master ~]$ hadoop fs -ls /user/hadoop
```

创建后可以看到当前路径下多出创建的文件夹 my_stuff，如图 4-6 所示。

```
[hadoop@master ~]$ hadoop fs -mkdir /user/hadoop/my_stuff
[hadoop@master ~]$ hadoop fs -ls /user/hadoop
Found 4 items
drwxr-xr-x   - hadoop supergroup          0 2018-04-23 15:09 /user/hadoop/QuasiMonteCarlo_1524467367134_1218466016
drwxr-xr-x   - hadoop supergroup          0 2018-04-23 15:41 /user/hadoop/QuasiMonteCarlo_1524469260966_1396029560
drwxr-xr-x   - hadoop supergroup          0 2018-04-23 16:30 /user/hadoop/my_stuff
drwxr-xr-x   - hadoop supergroup          0 2018-04-22 09:31 /user/hadoop/terasort
```

图 4-6　创建 my_stuff 文件夹

（4）使用 hadoop fs –copyFromLocal 将文件夹从本地复制到 HDFS 中。先在 Linux 家目录下创建测试文件夹 test，再将其复制到 HDFS 中，命令如下：

```
[hadoop@master ~]$ mkdir test
[hadoop@master ~]$ hadoop fs -copyFromLocal test /user/hadoop/
```

从图 4-7 中可以看出，已经从本地将 test 文件夹复制到 HDFS 中。

```
[hadoop@master ~]$ mkdir test
[hadoop@master ~]$ ls
data  hadoop  hadoop-2.9.0  java  software  test
[hadoop@master ~]$ hadoop fs -copyFromLocal test /user/hadoop/
[hadoop@master ~]$ hadoop fs -ls /user/hadoop/
Found 6 items
drwxr-xr-x   - hadoop supergroup          0 2018-04-23 15:09 /user/hadoop/QuasiMonteCarlo_1524467367134_1218466016
drwxr-xr-x   - hadoop supergroup          0 2018-04-23 15:41 /user/hadoop/QuasiMonteCarlo_1524469260966_1396029560
drwxr-xr-x   - hadoop supergroup          0 2018-04-23 17:11 /user/hadoop/QuasiMonteCarlo_1524474687401_842283761
drwxr-xr-x   - hadoop supergroup          0 2018-04-23 16:30 /user/hadoop/my_stuff
drwxr-xr-x   - hadoop supergroup          0 2018-04-22 09:31 /user/hadoop/terasort
drwxr-xr-x   - hadoop supergroup          0 2018-04-23 17:11 /user/hadoop/test
```

图 4-7　将文件夹从本地复制到 HDFS 中

（5）使用 hadoop fs -copyToLocal 将文件夹从 HDFS 复制到本地，命令如下：

```
[hadoop@master ~]$ hadoop fs -copyToLocal /user/hadoop/my_stuff /home/hadoop
```

从图 4-8 可以看出，本地环境多了一个 my_stuff 文件夹，这是从 HDFS 复制过来的。

```
[hadoop@master ~]$ hadoop fs -copyToLocal /user/hadoop/my_stuff /home/hadoop
[hadoop@master ~]$ ls
data  hadoop  hadoop-2.9.0  java  my_stuff  software  test
```

图 4-8　从 HDFS 复制文件夹到本地

（6）使用 hadoop fs –put 命令将文件夹从本地上传到 HDFS 的目录下，类似于命令 hadoop fs –copyFromLocal，命令如下：

```
[hadoop@master ~]$ mkdir test_put
[hadoop@master ~]$ hadoop fs -put /home/hadoop/test_put /user/hadoop/
[hadoop@master ~]$ hadoop fs -ls /user/hadoop/
```

上传后 HDFS 的目录下多了 test_put 文件夹，如图 4-9 所示。

```
[hadoop@master ~]$ mkdir test_put
[hadoop@master ~]$ hadoop fs -put /home/hadoop/test_put /user/hadoop/
[hadoop@master ~]$ hadoop fs -ls /user/hadoop/
Found 6 items
drwxr-xr-x   - hadoop supergroup          0 2018-04-23 15:09 /user/hadoop/QuasiMonteCarlo
drwxr-xr-x   - hadoop supergroup          0 2018-04-23 15:41 /user/hadoop/QuasiMonteCarlo
drwxr-xr-x   - hadoop supergroup          0 2018-04-23 16:30 /user/hadoop/my_stuff
drwxr-xr-x   - hadoop supergroup          0 2018-04-22 09:31 /user/hadoop/terasort
drwxr-xr-x   - hadoop supergroup          0 2018-04-23 17:11 /user/hadoop/test
drwxr-xr-x   - hadoop supergroup          0 2018-04-23 17:40 /user/hadoop/test_put
```

图 4-9　上传文件夹

（7）使用 hadoop fs –get 命令将文件夹从 HDFS 下载到本地，类似于命令 hadoop fs –copyToLocal，命令如下：

```
[hadoop@master ~]$ hadoop fs -get /user/hadoop/terasort /home/hadoop/
```

下载后，本地环境多了 terasort 文件夹，如图 4-10 所示。

```
[hadoop@master ~]$ hadoop fs -get /user/hadoop/terasort /home/hadoop/
[hadoop@master ~]$ ls
data  hadoop  hadoop-2.9.0  java  my_stuff  software  terasort  test  test_put
```

图 4-10　下载文件夹

（8）使用 hadoop fs –chmod 命令修改文件夹权限，命令如下：

```
[hadoop@master ~]$ hadoop fs -chmod 777 /user/hadoop/my_stuff
[hadoop@master ~]$ hadoop fs -ls /user/hadoop
```

修改后，my_stuff 文件夹权限由默认的 755 改为 777，如图 4-11 所示。

```
[hadoop@master ~]$ hadoop fs -chmod 777 /user/hadoop/my_stuff
[hadoop@master ~]$ hadoop fs -ls /user/hadoop
Found 6 items
drwxr-xr-x   - hadoop supergroup          0 2018-04-23 15:09 /user/hadoop/QuasiMont
drwxr-xr-x   - hadoop supergroup          0 2018-04-23 15:41 /user/hadoop/QuasiMont
drwxrwxrwx   - hadoop supergroup          0 2018-04-23 16:30 /user/hadoop/my_stuff
drwxr-xr-x   - hadoop supergroup          0 2018-04-22 09:31 /user/hadoop/terasort
drwxr-xr-x   - hadoop supergroup          0 2018-04-23 17:11 /user/hadoop/test
drwxr-xr-x   - hadoop supergroup          0 2018-04-23 17:40 /user/hadoop/test_put
```

图 4-11　修改文件夹权限

（9）使用 hadoop fs –rm 将 my_stuff 文件夹移动到回收站。当删除文件夹时，需要加参数-r；

如果需要直接删除，需要增加参数-skipTrash，命令如下：

```
[hadoop@master ~]$ hadoop fs -rm -r /user/hadoop/my_stuff
```

删除后，my_stuff 文件夹已经移动到回收站，HDFS 家目录中无法找到该文件夹，如图 4-12 所示。

```
[hadoop@master ~]$ hadoop fs -rm -r /user/hadoop/my_stuff
Deleted /user/hadoop/my_stuff
[hadoop@master ~]$ hadoop fs -ls /user/hadoop
Found 5 items
drwxr-xr-x   - hadoop supergroup          0 2018-04-23 15:09 /user/hadoop/QuasiMont
drwxr-xr-x   - hadoop supergroup          0 2018-04-23 15:41 /user/hadoop/QuasiMont
drwxr-xr-x   - hadoop supergroup          0 2018-04-22 09:31 /user/hadoop/terasort
drwxr-xr-x   - hadoop supergroup          0 2018-04-23 17:11 /user/hadoop/test
drwxr-xr-x   - hadoop supergroup          0 2018-04-23 17:40 /user/hadoop/test_put
```

图 4-12 删除文件夹

（10）使用 hadoop fs –help 查看命令的使用方法。在使用 HDFS 的过程中遇到不确定的命令时，要灵活利用 hadoop fs-help 命令查看使用方法，命令如下：

```
[hadoop@master ~]$ hadoop fs -help rm
```

使用 hadoop fs-help 命令能够查看命令的具体使用方法，图 4-13 显示了 rm 命令的使用方法。

```
[hadoop@master ~]$ hadoop fs -help rm
-rm [-f] [-r|-R] [-skipTrash] [-safely] <src> ... :
  Delete all files that match the specified file pattern. Equivalent to the Unix
  command "rm <src>"

  -f          If the file does not exist, do not display a diagnostic message or
              modify the exit status to reflect an error.
  -[rR]       Recursively deletes directories.
  -skipTrash  option bypasses trash, if enabled, and immediately deletes <src>.
  -safely     option requires safety confirmation, if enabled, requires
              confirmation before deleting large directory with more than
              <hadoop.shell.delete.limit.num.files> files. Delay is expected when
              walking over large directory recursively to count the number of
              files to be deleted before the confirmation.
```

图 4-13 rm 命令的使用方法

4.3 HDFS 数据平衡优化

HDFS 集群非常容易出现服务器与服务器之间磁盘利用率不平衡的情况。当数据不平衡时，Map 任务可能会被分配给没有存储数据的服务器，这会消耗网络带宽。当 HDFS 负载不均衡时，需要对 HDFS 进行数据的负载均衡调整，也就是对各节点服务器上数据的存储分布进行调整，即数据重分布。

容易引起磁盘利用率不均衡的情况包括集群内新增、删除节点，或者某个节点服务器内磁盘存储达到饱和。解决的方法是调整数据在各节点的分布，均衡 I/O 性能，防止热点的发生。Hadoop 中包含的 Balancer 程序可以让 HDFS 集群达到平衡状态，下面提供两种数据平衡方法。

（1）使用 hdfs balancer 命令自动平衡，命令如下：

```
[hadoop@master ~]$ hdfs balancer
```

使用 hdfs balancer 之后，HDFS 会按照默认参数进行数据平衡，如图 4-14 所示。

```
[hadoop@master ~]$ hdfs balancer
18/04/24 15:05:54 INFO balancer.Balancer: namenodes  = [hdfs://master:9000]
18/04/24 15:05:54 INFO balancer.Balancer: parameters = Balancer.BalancerPara
iteration = 5, #excluded nodes = 0, #included nodes = 0, #source nodes = 0,
18/04/24 15:05:54 INFO balancer.Balancer: included nodes = []
18/04/24 15:05:54 INFO balancer.Balancer: excluded nodes = []
18/04/24 15:05:54 INFO balancer.Balancer: source nodes = []
```

图 4-14　数据平衡

（2）在平衡时指定阈值，命令如下：

```
[hadoop@master ~]$ hdfs balancer -threshold 5
```

参数 5 表示每个 DataNode 的磁盘使用量必须在集群总磁盘使用量的 5%之内，平衡过程如图 4-15 所示。

```
[hadoop@master ~]$ hdfs balancer -threshold 5
18/04/24 15:25:24 INFO balancer.Balancer: Using a threshold of 5.0
18/04/24 15:25:24 INFO balancer.Balancer: namenodes  = [hdfs://mast
18/04/24 15:25:24 INFO balancer.Balancer: parameters = Balancer.Bal
iteration = 5, #excluded nodes = 0, #included nodes = 0, #source nc
18/04/24 15:25:24 INFO balancer.Balancer: included nodes = []
18/04/24 15:25:24 INFO balancer.Balancer: excluded nodes = []
18/04/24 15:25:24 INFO balancer.Balancer: source nodes = []
```

图 4-15　指定阈值进行数据平衡

4.3.1　编程原则

在开发 Balancer 程序的过程中，Hadoop 的开发人员遵循了以下 4 个原则。

（1）在执行数据重分布的过程中，严格保证数据不出现丢失的情况，不改变数据的副本数，不改变每个 Rack 中的 Block 数量。

（2）系统管理员可以通过一条命令启动或者停止数据重分布程序。

（3）Block 在移动的过程中，不能占用过多网络带宽等资源。

（4）在执行数据重分布程序的过程中，不能影响 NameNode 的正常工作。

4.3.2　平衡逻辑

数据平衡过程是一个迭代的过程。每一次迭代的最终目的是让高负载的服务器能够降低数据负载，所以数据平衡会最大限度上使用网络带宽。图 4-16 显示了数据平衡服务器（Rebalancing

Server）的交互情况。

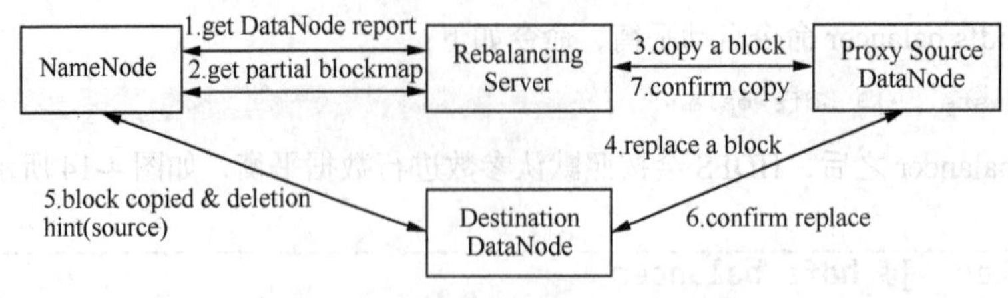

图 4-16　数据平衡服务器的交互情况

步骤分析如下。

（1）数据平衡服务器首先要求 NameNode 生成 DataNode 数据分布分析报告，获取每个 DataNode 磁盘使用情况。

（2）数据平衡服务器汇总需要移动的数据情况，计算具体数据块迁移线路，确保使用的是网络内最短的路径。

（3）开始数据块迁移任务，Proxy Source DataNode 复制一块需要移动的数据块。

（4）将第（3）步复制的数据块复制到目标 DataNode（Destination DataNode）上，进行数据迁移。

（5）数据迁移任务完成后，通过 NameNode 可以删除原始数据块。

（6）目标 DataNode 向 Proxy Source DataNode 确认该数据块迁移完成。

（7）Proxy Source DataNode 向数据平衡服务器确认本次数据块迁移完成，并继续执行上述步骤，直到集群达到数据平衡状态。

Hadoop 现有的这种 Balancer 程序工作方式适合绝大多数情况。

4.3.3　数据平衡案例

4.3.1 小节和 4.3.2 小节分别介绍了数据平衡的编程原则和平衡逻辑，本小节使用一个小案例进行讲解如何实现数据平衡。模拟如下场景。

HDFS 集群数据副本数为 3，集群中 Rack 数量为 2，并且这两个 Rack 中的磁盘型号不同。其中，第一个 Rack 中每台服务器的磁盘空间大小为 1TB，第二个 Rack 中每台服务器的磁盘空间大小为 10TB，第一个 Rack 存储着当前集群大部分备份数据。

在上述场景中，HDFS 集群中的数据明显不平衡。为解决数据不平衡问题，尝试运行 Balancer 程序。但是程序运行结束之后，整个 HDFS 集群中的数据依旧不平衡，可以发现第一个 Rack 中的磁盘剩余空间远远小于第二个 Rack。出现这种现象的原因在于 Balancer 程序的编程原则（1），在执行 Balancer 程序的时候，不会将数据从一个 Rack 迁移到另一个 Rack 中，所以导致了 Balancer 程序永远无法平衡 HDFS 集群的情况。

为解决上述问题，可采用以下两种方案。

方案一：对 Balancer 程序进行修改，允许改变每一个 Rack 中 Block 的数量，减少磁盘空间告急的 Rack 中的 Block 数量。

方案二：不修改 Balancer 程序，但是修改 Rack 中的服务器分布，将磁盘空间小的服务器分到不同的 Rack 中。

4.4　HDFS API 的使用方法

4.2 节和 4.3 节介绍了 HDFS 的基本命令和数据平衡优化，在本节中，将介绍 HDFS API 的使用方法。

Hadoop 中文件操作类 API 大部分可以在“org.apache.hadoop.fs”包中找到，这些 API 能够支持打开文件、读/写文件、修改文件、删除文件等基本操作。

Hadoop 类库最终面向用户提供 FileSystem 接口类，该类是抽象类，因此想要获取具体类只能通过该类的 get( )方法实现。get( )方法常用重载版本如下：

```
static FileSystem get(Configuration conf);
```

其获取的具体类封装了对文件的基本操作，对文件进行操作的程序框架大致如下：

```
operator(){
    获取 Configuration 对象
    获取 FileSystem 对象
    对文件进行相应操作
}
```

下面列出常见的 API 操作，包括上传本地文件、创建 HDFS 文件、创建 HDFS 目录、重命名 HDFS 文件、删除 HDFS 文件和删除 HDFS 目录等。

1. 上传本地文件

将本地文件上传到 HDFS 的指定位置上，具体实现如下：

```
1. package com.hdfs;
2. import org.apache.hadoop.conf.Configuration;
3. import org.apache.hadoop.fs.FileStatus;
4. import org.apache.hadoop.fs.FileSystem;
5. import org.apache.hadoop.fs.Path;
6. public class CopyFile {
7.     public static void main(String[] args) throws Exception {
8.          Configuration conf=new Configuration();
9.          FileSystemhdfs=FileSystem.get(conf);
10.         //本地文件
11.         Path src =new Path("/home/hadoop/CopyFile.txt ");
12.         //HDFS 的指定位置
13.         Path dst =new Path("/");
14.         hdfs.copyFromLocalFile(src, dst);
15.         System.out.println("Upload to"+conf.get("fs.default.name"));
16.         FileStatus files[]=hdfs.listStatus(dst);
17.         for(FileStatus file:files){
18.         System.out.println(file.getPath());
19.                     }
20.             }
21. }
```

编译上述 Java 代码，导出为 CopyFile.jar 文件，上传到集群中并在文件所在位置执行命令，命令如下：

```
[hadoop@master lib]$ hadoop jar CopyFile.jar com.hdfs.CopyFile
```

命令执行后可以看到在 hdfs 根目录下生成了 CopyFile.txt 文件，如图 4-17 所示。

```
[hadoop@master lib]$ hadoop jar CopyFile.jar com.hdfs.CopyFile
18/08/30 10:50:58 INFO Configuration.deprecation: fs.default.name is deprecated. Instea
Upload tohdfs://master:9000
hdfs://master:9000/CopyFile.txt
hdfs://master:9000/URLcat
hdfs://master:9000/URLcat.txt
[hadoop@master lib]$ hadoop fs -ls /
Found 3 items
-rw-r--r--   2 hadoop supergroup          0 2018-08-30 10:50 /CopyFile.txt
drwxr-xr-x   - hadoop supergroup          0 2018-08-29 11:40 /URLcat
-rw-r--r--   2 hadoop supergroup         21 2018-08-29 11:41 /URLcat.txt
```

图 4-17　上传本地文件

2. 创建 HDFS 文件

使用如下代码可在 HDFS 上创建文件，其中 f 为文件的完整路径。

```
Public FSDataOutputStream create(Path f)
```

具体实现如下：

```
1. package com.hdfs;
2. import org.apache.hadoop.conf.Configuration;
3. import org.apache.hadoop.fs.FSDataOutputStream;
4. import org.apache.hadoop.fs.FileSystem;
5. import org.apache.hadoop.fs.Path;
6. public class CreateFile {
7.    public static void main(String[] args) throws Exception {
8.        Configuration conf=new Configuration();
9.        FileSystemhdfs=FileSystem.get(conf);
10.        byte[] buff="hello hadoop world!\n".getBytes();
11.        Path dfs=new Path("/test");
12.        FSDataOutputStream outputStream=hdfs.create(dfs); //调用create()方法，创建文件
13.        outputStream.write(buff,0,buff.length);
14.    }
15. }
```

编译上述 Java 代码，导出为 CreateFile.jar 文件，上传到集群中并在文件所在位置执行命令，命令如下：

```
[hadoop@master lib]$ hadoop jar CreateFile.jar com.hdfs.CreateFile
```

命令执行后可以看到在 hdfs 根目录下创建了 test 文件，如图 4-18 所示。

```
[hadoop@master lib]$ hadoop jar CreateFile.jar com.hdfs.CreateFile
[hadoop@master lib]$ hadoop fs -ls /
Found 3 items
-rw-r--r--   2 hadoop supergroup          0 2018-08-30 10:50 /CopyFile.txt
-rw-r--r--   2 hadoop supergroup         21 2018-08-30 11:09 /URLcat1
-rw-r--r--   2 hadoop supergroup         20 2018-08-30 11:14 /test
```

图 4-18　创建 HDFS 文件

3. 创建 HDFS 目录

使用如下代码可在 HDFS 上创建目录，其中 f 为文件夹的完整路径。

```
public boolean mkdirs(Path f)
```

具体实现如下：

```
1. package com.hdfs;
2. import org.apache.hadoop.conf.Configuration;
3. import org.apache.hadoop.fs.FileSystem;
4. import org.apache.hadoop.fs.Path;
5. public class CreateDir {
6.     public static void main(String[] args) throws Exception{
7.         Configuration conf=new Configuration();
8.         FileSystemhdfs=FileSystem.get(conf);
9.         Path dfs=new Path("/TestDir");
10.        //调用mkdirs()方法，创建目录
11.        hdfs.mkdirs(dfs);
12.    }
13. }
```

编译上述 Java 代码，导出为 CreateDir.jar 文件，上传到集群中并在文件所在位置执行命令，命令如下：

```
[hadoop@master lib]$ hadoop jar CreateDir.jar com.hdfs.CreateDir
```

命令执行后可以看到在 hdfs 根目录下创建了 TestDir 目录，如图 4-19 所示。

```
[hadoop@master lib]$ hadoop jar CreateDir.jar com.hdfs.CreateDir
[hadoop@master lib]$ hadoop fs -ls /
Found 4 items
-rw-r--r--   2 hadoop supergroup          0 2018-08-30 10:50 /CopyFile.txt
drwxr-xr-x   - hadoop supergroup          0 2018-08-30 11:24 /TestDir
-rw-r--r--   2 hadoop supergroup         21 2018-08-30 11:09 /URLcat1
-rw-r--r--   2 hadoop supergroup         20 2018-08-30 11:14 /test
```

图 4-19　创建 HDFS 目录

4. 重命名 HDFS 文件

使用如下代码可为指定的 HDFS 文件重命名，其中 src 和 dst 均为文件的完整路径。

```
public abstract boolean rename(Path src, Path dst)
```

具体实现如下：

```
1. package com.hdfs;
2. import org.apache.hadoop.conf.Configuration;
3. import org.apache.hadoop.fs.FileSystem;
4. import org.apache.hadoop.fs.Path;
5. public class Rename{
6.     public static void main(String[] args) throws Exception {
7.         Configuration conf=new Configuration();
8.         FileSystemhdfs=FileSystem.get(conf);
9.         Path frpaht=new Path("/test");     //旧的文件名
10.        Path topath=new Path("/test1");    //新的文件名
11.        //调用rename()方法，修改文件名
```

```
12.         boolean flag=hdfs.rename(frpaht, topath);
13.         System.out.println("文件重命名结果为: "+flag);
14.
15.     }
16. }
```

编译上述 Java 代码，导出为 Rename.jar 文件，上传到集群中并在文件所在位置执行命令，命令如下：

```
[hadoop@master lib]$ hadoop jar Rename.jar com.hdfs.Rename
```

命令执行后可以看到 hdfs 根目录下的 test 文件重命名为 test1，如图 4-20 所示。

```
[hadoop@master lib]$ hadoop jar Rename.jar com.hdfs.Rename
文件重命名结果为: true
[hadoop@master lib]$ hadoop fs -ls /
Found 4 items
-rw-r--r--   2 hadoop supergroup          0 2018-08-30 10:50 /CopyFile.
txt
drwxr-xr-x   - hadoop supergroup          0 2018-08-30 11:24 /TestDir
-rw-r--r--   2 hadoop supergroup         21 2018-08-30 11:09 /URLcat1
-rw-r--r--   2 hadoop supergroup         20 2018-08-30 11:14 /test1
```

图 4-20　重命名 HDFS 文件

5. 删除 HDFS 文件

使用如下代码可删除指定的 HDFS 文件，其中 f 为需要删除文件的完整路径，recursive 表示确定是否进行递归删除。

```
public abstract boolean delete(Path f,boolean recursive)
```

具体实现如下：

```
1. package com.hdfs;
2. import org.apache.hadoop.conf.Configuration;
3. import org.apache.hadoop.fs.FileSystem;
4. import org.apache.hadoop.fs.Path;
5. public class DeleteFile {
6.    public static void main(String[] args) throws Exception {
7.       Configuration conf=new Configuration();
8.       FileSystemhdfs=FileSystem.get(conf);
9.       Path delef=new Path("/test1");
10.      boolean isDeleted=hdfs.delete(delef,false);
11.      //递归删除
12.      boolean isDeleted=hdfs.delete(delef,true);
13.      System.out.println("Delete?"+isDeleted);
14.    }
15. }
```

编译上述 Java 代码，导出为 DeleteFile.jar 文件，上传到集群中并在文件所在位置执行命令，命令如下：

```
[hadoop@master lib]$ hadoop jar DeleteFile.jar com.hdfs.DeleteFile
```

命令执行后可以看到 hdfs 根目录下的 test1 文件被删除了，如图 4-21 所示。

```
[hadoop@master lib]$ hadoop jar DeleteFile.jar com.hdfs.DeleteFile
Delete?true
[hadoop@master lib]$ hadoop fs -ls /
Found 3 items
-rw-r--r--   2 hadoop supergroup          0 2018-08-30 10:50 /CopyFile.txt
drwxr-xr-x   - hadoop supergroup          0 2018-08-30 11:24 /TestDir
-rw-r--r--   2 hadoop supergroup         21 2018-08-30 11:09 /URLcat1
```

图 4-21　删除 HDFS 文件

6. 删除 HDFS 目录

如下代码同删除文件 HDFS 代码一样，只是将待删除文件路径换成待删除目录路径。如果目录下有文件，要进行递归删除。

```
public abstract boolean delete(Path f,boolean recursive)
```

7. 查看 HDFS 文件是否存在

通过如下代码可查看 HDFS 文件是否存在，其中 f 为文件的完整路径。

```
public boolean exists(Path f)
```

具体实现如下：

```
1. package com.hdfs;
2. import org.apache.hadoop.conf.Configuration;
3. import org.apache.hadoop.fs.FileSystem;
4. import org.apache.hadoop.fs.Path;
5. public class CheckFile {
6.    public static void main(String[] args) throws Exception {
7.        Configuration conf=new Configuration();
8.        FileSystemhdfs=FileSystem.get(conf);
9.        Path findf=new Path("/test1");
10.       //调用 exists()方法，查看文件是否存在
11.       boolean isExists=hdfs.exists(findf);
12.       System.out.println("Exist?"+isExists);
13.    }
14. }
```

编译上述 Java 代码，导出为 CheckFile.jar 文件，上传到集群中并在文件所在位置执行命令，命令如下：

```
[hadoop@master lib]$ hadoop jar CheckFile.jar com.hdfs.CheckFile
```

命令执行后可以查看 hdfs 根目录下是否存在 test1 文件。由于前面删除了 test1，因此显示 false，如图 4-22 所示。

```
[hadoop@master lib]$ hadoop jar CheckFile.jar com.hdfs.CheckFile
Exist?false
[hadoop@master lib]$ hadoop fs -ls /
Found 3 items
-rw-r--r--   2 hadoop supergroup          0 2018-08-30 10:50 /CopyFile.txt
drwxr-xr-x   - hadoop supergroup          0 2018-08-30 11:24 /TestDir
-rw-r--r--   2 hadoop supergroup         21 2018-08-30 11:09 /URLcat1
```

图 4-22　查看 HDFS 文件是否存在

8. 查看 HDFS 文件的状态

通过如下代码可查看 HDFS 文件的各种状态，如修改时间。

```
public abstract FileStatus getFileStatus(Path f)
```

具体实现如下：

```
1. package com.hdfs;
2. import org.apache.hadoop.conf.Configuration;
3. import org.apache.hadoop.fs.FileStatus;
4. import org.apache.hadoop.fs.FileSystem;
5. import org.apache.hadoop.fs.Path;
6. public class GetLTime {
7.    public static void main(String[] args) throws Exception {
8.        Configuration conf=new Configuration();
9.        FileSystemhdfs=FileSystem.get(conf);
10.       //指定查看的文件为 CopyFile.txt
11.       Path fpath =new Path("/CopyFile.txt");
12.       FileStatusfileStatus=hdfs.getFileStatus(fpath);
13.       //调用 getModificationTime()方法，查看修改时间
14.       long modiTime=fileStatus.getModificationTime();
15.       System.out.println("file1.txt 的修改时间是"+modiTime);
16.   }
17. }
```

编译上述 Java 代码，导出为 GetLTime.jar 文件，上传到集群中并在文件所在位置执行命令，命令如下：

```
[hadoop@master lib]$ hadoop jar GetLTime.jar com.hdfs.GetLTime
```

命令执行后可以查看 CopyFile.txt 文件的修改时间，如图 4-23 所示。

```
[hadoop@master lib]$ hadoop jar GetLTime.jar com.hdfs.GetLTime
CopyFile.txt的修改时间是1535597458232
[1]+  Done                    cd ..  (wd: ~/GetLTime)
(wd now: ~/GetLTime/build/lib)
```

图 4-23　查看 HDFS 文件的状态

9. 读取 HDFS 目录下的所有文件

通过如下代码可读取 HDFS 目录下的所有文件。

```
public Path getPath()
```

具体实现如下：

```
1. package com.hdfs;
2. import org.apache.hadoop.conf.Configuration;
3. import org.apache.hadoop.fs.FileStatus;
4. import org.apache.hadoop.fs.FileSystem;
5. import org.apache.hadoop.fs.Path;
6. public class ListAllFile {
7.    public static void main(String[] args) throws Exception {
8.        Configuration conf=new Configuration();
9.        FileSystemhdfs=FileSystem.get(conf);
10.       //指定读取目录为/File
11.       Path listf =new Path("/File");
12.       FileStatus stats[]=hdfs.listStatus(listf);
13.       //依次读取/File 目录中的文件
```

```
14.        for(int i = 0; i < stats.length; ++i){
15.            System.out.println(stats[i].getPath().toString());
16.        }
17.        hdfs.close();  //关闭文件系统
18.    }
19. }
```

编译上述Java代码，导出为ListAllFile.jar文件，上传到集群中并在文件所在位置执行命令，命令如下：

```
[hadoop@master lib]$ hadoop jar ListAllFile.jar com.hdfs.ListAllFile
```

命令执行后可以查看File目录下的所有文件，如图4-24所示。

```
[hadoop@master lib]$ hadoop jar ListAllFile.jar com.hdfs.ListAllFile
hdfs://master:9000/File/data1
hdfs://master:9000/File/data2
hdfs://master:9000/File/data3
```

图4-24 读取HDFS目录下的所有文件

习 题

1. 下面哪个程序负责HDFS数据存储？（　　）
   A. NameNode　　B. DataNode　　C. SecondaryNameNode
2. HDFS中的Block默认保存几份？（　　）
   A. 3份　　B. 2份　　C. 1份　　D. 不确定
3. 下列哪个程序通常与NameNode在同一个节点启动？（　　）
   A. SecondaryNameNode　　B. DataNode　　C. TaskTracker　　D. JobTracker
4. HDFS默认的Block大小是多少？（　　）
   A. 32MB　　B. 64MB　　C. 128MB
5. 关于SecondaryNameNode哪项是正确的？（　　）
   A. 它是NameNode的热备　　B. 它对内存没有要求
   C. 它的目的是帮助NameNode合并编辑日志，减少NameNode启动时间
   D. SecondaryNameNode应与NameNode部署到同一个节点
6. 配置机架感知时，下面哪项是正确的？（　　）
   A. 如果一个机架出现问题，数据读/写就会受到影响
   B. 写入数据的时候优先填满一个DataNode
   C. MapReduce会根据机架获取离自己比较近的网络数据
7. 客户端上传文件的时候，下列哪项是正确的？（　　）
   A. 数据经过NameNode传递给DataNode
   B. 客户端将文件以Block为单位划分，再以管道方式依次传到DataNode
   C. 客户端只上传数据到一个DataNode，然后由NameNode负责Block复制工作
   D. 当某个DataNode失败时，客户端将停止文件的上传

第 5 章 MapReduce 分布式编程

第 4 章介绍了 HDFS 的体系结构以及使用方法，这解决了大数据资产存储问题。从存储的大数据中快速抽取信息，进一步挖掘数据的价值，则需要大数据的分布式计算技术进行支持。本章重点介绍 Hadoop 的 MapReduce 分布式编程操作。

5.1 MapReduce 简介

MapReduce 是一个分布式计算的编程框架，用于大规模数据集的并行处理。MapReduce 将一个数据处理过程拆分为 Map 和 Reduce 两部分：Map 表示映射，负责数据的过滤、分发；Reduce 表示规约，负责数据的计算、归并。开发人员只需编写 map( )方法和 reduce( )方法，不需要考虑分布式计算框架内部的运行机制，即可在 Hadoop 集群上实现分布式运算。MapReduce 框架引入后，工程人员可将精力集中在业务逻辑的开发上，复杂的分布式计算交由框架来处理。

MapReduce 把对数据集的大规模操作分发到计算节点，计算节点会周期性地返回其工作的最新状态和结果。如果计算节点沉默时间超过预设时间，主节点则默认该节点为宕机状态，把已分配给该节点的数据分发到其他节点重新计算，从而实现数据处理任务的自动调度。

Hadoop 支持多种语言进行 MapReduce 分布式编程，包括 Java、Ruby、Python 和 C++等。Hadoop 原生支持 Java，因此本书采用 Java 语言介绍 MapReduce 分布式编程。在 Hadoop 平台上运行 MapReduce 程序，主要任务是将 HDFS 存储的大文件数据分发到多个计算节点上的 Map 程序进行处理，然后由计算节点上的 Reduce 程序合并或进一步处理多个计算节点上的结果。从程序员的角度看，采用 Java 语言进行的 MapRedcue 分布式编程流程如图 5-1 所示。

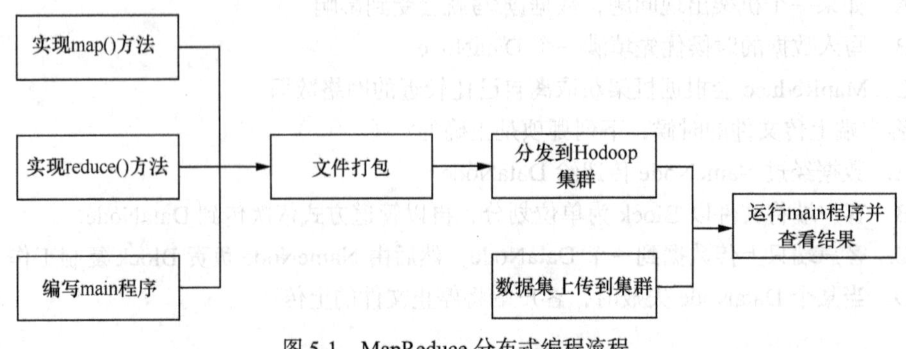

图 5-1　MapReduce 分布式编程流程

MapReduce 分布式编程流程如下：

（1）编写 Hadoop 中 org.apache.hadoop.mapreduce.Mapper 类的子类，并实现 map( )方法；

（2）编写 Hadoop 中 org.apache.hadoop.mapreduce.Reducer 类的子类，并实现 reduce( )方法；

（3）编写 main 程序，设置 MapReduce 程序的配置，并指定任务的 Map 程序类（第（1）步的 Java 类）、Reduce 程序类（第（2）步的 Java 类）等，指定输入文件和输出文件及格式，提交任务等；

（4）将前面步骤的类文件与 Hadoop 自带的包打包为.jar 文件，分发到 Hadoop 集群的任意节点；

（5）运行 main 程序，任务自动在 Hadoop 集群上运行；

（6）到指定文件夹查看计算结果。

Map 程序和 Reduce 程序的输入和输出都是以键值对的形式出现的，map( )方法的输出和 reduce( )方法的输入的键值对格式必须一致，MapReduce 的调度程序完成 Map 程序和 Reduce 程序间的数据传递。

5.2　词频统计编程实例

通过前文的学习，我们已经了解了分布式编程的整个流程，接下来，我们将在 Hadoop 集群运行一个 WordCount 程序来统计输入文件中单词出现的次数（词频）。WordCount 程序包含 3 个部分，分别为：

（1）编译一个 WordCount.java 程序，该程序将在 Hadoop 集群中运行；

（2）把在 HDFS 中处理的文件进行上传；

（3）运行 WordCount.java 程序并查看结果。

具体实践步骤如下。

（1）下载本实例源码 ch05/WordCount/src/WordCount.java，将源码以及数据集 YoutubeDataSets.txt 上传到集群的/home/hadoop/data/wordcount 目录，可通过 Linux 命令 scp 实现，或者借助 FileZilla 工具实现。

（2）打开命令行界面，切换到 wordcount 目录，命令如下：

```
[hadoop@master~]$ cd /home/hadoop/data/wordcount
```

wordcount 目录如图 5-2 所示。

```
[hadoop@master wordcount]$ ll
总用量 952
-rw-r--r-- 1 hadoop hadoop   3099 4月   24 09:01 WordCount.java
-rw-r--r-- 1 hadoop hadoop 969389 4月   24 09:02 YoutubeDataSets.txt
[hadoop@master wordcount]$
```

图 5-2　wordcount 目录

（3）查看在 Hadoop 集群上运行 MapReduce 的程序 WordCount.java，命令如下：

```
[hadoop@master wordcount]$gedit WordCount.java
```

WordCount.java 代码如下：

```
1.public class WordCount {
2.public static class MyMapper extends Mapper<Object, Text, Text, IntWritable> {
```

```
3.      private final static IntWritable one = new IntWritable(1);
4.      private Text word = new Text();
5.      public void map(Object key, Text value, Context context) throws IOException,
 InterruptedException {
6.          StringTokenizer itr = new
 StringTokenizer(value.toString(),"\t");//以制表符分隔一行文本
7.          while (itr.hasMoreTokens()) {
8.              word.set(itr.nextToken());
9.              context.write(word, one);
10.         }
11.     }
12. }
13.
14. public static class MyReducer extends Reducer<Text, IntWritable, Text,
  IntWritable> {
15.     public void reduce(Text key, Iterable<IntWritable> values, Context context)
16.             throws IOException, InterruptedException {
17.         int sum = 0;
18.         for (IntWritable value : values) {//相同的 key 累加计数
19.             sum += value.get();
20.         }
21.         context.write(key, new IntWritable(sum));
22.     }
23. }
24. public static void main(String[] args) throws Exception {
25.     Configuration conf = new Configuration();
26.     String[] otherArgs = new GenericOptionsParser(conf,
  args).getRemainingArgs();
27.     Job job = new Job(conf, "WordCount");
28.     job.setJarByClass(WordCount.class);
29.     job.setMapperClass(MyMapper.class);//指定 Map 任务类
30.     job.setReducerClass(MyReducer.class);//指定 Reduce 任务类
31.     job.setOutputKeyClass(Text.class);
32.     job.setOutputValueClass(IntWritable.class);
33.     FileInputFormat.addInputPath(job, new Path(otherArgs[0]));//指定文件输入路径
34.     FileOutputFormat.setOutputPath(job, new Path(otherArgs[1]));//指定文件输出路径
35.     System.exit(job.waitForCompletion(true) ? 0 : 1);
36. }
37.
38.
39.}
40.
```

WordCount.java 程序中第 2～13 行实现了 Mapper 接口，第 14～23 行实现了 Reducer 接口，第 24～36 行实现了 main( )方法。

第 2 行定义了 MyMapper 类继承泛型类 Mapper，类 Mapper<Object, Text, Text, IntWritable>指定的前两个泛型参数对应 map( )方法接收的输入 key 和 value 的类型，后两个泛型参数对应 map( )方法输出 key、value 的类型，也是后续 reduce( )函数要求的输入 key、value 的类型。

第 3～4 行声明用于存放单词的变量 one、word。

第 5 行定义 map( )方法，其中每一行的起始偏移量作为 key，每一行的文本内容作为 value，注意此处 map( )方法的前两个参数对应第 2 行泛型类的前两个参数。

第 6 行生成一个 StringTokenizer 对象。StringTokenizer 是 Java 中 Object 类的一个子类，实现了 Enumeration 接口。StringTokenizer 对象根据分隔符把字符串分割成标记（Token），然后按照请求返回各个标记。

第 7～10 行使用 StringTokenizer 类的 nextToken( )方法对字符串进行分割，通过 write( )方法将单词写入变量 word 中，最后 write( )方法将(word,one)形式的二元组存入 context 中。

第 14 行定义了 MyReducer 类继承泛型类 Reducer，Text、IntWritable 均是 Hadoop 中实现的用于封装 Java 数据类型的类，这些类实现了 WritableComparable 接口。Reducer<Text, IntWritable,Text, IntWritable> 的 4 个泛型参数中前两个参数对应 reduce( )方法输入需要的参数类型(key，values)和 reduce( )方法输出需要的 key、value 类型(key,value)。

第 15 行定义 reduce( )方法，map( )方法输出的一组键值对中的 key，作为 reduce( )方法输入参数中的 key，values 为一组键值对中所有 value 的迭代器。

第 18～20 行通过迭代器，遍历每一组键值对中所有的 value，进行累加计算。

第 21 行 依然将 key 作为键，同一个 key 的累加值作为值，产生新的键值对写到 context 中作为 Hadoop 框架输出使用。

第 25 行生成一个 Configuration 对象，程序运行后会加载 Hadoop 默认的一些配置，是 Job 必不可少的组件。

第 26 行中的 GenericOptionsParser( )用来解析输入参数中的 Hadoop 参数，并做相应处理，比如设置 NameNode 等；后面的 getRemainingArgs( )返回 Hadoop 参数以外的部分，留给具体的应用处理。

第 27～32 行产生 Job 类的实例，并设置相关属性值。Job 任务需要加载 Hadoop 的一些配置，并给这个 Job 命名为 “WordCount” 。第 28 行使用了 WordCount.class 的类加载器来寻找包含该类的.jar 包，然后设置该.jar 包为 job 所用的.jar 包。第 29～32 行设置 job 实例属性，其参数使用 Java 反射机制。

第 33～34 行设置程序输入和输出的路径，本实例是从命令行中接收参数的，第一个参数为输入路径，第二个参数为输出路径。

第 35 行通过 waitForCompletion( )方法提交 Hadoop 任务。

（4）在/home/hadoop/data/wordcount 本地文件系统目录创建一个新的目录 wordcount_classes，命令如下：

```
[hadoop@master wordcount]$mkdir wordcount_classes
```

（5）编译 WordCount.java 程序。输入下面的命令编译 WordCount.java 程序并且设置正确的路径及输出目录。用-d（directory，目录）选项指定编译结果的.class 文件的存放目录：

```
[hadoop@master wordcount]$javac -cp /home/hadoop/hadoop-2.9.0/share/hadoop/*
-d wordcount_classes WordCount.java
```

运行结果如图 5-3 所示。

```
[hadoop@master wordcount_classes]$ ll
总用量 12
-rw-r--r-- 1 hadoop hadoop 1922 4月  24 10:26 WordCount.class
-rw-r--r-- 1 hadoop hadoop 2302 4月  24 10:26 WordCount$MyMapper.class
-rw-r--r-- 1 hadoop hadoop 2340 4月  24 10:26 WordCount$MyReducer.class
[hadoop@master wordcount_classes]$
```

图 5-3 编译后的.class 文件

（6）为编译的 wordcount 目录创建一个.jar 文件。这个.jar 文件是必需的，因为需要将该.jar 文件发送到集群的其他节点上并且同时运行，命令如下：

```
[hadoop@master wordcount]$jar -cvf WordCount.jar -C wordcount_classes/ .
```

运行结果如图 5-4 所示。

```
[hadoop@master wordcount]$ jar -cvf WordCount.java -C wordcount_classes/ .
已添加清单
正在添加: WordCount$MyMapper.class(输入 = 2302) (输出 = 930)(压缩了 59%)
正在添加: WordCount$MyReducer.class(输入 = 2340) (输出 = 890)(压缩了 61%)
正在添加: WordCount.class(输入 = 1922) (输出 = 969)(压缩了 49%)
[hadoop@master wordcount]$
```

图 5-4　创建.jar 文件

（7）创建 MapReduce 程序的输入文件/tmp/MR-WordCount，上传 YouTube 数据集当作 WordCount 程序的输入文件，命令如下：

```
[hadoop@master wordcount]$ hadoop fs -mkdir /tmp/MR-WordCount
```

（8）使用 Hadoop 的 put 命令把 YouTube 数据集从本地文件系统的/home/hadoop/data/wordcount 复制到 HDFS 的/tmp/ MR-WordCount 目录，命令如下：

```
[hadoop@master wordcount]$ hadoop fs -put YoutubeDataSets.txt /tmp/MR-WordCount/
```

（9）查看文件是否上传成功，命令如下：

```
[hadoop@master wordcount]$hadoop fs -ls /tmp/MR-WordCount/
```

运行结果如图 5-5 所示。

```
[hadoop@master wordcount]$ hadoop fs -ls /tmp/MR-WordCount/
Found 1 items
-rw-r--r--   2 hadoop supergroup     969389 2018-04-24 10:46 /tmp/MR-WordCount/Y
outubeDataSets.txt
[hadoop@master wordcount]$
```

图 5-5　查看文件

（10）运行 MapReduce 任务，命令如下：

```
[hadoop@master wordcount]$ hadoop jar WordCount.jar cn.hust.book.bigdata.
ch04.WordCount /tmp/MR-WordCount/YoutubeDataSets.txt /tmp/MR-WordCount/output
```

其中，cn.hust.book.bigdata.ch04.WordCount 为程序的主类名称，/tmp/MR-WordCount/output 为输出目录。

Hadoop 集群提供的运行信息显示在命令行上，帮助跟踪任务的执行状态。当任务完成后，命令提示符会出现。图 5-6 所示的高亮的信息显示了 MapReduce 的任务 ID，如果需要，该 ID 可以用于杀死正在运行的任务。图 5-7 显示了任务输出的计数器信息，其中包括数据集大小信息、指定的 Map/Reduce 任务数、Map/Reduce 读取记录数等信息，对了解整个 MapReduce 任务运行机制有很大的帮助。

```
18/04/24 11:03:52 INFO client.RMProxy: Connecting to ResourceManager at master/1
92.168.2.176:8032
18/04/24 11:03:53 INFO input.FileInputFormat: Total input files to process : 1
18/04/24 11:03:53 INFO mapreduce.JobSubmitter: number of splits:1
18/04/24 11:03:53 INFO Configuration.deprecation: yarn.resourcemanager.system-me
trics-publisher.enabled is deprecated. Instead, use yarn.system-metrics-publishe
r.enabled
18/04/24 11:03:53 INFO mapreduce.JobSubmitter: Submitting tokens for job: job_15
24202612343_0027
18/04/24 11:03:53 INFO impl.YarnClientImpl: Submitted application application_15
24202612343_0027
18/04/24 11:03:53 INFO mapreduce.Job: The url to track the job: http://master:80
88/proxy/application_1524202612343_0027/
18/04/24 11:03:53 INFO mapreduce.Job: Running job: job_1524202612343_0027
18/04/24 11:04:00 INFO mapreduce.Job: Job job_1524202612343_0027 running in uber
 mode : false
18/04/24 11:04:00 INFO mapreduce.Job:  map 0% reduce 0%
18/04/24 11:04:07 INFO mapreduce.Job:  map 100% reduce 0%
18/04/24 11:04:13 INFO mapreduce.Job:  map 100% reduce 100%
18/04/24 11:04:13 INFO mapreduce.Job: Job job_1524202612343_0027 completed succe
ssfully
18/04/24 11:04:13 INFO mapreduce.Job: Counters: 49
        File System Counters
```

图 5-6　MapReduce 的任务 ID

```
18/04/24 11:04:13 INFO mapreduce.Job: Counters: 49
        File System Counters
                FILE: Number of bytes read=832799
                FILE: Number of bytes written=2069111
                FILE: Number of read operations=0
                FILE: Number of large read operations=0
                FILE: Number of write operations=0
                HDFS: Number of bytes read=969509
                HDFS: Number of bytes written=642323
                HDFS: Number of read operations=6
                HDFS: Number of large read operations=0
                HDFS: Number of write operations=2
        Job Counters
                Launched map tasks=1
                Launched reduce tasks=1
                Data-local map tasks=1
                Total time spent by all maps in occupied slots (ms)=3610
                Total time spent by all reduces in occupied slots (ms)=4205
                Total time spent by all map tasks (ms)=3610
                Total time spent by all reduce tasks (ms)=4205
                Total vcore-milliseconds taken by all map tasks=3610
                Total vcore-milliseconds taken by all reduce tasks=4205
                Total megabyte-milliseconds taken by all map tasks=3696640
                Total megabyte-milliseconds taken by all reduce tasks=4305920
        Map-Reduce Framework
                Map input records=4100
                Map output records=100258
                Map output bytes=1366193
                Map output materialized bytes=832799
                Input split bytes=120
                Combine input records=100258
                Combine output records=47920
                Reduce input groups=47920
                Reduce shuffle bytes=832799
                Reduce input records=47920
                Reduce output records=47920
                Spilled Records=95840
                Shuffled Maps =1
                Failed Shuffles=0
                Merged Map outputs=1
                GC time elapsed (ms)=184
                CPU time spent (ms)=3140
                Physical memory (bytes) snapshot=499556352
                Virtual memory (bytes) snapshot=4259733504
                Total committed heap usage (bytes)=322437120
```

图 5-7　任务输出的计数器信息

也可以使用浏览器登录 YARN 的 Web 图形用户界面（Graphical User Interface，GUI），更直观地查看本次任务的详细信息，此处不做详细介绍。

（11）进入 HDFS，查看执行结果，命令如下：

```
[hadoop@master wordcount]$ hadoop fs -ls /tmp/MR-WordCount/output
```

任务执行结果如图 5-8 所示。

```
[hadoop@master wordcount]$ hadoop fs -ls /tmp/MR-WordCount/output
Found 2 items
-rw-r--r--   2 hadoop supergroup          0 2018-04-24 11:04 /tmp/MR-WordCount/o
utput/_SUCCESS
-rw-r--r--   2 hadoop supergroup     642323 2018-04-24 11:04 /tmp/MR-WordCount/o
utput/part-r-00000
[hadoop@master wordcount]$
```

图 5-8　任务执行结果

（12）使用 Hadoop 的 cat 命令显示 part-r-00000 结果文件的内容，命令如下：

```
[hadoop@master wordcount]$hadoop fs -cat /tmp/MR-WordCount/output/part-r-00000
```

词频统计结果如图 5-9 所示。

```
zwsH5j7s2io     7
zwsH_a_so-w     1
zwxljUHq1n4     5
zx1XFUaik2Q     1
zx2LzKcKYxE     1
zx5w2ditREo     1
zxK6vkSSARs     1
zxNUWw533pU     1
zxX0DKE3oqw     2
zxppd6t6GAg     1
zxsCAyk5pDY     1
zyAkuFTSnQE     1
zyBAn9gi5Zc     1
zyMCjLBpm6g     1
zyVn9k6d1og     1
zydZq2lVnns     1
zyilSPsYdW8     4
zylHHp9k7_c     1
zylqaZISiNc     1
zyqBRZ73WHw     1
zyyCcjbrWOM     1
zzCUvBWsgCk     5
zzFjdGcQNWk     1
zzIMbjA5m8s     11
zzeKqnU7lDg     1
zzfDGK8VWCk     1
zzlDKaXAIgE     19
zznQtqa6Obs     4
zzpCV0qgxNg     1
zzyrHsYTveE     1
zzzzzzzzap      1
[hadoop@master wordcount]$
```

图 5-9　词频统计结果

（13）再次执行 MapReduce 任务，命令如下：

```
[hadoop@master wordcount]$ hadoop jar WordCount.jar cn.hust.book.bigdata.ch04.
WordCount /tmp/MR-WordCount/YoutubeDataSets.txt /tmp/MR-WordCount/output
```

输出信息如图 5-10 所示。

```
[hadoop@master wordcount]$ hadoop jar WordCount.jar /tmp/MR-WordCount/YoutubeDat
aSets.txt /tmp/MR-WordCount/output
18/04/24 11:42:51 INFO client.RMProxy: Connecting to ResourceManager at master/1
92.168.2.176:8032
Exception in thread "main" org.apache.hadoop.mapred.FileAlreadyExistsException:
Output directory hdfs://master:9000/tmp/MR-WordCount/output already exists
        at org.apache.hadoop.mapreduce.lib.output.FileOutputFormat.checkOutputSp
ecs(FileOutputFormat.java:146)
        at org.apache.hadoop.mapreduce.JobSubmitter.checkSpecs(JobSubmitter.java
:279)
        at org.apache.hadoop.mapreduce.JobSubmitter.submitJobInternal(JobSubmitt
er.java:145)
        at org.apache.hadoop.mapreduce.Job$11.run(Job.java:1570)
        at org.apache.hadoop.mapreduce.Job$11.run(Job.java:1567)
        at java.security.AccessController.doPrivileged(Native Method)
        at javax.security.auth.Subject.doAs(Subject.java:422)
        at org.apache.hadoop.security.UserGroupInformation.doAs(UserGroupInforma
tion.java:1886)
        at org.apache.hadoop.mapreduce.Job.submit(Job.java:1567)
        at org.apache.hadoop.mapreduce.Job.waitForCompletion(Job.java:1588)
        at bigdata.ch04.WordCount.main(WordCount.java:41)
        at sun.reflect.NativeMethodAccessorImpl.invoke0(Native Method)
        at sun.reflect.NativeMethodAccessorImpl.invoke(NativeMethodAccessorImpl.
java:62)
```

图 5-10　执行 MapReduce 任务

这时会出现一个错误，不能在结果目录更新数据，因为 Hadoop 不允许更新数据文件，只有读和写操作是允许的。这样，执行的时候没有地方存放输出结果。要想再次执行 MapReduce 任务，需要指定另外一个输出目录，或者删除现有的输出目录。

（14）删除输出目录，命令如下：

```
[hadoop@master wordcount]$hadoop fs -rm -r /tmp/MR-WordCount/output
```

执行结果如图 5-11 所示。

```
[hadoop@master wordcount]$ hadoop fs -rm -r /tmp/MR-WordCount/output
Deleted /tmp/MR-WordCount/output
[hadoop@master wordcount]$
```

图 5-11　删除输出目录

（15）再次执行 MapReduce 任务，任务成功执行，命令如下：

```
[hadoop@master wordcount]$ hadoop jar WordCount.jar cn.hust.book.bigdata.ch04.
WordCount /tmp/MR-WordCount/YoutubeDataSets.txt /tmp/MR-WordCount/output
```

5.3　MapReduce Shuffle 过程开发

通过前文的学习我们知道，Hadoop 框架使用 Mapper 将数据处理成键值对，然后在网络节点间对其进行整理，通过 Reducer 处理后最终输出。数据整理的中间过程即分拣（Shuffle）过程，主要涉及分区（Partition）、排序（Sort）、合并（Combine）。通过 Shuffle 过程，可以有效地降低在网络间传输的数据量，提高程序的执行效率。

5.3.1 MapReduce 数据类型

MapReduce 运算将完成的 Mapper 的计算结果发送给 Reducer，Reducer 收到任务后进行规约计算。Map 任务和 Reduce 任务多分布在不同的计算节点上，这就要求在网络上传递可序列化的 Java 对象。对象序列化是指把 Java 对象转化成字节序列的过程，反序列化是指把字节序列转化成 Java 对象的过程。Hadoop 重新定义了 Java 中常用的数据类型，如表 5-1 所示，并让它们具有序列化的特点。

表 5-1　Hadoop 定义的 MapReduce 数据类型与 Java 基本类型的对照

Java 数据类型	Hadoop 封装类型	说明
Byte	ByteWritable	单字节数值
Int	IntWritable	整型数值
Long	LongWritable	长整型数值
Float	FloatWritable	浮点型数值
Double	DoubleWritable	双字节数值
Boolean	BooleanWritable	标准布尔型数值
String	Text	UTF-8 格式存储的文本

MapReduce 除了常用的数据类型外，有时针对特定的应用场景，还可通过实现 org.apache.hadoop.io.Writable 接口自定义数据类型。

5.3.2 Partitioner 负载平衡编程

在 Hadoop 的 MapReduce 过程中，Mapper 完成数据处理后，会把数据发送到 Partitioner，由 Partitioner 来决定每条记录应该送往哪个 Reducer 节点。MapReduce 默认的 Partitioner 是 HashPartitioner。

```
1.public int getPartition(K2 key, V2 value,int numReduceTasks) {
2.    return (key.hashCode() & Integer.MAX_VALUE) % numReduceTasks;
3. }
```

getPartition( )的输入参数就是 Mapper 输出的 key 和 value，然后针对这样的输入，Partitioner 先计算 key 的散列值，然后根据 Reducer 个数执行取模运算。

假如，事前已经对数据以及 Mapper 处理后的输出数据都有一个很好的了解，那么可以控制 Mapper 结果送往某个 Reducer 进行处理。这样方便我们采取某种策略，使 Reducer 处理的数据量基本相同，达到一种均衡的效果。这样，数据处理的效率也会有很大的提高。当然，这种策略需要对数据的了解程度比较高。

接下来，我们通过完善词频统计的实例，实现 Partitioner 负载平衡编程。

```
1.public class WordCountWithPartition {
2.   public static class MyMapper extends Mapper<Object, Text, Text, IntWritable> {
3.        private final static IntWritable one = new IntWritable(1);
4.        private Text word = new Text();
5.
6.        public void map(Object key, Text value, Context context) throws IOException,
InterruptedException {
```

```
7.            StringTokenizer itr = new StringTokenizer(value.toString(), "\t");
8.            while (itr.hasMoreTokens()) {
9.                word.set(itr.nextToken());
10.               context.write(word, one);
11.           }
12.       }
13. }
14.
15. public static class MyReducer extends Reducer<Text, IntWritable, Text,
  IntWritable> {
16.       public void reduce(Text key, Iterable<IntWritable> values, Context
17.              context)throws IOException, InterruptedException {
18.           int sum = 0;
19.           for (IntWritable value : values) {
20.               sum += value.get();
21.           }
22.           context.write(key, new IntWritable(sum));
23.       }
24. }
25.
26. public static void main(String[] args) throws Exception {
27.       Configuration conf = new Configuration();
28.       String[] otherArgs = new GenericOptionsParser(conf, args).getRemainingArgs();
29.       Job job = new Job(conf, "WordCount");
30.       job.setJarByClass(WordCountWithPartition.class);
31.       job.setMapperClass(MyMapper.class);//指定 Map 任务类
32.       Job.setReducerClass(MyReducer.class);//指定 Reduce 任务类
33.       job.setOutputKeyClass(Text.class);
34.       job.setOutputValueClass(IntWritable.class);
35.       FileInputFormat.addInputPath(job, new Path(otherArgs[0]));
36.       FileOutputFormat.setOutputPath(job, new Path(otherArgs[1]));
37.       job.setPartitionerClass(MyPartitioner.class);//指定自定义 Partitioner 类
38.       job.setNumReduceTasks(2);//指定 Reduce 任务数
39.       System.exit(job.waitForCompletion(true) ? 0 : 1);
40. }
41.}
42.class MyPartitioner<K1, V1> extends Partitioner<K1, V1>{//自定义 Partitioner 类，重
  写 getPartition()方法
43. @Override
44. public int getPartition(K1 key, V1 value, int numPartitions) {
45.       String tmpValue = value.toString();
46. if(tmpValue!=null&&(tmpValue.indexOf("A")==0||tmpValue.indexOf("a")==0)){
47. return 0;
48.       }
49. return 1;
50. }
51.}
```

在上述代码中，我们并没有修改 Mapper 类和 Reducer 类，只是自定义了 MyPartitioner 类继承 Partitioner 基类，重写了 getPartition( )方法。第 37 行设置自定义 Partitioner。第 44～50 行重写 getPartition( )方法：如果 value 值以 A 开头，就把数据发送到其中一个 Reducer，否则发送到另外一个。

5.3.3 Sort 排序编程

Mapper 的输出结果在写入磁盘前，会进行二次快速排序。首先根据数据所属的 Partitioner 排序，然后每个 Partitioner 按 key 排序。

如果 key 的类型是用户自定义的类型，并没有默认的比较函数时，就需要自己定义 key 的比较函数，也就是继承 WritableComparator。

接下来，我们通过使用词频统计实例的程序，进行 Sort 排序的再编程。

```
1.public class WordCountWithSort {
2.
3.  public static class MyMapper extends Mapper<Object, Text, Text, IntWritable> {
4.       private final static IntWritable one = new IntWritable(1);
5.       private Text word = new Text();
6.       public void map(Object key, Text value, Context context) throws IOException,
InterruptedException {
7.           StringTokenizer itr = new StringTokenizer(value.toString());
8.           while (itr.hasMoreTokens()) {
9.               word.set(itr.nextToken());
10.              context.write(word, one);
11.          }
12.      }
13. }
14. public static class MyReducer extends Reducer<Text, IntWritable, Text,
  IntWritable> {
15.      public void reduce(Text key, Iterable<IntWritable> values, Context context)
16.              throws IOException, InterruptedException {
17.          int sum = 0;
18.          for (IntWritable value : values) {
19.              sum += value.get();
20.          }
21.          context.write(key, new IntWritable(sum));
22.      }
23. }
24.
25. public static void main(String[] args) throws Exception {
26.      Configuration conf = new Configuration();
27.      String[] otherArgs = new GenericOptionsParser(conf,
  args).getRemainingArgs();
28.      Job job = new Job(conf, "WordCount");
29.      job.setJarByClass(WordCountWithSort.class);
30.      job.setMapperClass(MyMapper.class);
31.      job.setReducerClass(MyReducer.class);
32.      job.setOutputKeyClass(Text.class);
33.      job.setOutputValueClass(IntWritable.class);
34.      FileInputFormat.addInputPath(job, new Path(otherArgs[0]));
35.      FileOutputFormat.setOutputPath(job, new Path(otherArgs[1]));
36.      job.setSortComparatorClass(TextComparator.class);//指定自定义 Sort 类
37.      job.setGroupingComparatorClass(TextComparator.class);
38.      System.exit(job.waitForCompletion(true) ? 0 : 1);
39. }
40.}
41.class TextComparator extends WritableComparator{//自定义 Sort 类，重写 compare()方法
```

```
42. public TextComparator(){
43.     super(Text.class, true);
44. }
45. @Override
46. public int compare(WritableComparable a, WritableComparable b) {
47.     Text o1 = (Text) a;
48.     Text o2 = (Text) b;
49.     return -o1.compareTo(o2);
50. }
51.}
52.
```

通过上述代码我们发现，自定义 TextComparator 类继承 WritableComparator 基类，重写了 compare( )方法。第 36、37 行在主函数中设置自定义 Sort 类，第 41～51 行的 compare( )方法实现了 Mapper 的输出数据按照 key 降序排列。

5.3.4 Combiner 减少中间数据编程

每一个 Mapper 都可能产生大量的本地输出，Combiner 的作用是对 Mapper 的输出先做一次合并，以减少在 Mapper 和 Reducer 节点之间的数据传输量，提高网络 I/O 性能。使用 Combiner 减少数据是 MapReduce 的一种优化手段。

通过使用 Combiner 可以使 MapReduce 整体性能得到提升。Combiner 可以从根本上优化和最小化键值对的数量。因为键值对是通过网络在 Mapper 和 Reducer 之间传输的，所以 Combiner 可以降低网络传输负载。

Combiner 在词频统计中的使用十分简单，我们并不需要修改之前的词频统计源码，只需在主方法 main( )中添加如下代码：

```
job.setCombinerClass(MyReducer.class);
```

即通过 job 设置 Combiner 类，其逻辑是可以直接使用 Reducer。

添加 Combiner 类后，我们可以通过 Hadoop 计数器查看合并效果，此处我们运行一项作业，Hadoop 计数器部分输出结果如图 5-12 所示。

```
Map-Reduce Framework
        Map input records=4100
        Map output records=100258
        Map output bytes=1366193
        Map output materialized bytes=832799
        Input split bytes=112
        Combine input records=100258
        Combine output records=47920
        Reduce input groups= =47920
        Reduce shuffle bytes=832799
        Reduce input records=47920
        Reduce output records= 47920
```

图 5-12　Hadoop 计数器部分输出结果

由图 5-12 可知，Map 输出记录数为 100258，Combiner 将 100258 作为输入值，经过合并后输出记录数为 47920，47920 作为 Reduce 的输入值，最终将结果写入 HDFS 中。

5.4 MapReduce 的性能优化

影响 MapReduce 性能的因素很多，根据我们的经验和观察，主要因素有以下几个：硬件（或者资源）因素，如 CPU 时钟、磁盘 I/O、网络带宽和内存大小；底层存储系统；输入数据、Shuffle 数据以及输出数据的大小，这与作业的运行时间紧密相关；作业算法，如分区、合并和压缩算法等。针对以上因素，我们将深入探讨 MapReduce 性能的优化方案。

5.4.1 Hadoop 配置参数调优

mapred-site.xml 是优化 Hadoop MapReduce 性能的关键配置文件，该文件包含了与 CPU、内存、网络和磁盘 I/O 相关的参数。

表 5-2 所示的两个参数与 CPU 利用率关联最密切，通过修改这些变量，可以指定并发运行的 Map/Reduce 任务的最大数目。理论上，增加上述默认值有助于提高 CPU 利用率，从而提高性能，但同时需要考虑集群资源的设置，比如 CPU 和可用内存；否则，可能会导致 MapReduce 作业和集群性能整体降级的风险。

表 5-2　　与 CPU 利用率关联最密切的参数

配置变量	含义描述	默认值
mapred.tasktracker.map.tasks.maximum	设置可同时运行的 Map 任务最大数目	2
mapred.tasktracker.reduce.tasks.maximum	设置可同时运行的 Reduce 任务最大数目	2

内存资源非常重要，精确分配内存资源可以有效地避免内存交换，并优化 Hadoop 作业的运行性能。表 5-3 列出了与内存相关的常用参数。mapred.child.ulimit 默认值未指定，若要设置此参数，其值应大于 2* mapred.child.java.opts，否则，Java 虚拟机（Java Virtual Machine，JVM）可能无法启动。增加 io.sort.mb 和 io.sort.factor 的值会为排序与合并操作分配更多的内存，这会减少对磁盘的流量冲击，从而减 Mapper 和 Reducer 任务的 I/O 时间。为避免任务耗尽内存现象的出现，io.sort.mb 的值应该大于 0.25* mapred.child.java.opts 且小于 0.5* mapred.child.java.opts。通过增加 mapred.job.reduce.input.buffer.percent 的值可以获得更多的内存缓冲区，从而在 Reduce 阶段减少本地磁盘的 I/O 时间。

表 5-3　　与内存相关的常用参数

配置变量	含义描述	默认值
mapred.child.java.opts	设置每个 JVM 任务的可用内存，默认值是为 MapReduce 作业保留 200MB 内存	-Xmx200m
mapred.child.ulimit	设置分配给每个 MapReduce 作业的虚拟内存的极限值	
io.sort.mb	指定对流进行排序时使用的缓冲区大小	100MB
io.sort.factor	设置对文件进行排序时一次合并的 Map 输出分区的数量	10
mapred.job.reduce.input.buffer.percent	设置 Reduce 阶段内存相对于保留 Map 输出的最大堆大小的百分比	0.0

表 5-4 所示为与网络相关的常用参数，增加 mapred.tasktracker.parallel.copies 的值会增加网络流量，加速 Map 输出的复制过程，但需要消耗更多的内存。建议只在 Mapper 任务产生的输出量非常大的时候考虑增加这一参数的值。

表 5-4 与网络相关的常用参数

配置变量	含义描述	默认值
mapred.tasktracker.parallel.copies	设置在 Shuffle 阶段获取 Map 输出的并行传输数量	5

为了优化磁盘 I/O 操作，可使用数据压缩（compress），默认情况下它处于关闭状态，可通过修改控制压缩的参数默认值来启用数据压缩，如表 5-5 所示。启用 Map 任务的输出压缩可有效减少写入存储系统的中间数据量。实践表明，只有当输入数据量很大且容易拆分时，才应该使用压缩，否则会造成集群性能的降级。

表 5-5 与压缩相关的常用参数

配置变量	含义描述	默认值
mapred.compress.map.output	设置 Map 任务的输出是否使用 SequenceFile 编码器解码	false
mapred.output.compress	设置是否启用对作业输出进行压缩	false
mapred.map.output.compression.codec	设置 Map 输出进行压缩时的编/解码方式	org.apache.hadoop.io.compress.DefaultCodec
mapred.local.dir	设置存储 Map 中间文件的本地目录，可定义多个目录，每个目录应该对应不同的物理设备	${hadoop.tmp.dir}/mapred/local
dfs.data.dir	设置 DataNode 在本地文件系统上存储数据块的目录，该参数在 hdfs-site.xml 文件中	${hadoop.tmp.dir}/dfs/data

要获得磁盘 I/O 的平衡并大幅提升 I/O 性能，可以使用向多个位置写入的功能，把数据写到每个节点的所有磁盘上。mapred.local.dir 指定中间 Map 输出数据的存储位置，dfs.data.dir 指定 HDFS 数据的存储位置。

5.4.2 使用合适的数据类型

初次编写 MapReduce 的开发人员，经常在不必要的时候使用 Text 对象。虽然 Text 对象使用起来很方便，但它在由数值转换到文本或由 UTF-8 字符串转换到文本时都是低效的，且会消耗大量的 CPU 时间。当处理那些非文本的数据时，可以使用二进制的 Writable 类型，如 IntWritable、FloatWritable 等。

另外，很多 MapReduce 开发人员常犯的一个错误是，在 map( )/reduce( )方法中为每个输出都创建 Writable 对象。例如，WordCount 的 Mapper 如下：

```
1. class MyMapper extends Mapper<…> {
2.       …
3.     public void map(…) {
4.         For (String word : words){
5.         output.collect(new Text(word),new IntWritable(1));
6.     }
7.}
```

这样的写法会导致程序被分配若干个短周期的对象，Java 垃圾回收器就要为此做很多的工作。更有效的写法如下：

```
1.class MyMapper extends Mapper<…> {
2.     Text text= new Text();
3.    IntWritable one = new IntWritable(1);
4.     …
5.   public void map(…) {
6.          for (String word: words) {
7.                text.set(word);
8.                output.collect(text, one);
9.          }
10.      }
11.}
```

按照上面的写法修改 WordCount 例子后， job 刚运行时与修改之前没有任何不同。这是因为集群默认为每个任务分配一个 1GB 的堆，所以 Java 垃圾回收机制没有启动。当重新设置参数时，比如为每个任务只分配 100MB 的堆时，没有重用 Writable 对象的方法，计算效率明显下降。重用 Writable 对象的方法，在设置更小的堆时还是保持相同的执行速度。因此重用 Writable 对象是提升效率的一个常用方法。

5.4.3 基准性能测试工具

测试对验证系统的正确性和分析系统的性能来说非常重要，但往往容易被忽视。Hadoop 自带了几个基准性能测试工具，被打包在.jar 包中，如 hadoop-mapreduce–client- jobclient-×××-tests.jar 和 hadoop-mapreduce-examples-×××.jar，在 Hadoop 环境中可以很方便地运行测试实例。本小节将介绍 Hadoop 自带的基准性能测试工具，帮助我们对系统有更全面的了解、发现系统的瓶颈、对系统性能做出更好的改进。

1. TestDFSIO

TestDFSIO 主要用于测试 HDFS 的 I/O 性能。其测试方法是：通过一个 MapReduce 作业来并发地执行读/写操作，每个 Map 任务用于读或写每个文件，Map 的输出用于收集与处理文件相关的统计信息，Reduce 用于累积统计信息，并产生结果。

执行步骤如下。

（1）登录 Hadoop 集群，先删除历史测试数据，命令如下：

```
[hadoop@master~]$hadoop jar /home/hadoop/hadoop-2.9.0/share/hadoop/mapreduce/
hadoop-mapreduce-client-jobclient-2.9.0-tests.jar TestDFSIO -clean
```

（2）测试写性能，生成 5 个文件，每个文件为 20MB 大小，命令如下：

```
[hadoop@master~]$hadoop jar /home/hadoop/hadoop-2.9.0/share/hadoop/mapreduce/
hadoop-mapreduce-client-jobclient-2.9.0-tests.jar TestDFSIO -write -nrFiles 5
-fileSize 20
```

（3）查看结果，测试结果会保存到 TestDFSIO_results.log，命令如下：

```
[hadoop@master mapreduce]$ cat TestDFSIO_results.log
```

输出结果如图 5-13 所示。

```
[hadoop@master mapreduce]$ cat TestDFSIO_results.log
----- TestDFSIO ----- : write
           Date & time: Sat Apr 21 18:36:53 CST 2018
       Number of files: 5
Total MBytes processed: 100.0
     Throughput mb/sec: 107.18113612004288
Average IO rate mb/sec: 109.38374328613281
 IO rate std deviation: 15.724167357554894
    Test exec time sec: 19.423

[hadoop@master mapreduce]$
```

图 5-13 I/O 写性能测试结果

（4）结果说明如下：

- Number of files：生成的文件数目。
- Total MBytes processed：总共需要写入的数据量。
- Throughput mb/sec：总共需要写入的数据量/每个 Map 任务写入数据的时间之和，即 100/（Map1 写时间+Map2 写时间+…），表示数据写入速率。
- Average IO rate mb/sec：（每个 map 写入的数据量/每个 Map 任务写入数据的时间）之和，即 20/Map1 写时间+20/Map2 写时间+…。
- IO rate std deviation：上一个值的标准差。
- Test exec time sec：整个 job 的执行时间。

（5）测试读性能，读取之前写入的文件，命令如下：

```
[hadoop@master~]$hadoop jar /home/hadoop/hadoop-2.9.0/share/hadoop/mapreduce/
hadoop-mapreduce-client-jobclient-2.9.0-tests.jar TestDFSIO -read -nrFiles 5
-fileSize 20
```

（6）查看结果，测试结果追加保存到 TestDFSIO_results.log，命令如下：

```
[hadoop@master mapreduce]$ cat TestDFSIO_results.log
```

输出结果如图 5-14 所示。

```
----- TestDFSIO ----- : read
           Date & time: Sat Apr 21 19:25:41 CST 2018
       Number of files: 5
Total MBytes processed: 100.0
     Throughput mb/sec: 452.4886877828054
Average IO rate mb/sec: 457.4574279785156
 IO rate std deviation: 44.35173713938598
    Test exec time sec: 19.524
```

图 5-14 I/O 读性能测试结果

结果中各选项含义与写入相同，但是读速率比写快很多，总执行时间相差不大。

2. TeraSort

排序通常用于衡量分布式数据处理框架的数据处理能力。TeraSort 是 Hadoop 中的一个排序作业，通过调节不同的 Map 任务数量和 Reduce 任务数量，测试对 Hadoop 性能的影响。TeraSort 测试主要流程：产生随机数据→排序→验证排序结果。

执行步骤如下。

（1）使用 Teragen 产生随机数据，本次我们产生 1GB 数据，命令如下：

```
[hadoop@master~]$cd /home/hadoop/hadoop-2.9.0/share/hadoop/mapreduce
```

```
[hadoop@master mapreduce]$hadoop jar hadoop-mapreduce-examples-2.9.0.jar teragen
-Dmapred.map.tasks=10 10737418 terasort/1T-input
```

其中，-Dmapred.map.tasks 用于指定 Map 任务数，Teragen 默认开启 2 个 Map 任务；10737418 表示要产生的数据的行数，Teragen 每行的数据大小是 100B，要产生 1GB 的数据，需要的行数=1024×1024×1024÷100=10737418；terasort/1T-input 用于指定数据存放的位置，数据默认存放在 HDFS 中。

（2）查看产生的数据，命令如下：

```
[hadoop@master mapreduce]$hadoop fs -ls /user/hadoop/terasort/1T-input
```

输出结果如图 5-15 所示。

```
[hadoop@master mapreduce]$ hadoop fs -ls /user/hadoop/terasort/1T-input
Found 12 items
-rw-r--r--   1 hadoop supergroup          0 2018-04-22 01:34 /user/hadoop/terasort
/1T-input/_SUCCESS
drwxr-xr-x   - hadoop supergroup          0 2018-04-22 01:34 /user/hadoop/terasort
/1T-input/_temporary
-rw-r--r--   1 hadoop supergroup  107374200 2018-04-22 01:34 /user/hadoop/terasort
/1T-input/part-m-00000
-rw-r--r--   1 hadoop supergroup  107374200 2018-04-22 01:34 /user/hadoop/terasort
/1T-input/part-m-00001
-rw-r--r--   1 hadoop supergroup  107374200 2018-04-22 01:34 /user/hadoop/terasort
/1T-input/part-m-00002
-rw-r--r--   1 hadoop supergroup  107374200 2018-04-22 01:34 /user/hadoop/terasort
/1T-input/part-m-00003
-rw-r--r--   1 hadoop supergroup  107374100 2018-04-22 01:34 /user/hadoop/terasort
/1T-input/part-m-00004
-rw-r--r--   1 hadoop supergroup  107374200 2018-04-22 01:34 /user/hadoop/terasort
/1T-input/part-m-00005
-rw-r--r--   1 hadoop supergroup  107374200 2018-04-22 01:34 /user/hadoop/terasort
/1T-input/part-m-00006
-rw-r--r--   1 hadoop supergroup  107374200 2018-04-22 01:34 /user/hadoop/terasort
/1T-input/part-m-00007
-rw-r--r--   1 hadoop supergroup  107374200 2018-04-22 01:34 /user/hadoop/terasort
/1T-input/part-m-00008
-rw-r--r--   1 hadoop supergroup  107374100 2018-04-22 01:34 /user/hadoop/terasort
/1T-input/part-m-00009
[hadoop@master mapreduce]$
```

图 5-15　Teragen 产生的数据结果

（3）使用 TeraSort 对 Teragen 产生的数据进行排序，排序结果输出到/user/hadoop/terasort/1T-output，命令如下：

```
[hadoop@master mapreduce]$ hadoop jar hadoop-mapreduce-examples-2.9.0.jar
terasort -Dmapred.reduce.tasks=20 /user/hadoop/terasort/1T-input /user/hadoop/
terasort/1T-output
```

其中，-Dmapred.reduce.tasks 用于指定 Reduce 任务数，默认的 Reduce 任务数是 1，会导致任务运行缓慢。此处指定 Reduce 任务数为 20，会输出 20 个排序后的文件，整个排序花费的时间如图 5-16 所示。

```
[hadoop@master mapreduce]$ hadoop jar hadoop-mapreduce-examples-2.9.0.jar teraso
rt -Dmapred.reduce.tasks=20 /user/hadoop/terasort/1T-input /user/hadoop/terasort
/1T-output
18/04/22 09:26:16 INFO terasort.TeraSort: starting
18/04/22 09:26:17 INFO input.FileInputFormat: Total input files to process : 10
Spent 143ms computing base-splits.
Spent 3ms computing TeraScheduler splits.
Computing input splits took 147ms
Sampling 10 splits of 10
Making 20 from 100000 sampled records
Computing paritiions took 1094ms
Spent 1243ms computing partitions.
```

图 5-16　TeraSort 排序时间

其中，Spent 143ms computing base-splits 表示花费了 143ms 完成基本数据分片操作；Spent 3ms computing TeraScheduler splits 表示花费了 3ms 完成资源调度工作；Computing input splits took 147ms 表示花费了 147ms 完成分片输入 map( )函数操作；Spent 1243ms computing partitions 表示花费了 1243ms 完成分区工作。

（4）使用 TeraValidate 对 TeraSort 结果进行验证，命令如下：

```
[hadoop@master mapreduce] hadoop jar hadoop-mapreduce-examples-2.9.0.jar
teravalidate /user/hadoop/terasort/1T-output /user/hadoop/terasort/terasort
-report
```

验证结果输出到/user/hadoop/terasort/terasort-report，如图 5-17 所示。

```
[hadoop@master mapreduce]$ hadoop fs -ls /user/hadoop/terasort/terasort-report
Found 2 items
-rw-r--r--   2 hadoop supergroup          0 2018-04-22 09:32 /user/hadoop/teraso
rt/terasort-report/_SUCCESS
-rw-r--r--   2 hadoop supergroup         24 2018-04-22 09:32 /user/hadoop/teraso
rt/terasort-report/part-r-00000
[hadoop@master mapreduce]$
```

图 5-17　TeraValidate 验证结果

TeraSort 巧妙地使用了 Hadoop 的 MapReduce 机制实现了排序的目的，我们可以在集群上利用 TeraSort 来测试 Hadoop，常用的测试场景如下：

- 在不同版本的 Hadoop 上运行 TeraSort，使用相同的配置参数，来进行正确性对比测试；
- 在不同版本的 Hadoop 上运行 TeraSort，使用相同的配置参数，来进行性能对比测试；
- 在相同版本的 Hadoop 上运行 TeraSort，使用不同的配置参数，进行性能对比测试；

3. MRbench

MRbench 主要检验小型作业的快速响应能力，MRbench 会多次重复执行一个小型作业，用于检查在集群上小型作业的运行是否可重复以及运行是否高效。

执行步骤如下。

（1）运行一个小型作业 10 次，命令如下：

```
[hadoop@master ~]cd /home/hadoop/hadoop-2.9.0/share/hadoop/mapreduce
[hadoop@master mapreduce] hadoop jar hadoop-mapreduce-client-jobclient-2.9.0-
tests.jar mrbench -numRuns 10
```

（2）查看运行结果，如图 5-18 所示。

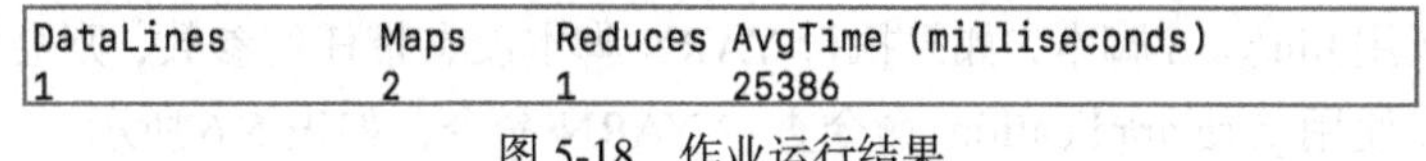

DataLines	Maps	Reduces	AvgTime (milliseconds)
1	2	1	25386

图 5-18　作业运行结果

由图 5-18 可看出，该作业运行 10 次花费了 25386ms，因为没有指定 Map、Reduce 任务数，所以此处全部采用默认值（Maps=2，Reduces=1）。若想指定 Map、Reduce 任务数，可通过选项 -maps、-reduces 设置。

5.5 YARN 数据处理框架

Apache Hadoop YARN 是一种新的 Hadoop 资源管理器。它是一个通用资源管理系统，可为上层应用提供统一的资源管理和调度。它的引入为集群在利用率、资源统一管理和数据共享等方面带来了巨大好处。

YARN 的执行模式比以前的 MapReduce 实现更通用。与 Apache Hadoop 的 MapReduce 的原始版本（也称为 MR1）相比，YARN 可以运行不遵循 MapReduce 模型的应用程序。此应用程序被称为 MR2（MapReduce 的第 2 版）。

YARN 的基本思想是将经典调度框架中 JobTracker 的资源管理和任务调度/监控功能分离成两个单独的组件，即一个全局的资源管理器 ResourceManager 和每个应用程序特有的 ApplicationMaster。ResourceManager 负责整个系统资源的管理和分配，而 ApplicationMaster 则负责单个应用程序的资源管理。

YARN 调度框架包括 ResourceManager、ApplicationMaster、NodeManager 及 Container 等组件概念。

ResourceManager 包含两个部分：Scheduler 和 ApplicationManager。Scheduler 负责将资源分配给各种正在运行的应用程序；ApplicationManager 主要负责接收 Job 的提交请求，为应用分配第一个 Container 来运行 ApplicationMaster，还负责监控 ApplicationMaster，在遇到失败时重启 ApplicationMaster 运行的 Container。

ApplicationMaster 负责向调度器申请、释放资源，请求 NodeManager 运行任务，跟踪应用程序的状态和监控它们的进程。

NodeManager 是 YARN 中单个节点的代理，负责与应用程序的 ApplicationMaster 和集群管理者 ResourceManager 交互；从 ApplicationMaster 上接收有关 Container 的命令并执行（例如，启动、停止 Container）；向 ResourceManager 汇报各个 Container 的执行状态和节点健康状况，并读取有关 Container 的命令；执行应用程序的 Container、监控应用程序的资源使用情况并且向 ResourceManager 调度器汇报。

Container 是 YARN 框架的计算单元，是具体执行应用任务（如 Map 任务、Reduce 任务）的基本单位。Container 和集群节点的关系：一个节点会运行多个 Container，但一个 Container 不会跨节点。

一个 Container 就是一组分配的系统资源，它封装了某个节点上的多维度资源，如内存、CPU、磁盘、网络等。YARN 会为每个任务分配一个 Container，且该任务只能使用该 Container 中的资源。

5.5.1 YARN 常用命令

YARN 命令调用 bin/yarn 脚本，如果执行 YARN 脚本没有带任何参数，则会输出 YARN 所有命令的描述。可以使用 yarn application 命令查看 YARN 命令，如表 5-6 所示。

表 5-6　　YARN 命令

命令选项	描述
-appStates <States>	与-list 命令一起使用，基于应用程序的状态来过滤应用程序。应用程序状态包括 ALL、NEW、NEW_SAVING、SUBMITTED、ACCEPTED、RUNNING、FINISHED、FAILED、KILLED
-appTypes <Types>	与-list 命令一起使用，基于应用程序类型来过滤应用程序
-list	输出 ResourceManager 返回的应用程序列表
-kill <ApplicationId>	终止指定的应用程序
-status <ApplicationId>	输出应用程序的状态

示例如下。

（1）本示例运行带有 10 个 Map 和 100000 个样本的 pi 实例，通过命令行查询运行状态，命令如下：

```
[hadoop@master ~]$cd /home/hadoop/hadoop-2.9.0/share/hadoop/mapreduce
[hadoop@master mspreduce]$ hadoop jar hadoop-mapreduce-examples-2.9.0.jar
pi 10 100000
```

（2）新打开一个命令行界面，查看当前是否有正在执行的应用，命令如下：

```
[hadoop@master ~]$ yarn application -list
```

输出结果如图 5-19 所示。

```
[hadoop@master mapreduce]$ yarn application -list
18/04/23 15:41:05 INFO client.RMProxy: Connecting to ResourceManager at master/192.168.2.176:8032
Total number of applications (application-types: [], states: [SUBMITTED, ACCEPTED, RUNNING] and tags: []):1
                Application-Id      Application-Name        Application-Type          User           Queue
           State             Final-State             Progress                        Tracking-URL
application_1524202612343_0022         QuasiMonteCarlo               MAPREDUCE        hadoop         default
        ACCEPTED               UNDEFINED                   0%                                   N/A
```

图 5-19　正在执行的应用列表

（3）输出当前运行程序的状态，命令如下：

```
[hadoop@master ~]$ yarn application -status application_1524202612343_0022
```

其中，application_1524202612343_0022 为当前运行程序的 Application-Id 值。

输出结果如图 5-20 所示。

```
[hadoop@master mapreduce]$ yarn application -status application_1524202612343_0022
18/04/23 15:41:13 INFO client.RMProxy: Connecting to ResourceManager at master/192.168.2.176:8032
Application Report :
        Application-Id : application_1524202612343_0022
        Application-Name : QuasiMonteCarlo
        Application-Type : MAPREDUCE
        User : hadoop
        Queue : default
        Application Priority : 0
        Start-Time : 1524469265501
        Finish-Time : 0
        Progress : 5%
        State : RUNNING
        Final-State : UNDEFINED
        Tracking-URL : http://slave:37230
        RPC Port : 39867
        AM Host : slave
        Aggregate Resource Allocation : 12021 MB-seconds, 5 vcore-seconds
        Aggregate Resource Preempted : 0 MB-seconds, 0 vcore-seconds
        Log Aggregation Status : DISABLED
        Diagnostics :
        Unmanaged Application : false
        Application Node Label Expression : <Not set>
        AM container Node Label Expression : <DEFAULT_PARTITION>
        TimeoutType : LIFETIME  ExpiryTime : UNLIMITED  RemainingTime : -1seconds
```

图 5-20　当前运行程序的状态

（4）停止当前正在运行的程序，命令如下：

```
[hadoop@master ~]$ yarn application -kill application_1524202612343_0022
```

输出结果如图 5-21 所示。

```
[hadoop@master mapreduce]$ yarn application -kill application_1524202612343_0022
18/04/23 15:41:25 INFO client.RMProxy: Connecting to ResourceManager at master/192.168.2.176:8032
Killing application application_1524202612343_0022
18/04/23 15:41:25 INFO impl.YarnClientImpl: Killed application application_1524202612343_0022
```

图 5-21　停止运行程序

5.5.2　使用 Web GUI 监控实例

本小节通过图解说明如何使用 Hadoop YARN 的 Web GUI 来监控和获取 YARN 作业的信息。YARN 的 Web 主界面（IP 地址可通过 yarn-site.xml 文件中的 yarn.resourcemanager.webapp.address 属性设置）如图 5-22 所示。

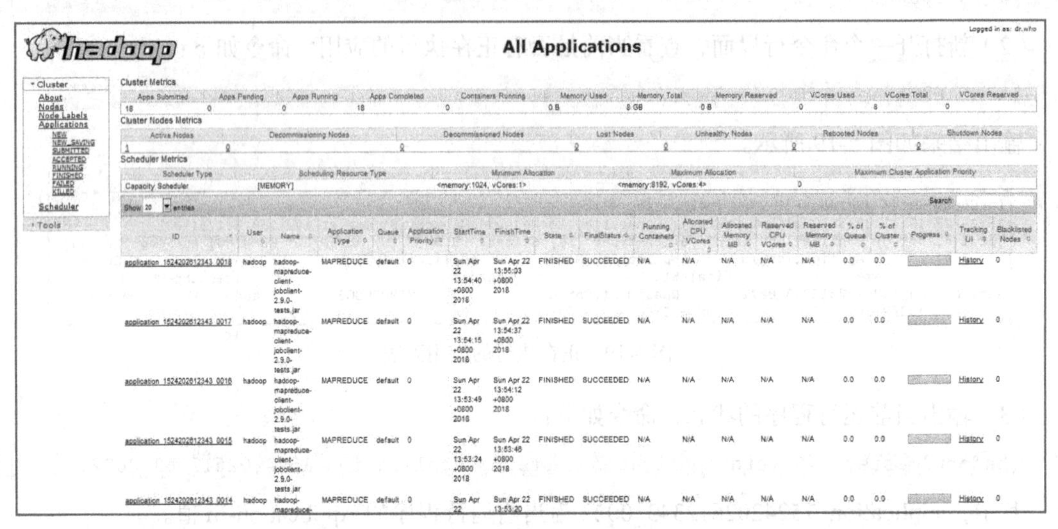

图 5-22　YARN 的 Web 主界面

通过集群的监控指标列表，我们会看到一些新的信息。首先，看到的不是 hadoop1.x 的“Map/Reduce Task Capacity”，而是运行中的 Container 的数量信息。如果 YARN 中运行着 MapReduce 作业，这些 Container 既用于 Map 任务也用于 Reduce 任务。如果单击指标列表 Nodes 链接，会看到该节点活动的概述。图 5-23 所示为 pi 应用在运行时一个节点活动的快照。

hadoop　Nodes of the cluster　Logged in as: dr.who

Cluster: About, Nodes, Node Labels, Applications (NEW, NEW_SAVING, SUBMITTED, ACCEPTED, RUNNING, FINISHED, FAILED, KILLED), Scheduler; Tools

Cluster Metrics

Apps Submitted	Apps Pending	Apps Running	Apps Completed	Containers Running	Memory Used	Memory Total	Memory Reserved	VCores Used	VCores Total	VCores Reserved
23	0	1	22	7	8 GB	8 GB	0 B	7	7	0

Cluster Nodes Metrics

Active Nodes	Decommissioning Nodes	Decommissioned Nodes	Lost Nodes	Unhealthy Nodes	Rebooted Nodes	Shutdown Nodes
1	0	0	0	0	0	0

Scheduler Metrics

Scheduler Type	Scheduling Resource Type	Minimum Allocation	Maximum Allocation	Maximum Cluster Application Priority
Capacity Scheduler	[MEMORY]	<memory:1024, vCores:1>	<memory:8192, vCores:4>	0

Show 20 entries　Search:

Node Labels	Rack	Node State	Node Address	Node HTTP Address	Last health-update	Health-report	Containers	Mem Used	Mem Avail	VCores Used	VCores Avail	Version
	/default-rack	RUNNING	slave:53611	slave:8042	Mon Apr 23 17:00:53 +0800 2018		7	8 GB	0 B	7	1	2.9.0

Showing 1 to 1 of 1 entries　First Previous 1 Next Last

图 5-23　节点活动的快照

其中，Apps Running 表示当前正在运行的 Job 数量；Containers Running 表示当前正在使用的 Container 数量；Memory Used 表示当前内存使用情况。

单击指标列表 Applications 链接回到主界面，单击刚才运行的 pi 应用的 ID，应用状态窗口会呈现出来，如图 5-24 所示。

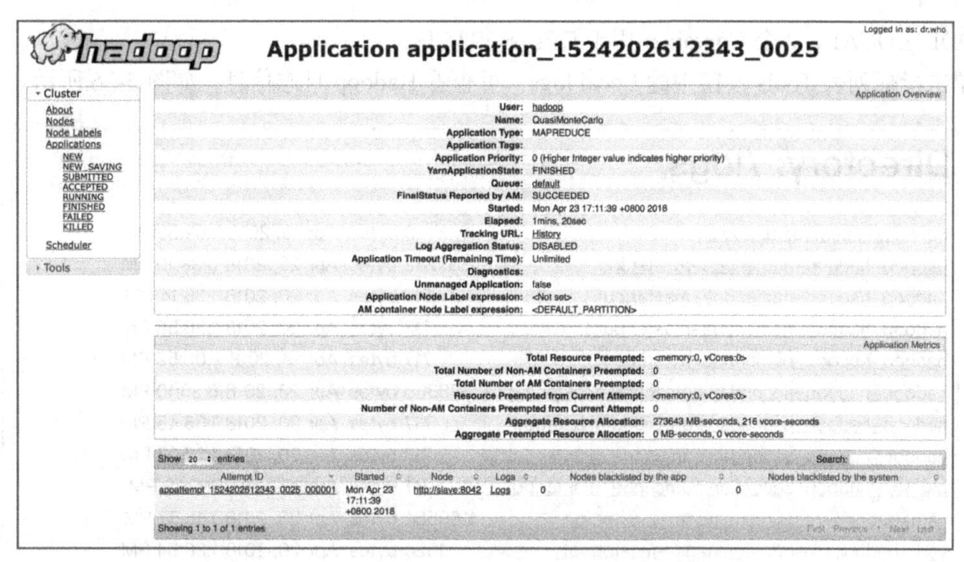

图 5-24 pi 应用的状态

这个窗口提供了一些被选中的作业的信息，其中包括作业提交者 User、作业的名字 Name（可在程序中设置）、作业最终执行状态 FinalStatus Reported by AM，以及作业最终花费时间 Elapsed。

单击图 5-24 中的 Attempt ID（在应用界面的最后一行），进入 Application Attempt 应用窗口，可以获取任务分配 Container 的详细信息，如图 5-25 所示。

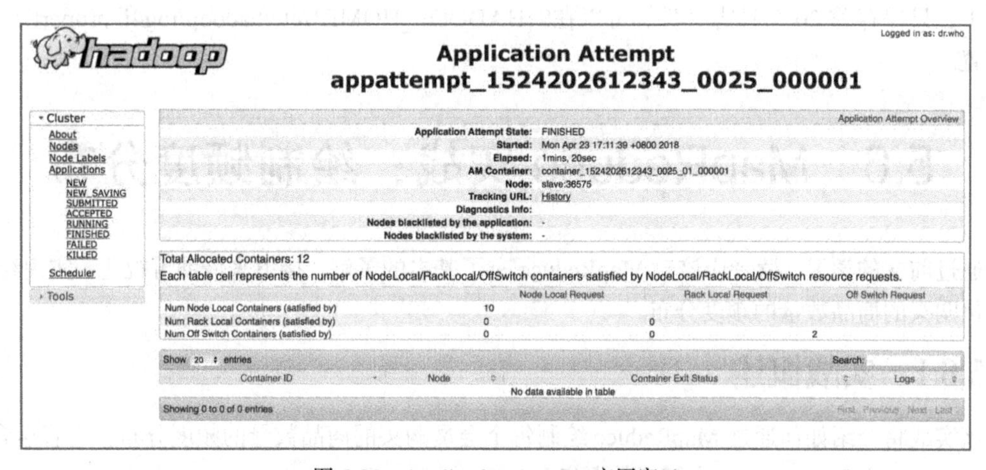

图 5-25 Application Attempt 应用窗口

在获取任务分配 Container 的详细信息前，这里有必要讲解一下 YARN 的本地化（Locality）。ApplicationMaster 客户端在提交应用的时候，有时候会对 Container 运行的位置提出限制。比如，由于某些数据在服务器 Node1 上，因此 ApplicationMaster 客户端希望用来处理这些数据的 Container 就运行在这个服务器上，这种要求运行在某个特定节点的本地化叫作 NODE_LOCAL；也可以要求运行在某个固定的机架 Rack1 上，这种机架的本地化叫作 RACK_LOCAL；或者没有任何要求，这种本地化叫作 OFF_SWITCH。因此，Resource Name 可以是某个节点的 IP（NODE_LOCAL），或者某个机架的 IP（RACK_LOCAL），或者是通配符*（OFF_SWITCH）。由该应用窗口提供的信息，总共分配了 12 个 Container 执行作业，其中，要求 10 个 Container 用于 NODE_LOCAL，2 个 Container 用于 OFF_SWITCH。

单击指标列表 Tools 按钮中的 Local logs，可查看 Hadoop 日志信息，如图 5-26 所示。

Directory: /logs/

SecurityAuth-hadoop.audit	0 bytes	Apr 19, 2018 9:09:58 PM
hadoop-hadoop-namenode-master.log	587232 bytes	Apr 23, 2018 6:38:10 PM
hadoop-hadoop-namenode-master.out	717 bytes	Apr 20, 2018 1:36:14 PM
hadoop-hadoop-namenode-master.out.1	717 bytes	Apr 20, 2018 1:26:56 PM
hadoop-hadoop-namenode-master.out.2	717 bytes	Apr 19, 2018 9:09:59 PM
hadoop-hadoop-secondarynamenode-master.log	398885 bytes	Apr 23, 2018 6:38:10 PM
hadoop-hadoop-secondarynamenode-master.out	717 bytes	Apr 20, 2018 1:36:45 PM
hadoop-hadoop-secondarynamenode-master.out.1	717 bytes	Apr 20, 2018 1:27:08 PM
hadoop-hadoop-secondarynamenode-master.out.2	717 bytes	Apr 19, 2018 9:10:12 PM
yarn-hadoop-resourcemanager-master.log	1169584 bytes	Apr 23, 2018 7:16:57 PM
yarn-hadoop-resourcemanager-master.out	1524 bytes	Apr 20, 2018 1:36:54 PM
yarn-hadoop-resourcemanager-master.out.1	1524 bytes	Apr 20, 2018 1:27:17 PM
yarn-hadoop-resourcemanager-master.out.2	1524 bytes	Apr 19, 2018 9:10:21 PM

图 5-26　Hadoop 日志信息

当日志达到一定的大小时，将会被切割出一个新的文件，切割出来的日志文件名类似 yarn-${USER}-resourcemanager-${hostname}.log.数字，后面的数字越大，代表日志文件越旧。在默认情况下，只保存前 20 个日志文件，可以在${HADOOP_HOME}/etc/hadoop/log4j. properties 文件中配置。

5.6　MapReduce 实战：绘制频度分布

通过前文的学习，读者应该对 MapReduce 有了基本的了解。接下来借助可视化，实现绘制每个会员购买的商品数量的频度分布。

5.6.1　实战概述

该实战将介绍如何通过 MapReduce 绘制每个会员购买的商品数量的频度分布，内容包含：

（1）自定义 InputFormat 来解析数据集；

（2）运行第 1 个 Job 来计算购买频度；

（3）运行第 2 个 Job 对第 1 个 Job 的结果进行排序，使用 Gnuplot（一个免费且强大的绘图程序）对 Job 的结果进行绘制。

5.6.2　实战步骤

（1）下载本示例源码 ch04/FrequencyAnalyzer/src，将源码、数据集 amazon-meta.txt 以及配置文件 build.xml 文件上传到集群的/home/hadoop/data/frequencyAnalyzer 目录，可通过 Linux 命令 scp 实现，或者借助 FileZilla 工具实现。源码目录结构如图 5-27 所示。

```
[[hadoop@master frequencyAnalyzer]$ ll
总用量 580
-rw-r--r-- 1 hadoop hadoop 588705 4月   24 16:15 amazon-meta.txt
-rw-r--r-- 1 hadoop hadoop   1416 4月   24 16:29 build.xml
drwxr-xr-x 2 hadoop hadoop    156 4月   24 16:28 src
[[hadoop@master frequencyAnalyzer]$ cd src
[[hadoop@master src]$ ll
总用量 28
-rw-r--r-- 1 hadoop hadoop 7229 4月   24 16:28 BuyerRecord.java
-rw-r--r-- 1 hadoop hadoop 2510 4月   24 16:28 BuyingFrequencyAnalyzer.java
-rw-r--r-- 1 hadoop hadoop  792 4月   24 16:28 ItemSalesDataFormat.java
-rw-r--r-- 1 hadoop hadoop 5059 4月   24 16:28 ItemSalesDataReader.java
-rw-r--r-- 1 hadoop hadoop 2835 4月   24 16:28 SimpleResultSorter.java
```

图 5-27　源码目录结构

（2）本实践通过 Ant 工具打包项目，请确保已经安装了 Ant 工具。进入/home/hadoop/ data/frequencyAnalyzer 目录，打包命令如下：

```
[hadoop@master frequencyAnalyzer]$ ant build
```

结果如图 5-28 所示。

```
[hadoop@master frequencyAnalyzer]$ ant build
Buildfile: /home/hadoop/data/frequencyAnalyzer/build.xml
init:
    [mkdir] Created dir: /home/hadoop/data/frequencyAnalyzer/build
    [mkdir] Created dir: /home/hadoop/data/frequencyAnalyzer/build/classes

compile:
    [javac] Compiling 5 source files to /home/hadoop/data/frequencyAnalyzer/buil
d/classes
    [javac] 注：某些输入文件使用或覆盖了已过时的 API。
    [javac] 注：有关详细信息，请使用 -Xlint:deprecation 重新编译。

build:
      [jar] Building jar: /home/hadoop/data/frequencyAnalyzer/build/lib/frequenc
yAnalyzer.jar

BUILD SUCCESSFUL
Total time: 2 seconds
[hadoop@master frequencyAnalyzer]$
```

图 5-28　Ant 打包项目

（3）进入目录/home/hadoop/data/frequencyAnalyzer/build/lib/，查看编译后的.jar 包，如图 5-29 所示。

```
[[hadoop@master frequencyAnalyzer]$ ll
总用量 580
-rw-r--r-- 1 hadoop hadoop 588705 4月   24 16:15 amazon-meta.txt
drwxrwxr-x 4 hadoop hadoop     30 4月   24 16:56 build
-rw-r--r-- 1 hadoop hadoop   1411 4月   24 16:54 build.xml
drwxr-xr-x 2 hadoop hadoop    156 4月   24 16:28 src
[[hadoop@master frequencyAnalyzer]$ cd build
[[hadoop@master build]$ ll
总用量 0
drwxrwxr-x 3 hadoop hadoop 15 4月   24 16:56 classes
drwxrwxr-x 2 hadoop hadoop 34 4月   24 16:56 lib
[[hadoop@master build]$ cd lib/
[[hadoop@master lib]$ ll
总用量 16
-rw-rw-r-- 1 hadoop hadoop 14799 4月   24 16:56 frequencyAnalyzer.jar
```

图 5-29　Ant 编译后的.jar 包

（4）切换到目录/home/hadoop/data/frequencyAnalyzer/，通过 put 命令上传 Amazon 数据集到 HDFS 的/tmp/MR-Frequency/目录中，命令如下：

```
[hadoop@master frequencyAnalyzer]$ hadoop fs -mkdir /tmp/MR-Frequency/
[hadoop@master frequencyAnalyzer]$ hadoop fs -put amazon-meta.txt  /tmp/MR-Frequency/
```

（5）查看上传到 HDFS 的数据集，命令如下：

```
[hadoop@master frequencyAnalyzer]$ hadoop fs -ls /tmp/MR-Frequency/
```

结果如图 5-30 所示。

```
[hadoop@master frequencyAnalyzer]$ hadoop fs -ls  /tmp/MR-Frequency/
Found 1 items
-rw-r--r--   2 hadoop supergroup     588705 2018-04-24 16:16 /tmp/MR-Frequency/a
mazon-meta.txt
```

图 5-30　上传到 HDFS 的数据集

（6）运行第 1 个 Job 来计算购买频度，命令如下：

```
[hadoop@master frequencyAnalyzer]$hadoop jar build/lib/frequencyAnalyzer.jar
cn.hust.book.bigdata.ch04.BuyingFrequencyAnalyzer /tmp/MR-Frequency/amazon-
meta.txt /tmp/MR-Frequency/output1
```

其中 cn.hust.book.bigdata.ch04.BuyingFrequencyAnalyzer 为第 1 个 Job 的主类名，/tmp/MR- Frequency/amazon-meta.txt 为数据集位置，/tmp/MR-Frequency/output1 为第 1 个 Job 的输出结果位置 。

（7）查看第 1 个 Job 的输出结果，命令如下：

```
[hadoop@master frequencyAnalyzer]$ hadoop fs -ls /tmp/MR-Frequency/output1
```

结果如图 5-31 所示。

```
[hadoop@master frequencyAnalyzer]$ hadoop fs -ls /tmp/MR-Frequency/output1
Found 2 items
-rw-r--r--   2 hadoop supergroup          0 2018-04-24 17:19 /tmp/MR-Frequency/o
utput1/_SUCCESS
-rw-r--r--   2 hadoop supergroup      60815 2018-04-24 17:19 /tmp/MR-Frequency/o
utput1/part-r-00000
[hadoop@master frequencyAnalyzer]$
```

图 5-31　第 1 个 Job 的输出结果

（8）运行第 2 个 Job，对第 1 个 Job 的结果进行排序，命令如下：

```
[hadoop@master frequencyAnalyzer]$ hadoop jar build/lib/frequencyAnalyzer.jar
cn.hust.book.bigdata.ch04.SimpleResultSorter /tmp/MR-Frequency/output1 /tmp/MR-
Frequency/output2
```

其中，cn.hust.book.bigdata.ch04.SimpleResultSorter 为第 2 个 Job 的主类名，/tmp/MR-Frequency/output1 为第 1 个 Job 的输出结果，作为第 2 个 Job 的输入数据集，/tmp/MR-Frequency/output2 为第 2 个 Job 的输出结果。

（9）查看第 2 个 Job 的输出结果，命令如下：

```
[hadoop@master frequencyAnalyzer]$ hadoop fs -ls /tmp/MR-Frequency/output2
```

结果如图 5-32 所示。

```
[hadoop@master frequencyAnalyzer]$ hadoop fs -ls /tmp/MR-Frequency/output2
Found 2 items
-rw-r--r--   2 hadoop supergroup          0 2018-04-24 18:01 /tmp/MR-Frequency/o
utput2/_SUCCESS
-rw-r--r--   2 hadoop supergroup      60815 2018-04-24 18:01 /tmp/MR-Frequency/o
utput2/part-r-00000
[hadoop@master frequencyAnalyzer]$
```

图 5-32　第 2 个 Job 的输出结果

（10）将第 2 个 Job 的输出结果通过 get 命令复制到/home/hadoop/data/frequencyAnalyzer 目录，输出结果重新命名为 1.data，命令如下：

```
[hadoop@master ~]$ cd /home/hadoop/data/frequencyAnalyzer
[hadoop@master frequencyAnalyzer]$ hadoop fs -get /tmp/MR-Frequency/output2/
part-r-00000 1.data
```

（11）安装 Gnuplot 绘图工具。

Gnuplot 是一个命令行的交互式绘图工具（Command-driven Interactive Function Plotting Program）。用户通过输入命令，可以设置绘图环境并以图形描述数据或函数，然后根据图形做更进一步的分析。

①切换到 root 用户，用 yum 命令安装软件，命令如下：

```
[root@master ~]# yum install gnuplot
```

②配置环境变量，编译/etc/profile.d/gnuplot.sh 文件，命令如下：

```
[root@master ~]# vim /etc/profile.d/gnuplot.sh
```

添加内容如图 5-33 所示。

```
export GNUTERM=dumb
export GNUPLOT=/home/hadoop/software/gnuplot
export PATH=/home/hadoop/gnuplot/bin:$PATH
export MANPATH=/home/hadoop/gnuplot/share/man/man1:$MANPATH
~
~
~
~
```

图 5-33　配置环境变量

③立即生效环境变量，命令如下：

```
[root@master ~]# source /etc/profile.d/gnuplot.sh
```

④切换到 hadoop 用户，启用环境变量，命令如下：

```
[root@master ~]#su - hadoop
[hadoop@master ~]$ source /etc/profile.d/gnuplot.sh
```

⑤进入运行界面，命令如下：

```
[hadoop@master ~]$ gnuplot
```

Gnuplot 运行界面如图 5-34 所示。

```
[hadoop@master ~]$ gnuplot

        G N U P L O T
        Version 4.6 patchlevel 2    last modified 2013-03-14
        Build System: Linux x86_64

        Copyright (C) 1986-1993, 1998, 2004, 2007-2013
        Thomas Williams, Colin Kelley and many others

        gnuplot home:     http://www.gnuplot.info
        faq, bugs, etc:   type "help FAQ"
        immediate help:   type "help"  (plot window: hit 'h')

Terminal type set to 'dumb'
gnuplot>
```

图 5-34　Gnuplot 运行界面

⑥测试 sin( )函数，命令如下：

```
gnuplot> plot sin(x)
```

运行结果如图 5-35 所示。

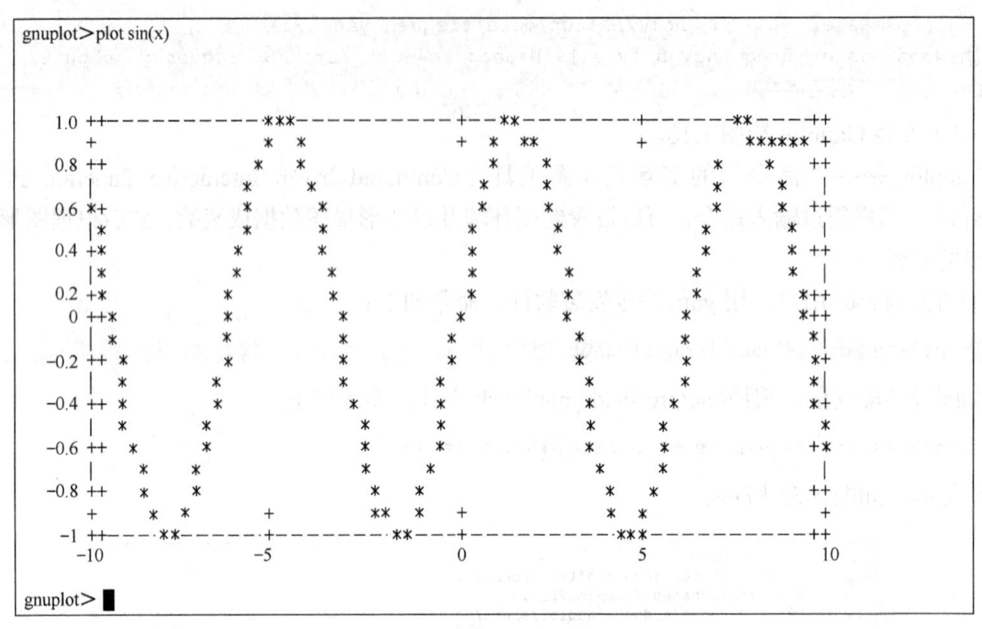

图 5-35　测试 sin()函数

⑦使用 Gnuplot 绘图工具。

切换到/home/hadoop/data/frequencyAnalyzer 目录，命令如下：

```
[hadoop@master ~]$ cd /home/hadoop/data/frequencyAnalyzer
```

新建 buyfreq.plot 文件，命令如下：

```
[hadoop@master frequencyAnalyzer]$ vim buyfreq.plot
```

添加如下内容：

```
1. set terminal png
2. set output "buyfreq.png"
3. set title "Frequency Distribution of Items Brought by Buyer";
4. set ylabel "Number of Items Brought";
5. set xlabel "Buyers Sorted by Items Count";
6. set key left top
7. set log y
8. set log x
9. plot "1.data" using 2 title "Frequency" with linespoints
```

其中，第 1～2 行定义了输出格式；第 3～6 行定义了坐标轴标签与标题；第 7～8 行定义了每个坐标轴的刻度，分布图中两个坐标轴的刻度值都使用了 log 函数；第 9 行定义了图形，Gnuplot 从 1.data 文件中读取数据，using 2 表示使用文件的第 2 列的数据，linespoints 表示使用直线进行绘制。

退出并保存 buyfreq.plot 文件，按 Esc 键跳转到命令模式，然后输入:wq，保存文件并退出。

（12）生成文件，命令如下：

```
[hadoop@master frequencyAnalyzer]$gnuplot buyfreq.plot
```

生成文件如图 5-36 所示。

```
[hadoop@master frequencyAnalyzer]$ gnuplot buyfreq.plot
[hadoop@master frequencyAnalyzer]$ ll
总用量 652
-rw-r--r-- 1 hadoop hadoop  60815 4月  24 18:12 1.data
-rw-r--r-- 1 hadoop hadoop 588705 4月  24 16:15 amazon-meta.txt
drwxrwxr-x 4 hadoop hadoop     30 4月  24 16:56 build
-rw-r--r-- 1 hadoop hadoop   1411 4月  24 16:54 build.xml
-rwxr-xr-x 1 hadoop hadoop    267 4月  24 18:34 buyfreq.plot
-rw-rw-r-- 1 hadoop hadoop   4835 4月  24 18:37 buyfreq.png
drwxr-xr-x 2 hadoop hadoop    156 4月  24 16:28 src
[hadoop@master frequencyAnalyzer]$
```

图 5-36　生成文件

（13）生成一个名为 buyfreq.png 的文件，如图 5-37 所示。

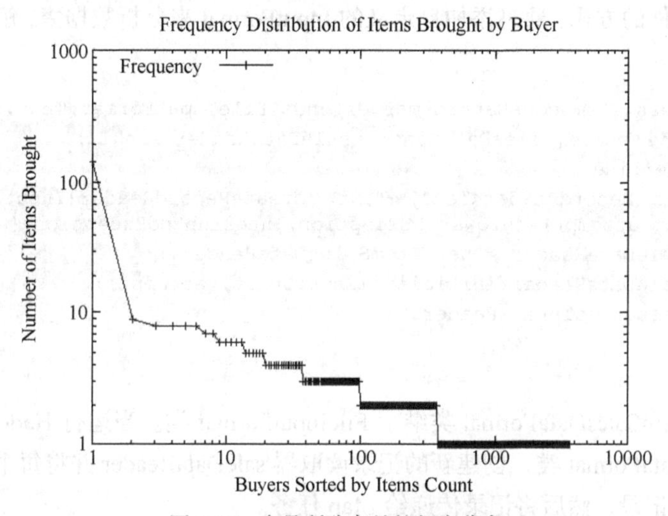

图 5-37　会员所购商品的频度分布

5.6.3　源码分析

打开示例源码 ch05/FrequencyAnalyzer/src 目录，我们会看到定义了 5 个类，如图 5-38 所示。

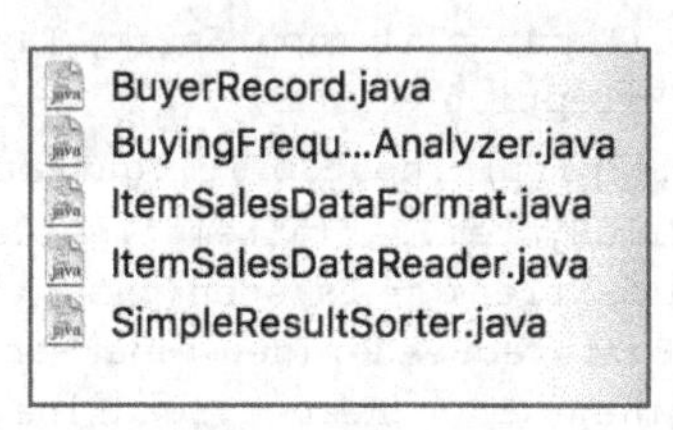

图 5-38　5 个类

其中，BuyerRecord 类、BuyingFrequencyAnalyzer 类用于计算购买频度，ItemSalesDataFormat 类、ItemSalesDataReader 类用于自定义 InputFormat，SimpleResultSorter 类用于对第 1 个 Job 的结果进行排序。

自定义 InputFormat。默认情况下，MapReducer 的 Job 会逐行读取输入文件并将每一行内容传送给 map( )函数。然而，有时一条数据记录会占据多行，比如本实战使用的 Amazon 数据集，我们查询前 10 行，会发现它们都属于第一条记录，如图 5-39 所示。

```
[hadoop@master frequencyAnalyzer]$ cat amazon-meta.txt | head -n 10
Id:   547020
ASIN: B0000694WE
  title: Contract Killer
  group: DVD
  salesrank: 21409
  similar: 5  B00005BJX9  B0000541U5  B00005OOQ1  6305801193  B00003W8NS
  categories: 16
   |[139452]|DVD[130]|Actors & Actresses[404278]|( L )[432148]|Leung, Gigi[43339
2]
   |[139452]|DVD[130]|Actors & Actresses[404278]|( L )[432148]|Li, Jet[433544]
   |[139452]|DVD[130]|Actors & Actresses[404278]|( S )[444858]|Sahara, Kenji[444
938]
[hadoop@master frequencyAnalyzer]$
```

图 5-39　Amazon 数据集一条记录的前 10 行内容

在这种情况下，编写一个 Job 将这些行放在一起进行处理是非常复杂的。但是，我们可以重写 Hadoop 读/写文件的方式，通过添加自定义的 InputFormat 来分析数据集，解决一条数据占据多行的问题。

```
1. public class ItemSalesDataFormat extends FileInputFormat<Text,Text> {
2.        private ItemSalesDataReader saleDataReader = null;
3.        @Override
4.      public RecordReader<Text, Text> createRecordReader(InputSplit inputSplit,
TaskAttemptContext attempt) throws IOException,InterruptedException {
5.          saleDataReader = new ItemSalesDataReader();
6.          saleDataReader.initialize(inputSplit, attempt);
7.          return saleDataReader;
8.      }
9. }
```

代码自定义 ItemSalesDataFormat 类继承 FileInputFormat 类。当运行 Hadoop 的 Job 时，首先会找到 ItemSalesDataFormat 类，创建新的记录读取器 saleDataReader 并将每个文件传递进去，从记录读取器中读取记录，然后将记录传递给 Map 任务。

如下代码展示了记录读取器：

```
1. public class ItemSalesDataReader extends RecordReader<Text, Text> {
2.   public ItemSalesDataReader() {
3.   }
4.   @Override
5.   public void initialize(InputSplit inputSplit, TaskAttemptContext attempt) throws
  IOException, InterruptedException {
6.       Path path =  ((FileSplit) inputSplit).getPath();//获取文件位置
7.       FileSystem fs = FileSystem.get(attempt.getConfiguration());
8.       FSDataInputStream fsStream = fs.open(path);
9.       inputReader = new BufferedReader(new InputStreamReader(fsStream), 1024*100);
10.       while ((currentLine = inputReader.readLine()) != null) {
11.          if(currentLine.startsWith("Id:")){//当前行以“Id”开头
12.              break;
13.          }
14.       }
15.    }
16.    @Override
17.    public boolean nextKeyValue() throws IOException, InterruptedException {
18.       currentLineData = new StringBuffer();
19.       inputCountSofar++;
20.       boolean readingreview = false;
```

```
21.        while ((currentLine = inputReader.readLine()) != null) {
22.            if(currentLine.trim().length() == 0){
23.                currentValue = new Text(currentLineData.toString());
24.                return true;
25.            }else{
26.                if(readingreview){
27.                   Matcher matcher = regexPattern.matcher(currentLine);//进行正则匹配
28.                    if(matcher.matches()){
29.                        //格式转化
30.                         currentLineData.append("review=")
31.                        .append(matcher.group(2)).append("|")
32.                        .append(matcher.group(3)).append("|")
33.                        .append(matcher.group(4)).append("|")
34.                        .append(matcher.group(5)).append("#");
35.                    }else{
36.                        System.out.println(currentLine + "不匹配");
37.                    }
38.                }else{
39.                    int indexOf = currentLine.indexOf(":");
40.                    if(indexOf > 0){
41.                        String key = currentLine.substring(0,indexOf).trim();
42.                        String value = currentLine.substring(indexOf+1).trim();
43.                        if(value == null || value.length() == 0){
44.                            continue;
45.                        }
46.                        if(value.indexOf("#") > 0){
47.                            value = value.replaceAll("#", "_");
48.                        }
49.
50.                        if(key.equals("ASIN") || key.equals("Id") ||
key.equals("title")
51.                             || key.equals("group") || key.equals("salesrank")){
52.                            if(key.equals("ASIN")){
53.                                this.currentKey = new Text(value);
54.                            }
55.                            currentLineData.append(key).append("=")
56.                            .append(value.replaceAll(",", "")).append("#");
57.                        }else  if(key.equals("similar")){
58.                            String[] tokens = value.split("\\s+");
59.                            if(tokens.length >= 2){
60.                                currentLineData.append(key).append("=");
61.                                for(int i=1;i<tokens.length;i++){
62.   currentLineData.append(tokens[i].trim())
63.                                                .append("|");
64.                                }
65.                                currentLineData.append("#");
66.                            }
67.                        }else  if( key.equals("reviews")){
68.                            readingreview = true;
69.                        }
70.                    }
71.                }
72.            }
73.        }
74.        System.out.println("数据处理中断");
```

```
75.        return false;
76.    }
```

Hadoop 调用 initialize( )方法并传递输入文件，第 5～15 行打开数据集，定位到记录的起始位置（本数据集记录的起始位置为“Id”）；第 17～72 行调用 nextKeyValue( )方法处理下一条记录，通过 java.util.regex 包下的 Pattern 和 Matcher 类将消费记录（如 1999-1-21 cutomer: A1WK56YQY6DS1K rating: 4 votes:1 helpful:1）转化为 review= A1WK56YQY6DS1K|4|1|1#的形式。

计算购买频度。BuyingFrequencyAnalyzer 类、BuyerRecord 类用于计算购买频度。BuyingFrequencyAnalyzer 类的关键源码如下：

```
1. public static class BuyingFrequencyMapper extends Mapper<Object, Text, Text, 
 IntWritable> {
2.   public void map(Object key, Text value, Context context) throws IOException,
 InterruptedException {
3.      List<BuyerRecord> records=BuyerRecord.parseAItemLine(value.toString());
4.            for (BuyerRecord record : records) {
5.               context.write(new Text(record.customerID), new
6.                                IntWritable(record.itemsBrought.size()));
7.            }
8.         }
9.     }
10.public static class BuyingFrequencyReducer extends Reducer<Text,  IntWritable,
 Text, IntWritable> {
11.    public void reduce(Text key, Iterable<IntWritable> values, Context context)
 throws
12.          IOException,InterruptedException {
13.            int sum = 0;
14.            for (IntWritable val : values) {
15.               sum += val.get();
16.            }
17.            System.out.println(key + "= "+ sum);
18.            context.write(key, new IntWritable(sum));
19.         }
20.     }
21.
```

第 2～6 行为 map( )函数，读取 BuyerRecord 类的 parseAItemLine( )方法传递过来的记录，会员 ID（customerID）作为 key、会员一次购买商品的数量（itemsBrought）作为 value 输出；第 10～18 行为 reduce( )函数，通过迭代器累加会员购买商品的数量，context 输出最终的统计结果。

如下代码展示了 BuyerRecord 类的关键源码：

```
1. public static List<BuyerRecord> parseAItemLine(String itemLine){
2.       List<BuyerRecord> list = new ArrayList<BuyerRecord>();
3.       String[] tokens = itemLine.toString().split("#");//重新清洗后的一条记录以“#”隔开
4.       String itemID = null;
5.       String title = null;
6.       String salesrank = null;
7.       String group = null;
8.       List<String> similarItems = new ArrayList<String>();
9.       for(String token: tokens){      //获取一条记录中的各个字段
10.          String[] keyValue = token.trim().split("=");//字段以“=”隔开
11.          if(keyValue.length < 2){
12.             continue;
```

```
13.             }
14.             if(keyValue[0].equals("ASIN")){
15.                itemID = keyValue[1];
16.                if(itemID.equals("B0000694WE")){
17.                   System.out.println();
18.                }
19.             }else if(keyValue[0].equals("title")){
20.                title = keyValue[1];
21.             }else if(keyValue[0].equals("group")){
22.                group = keyValue[1];
23.             }else if(keyValue[0].equals("salesrank")){
24.                salesrank = keyValue[1];
25.             }else if(keyValue[0].equals("similar")){
26.                String[] items = keyValue[1].split("\\|");
27.                for(String item:items){
28.                   similarItems.add(item);
29.                }
30.             }else if(keyValue[0].equals("review")){    //本次实战的重点字段
31.                BuyerRecord amazonCustomer = new BuyerRecord();
32.                String[] items = keyValue[1].split("\\|");
33.                String customerID = items[0];
34.                String rating = items[1];
35.                ItemData itemData = amazonCustomer.new ItemData();
36.                itemData.itemID = itemID;
37.                itemData.rating= Integer.valueOf(rating);
38.                itemData.title = title;
39.                itemData.salesrank = Integer.valueOf(salesrank);
40.                itemData.group = group;
41.                itemData.similarItems = similarItems;
42.                amazonCustomer.customerID = customerID;   //获取会员 ID
43.                amazonCustomer.itemsBrought.add(itemData); //会员购买商品的详细信息
44.                list.add(amazonCustomer);//添加会员信息到列表
45.             }else if(keyValue[0].equals("Id")){
46.             }else{
47.                throw new RuntimeException("Unknown token 2"+ token  +","+ keyValue
  + " from the line '"+itemLine + "'" );
48.             }
49.    }
50.       return list;
51.    }
```

parseAItemLine( )方法的传入参数为经过 ItemSalesDataReader 类处理后的记录，其中每条记录以“#”隔开，提取每条记录中的各个字段（其中包括了会员 ID、购买商品的数量等）添加到列表 list 中。SimpleResultSorter 类用于对第 1 个 Job 的结果进行排序，关键源码如下所示：

```
1. public static class SimpleKeyMapper extends Mapper<Object, Text, IntWritable, Text>
  {
2.     private static Pattern parsingPattern =Pattern.compile("([^\\s]+)\\s+([^\\s]+)");
3.     public void map(Object key, Text value, Context context) throws IOException,
4.          InterruptedException {
5.              Matcher matcher = parsingPattern.matcher(value.toString());
6.              if (matcher.find()) {
7.                 String propName = matcher.group(1);  //会员 ID
8.                 String propValue = matcher.group(2);  //购买商品的数量
```

```
9.                  context.write(new
10.            IntWritable(-1*Integer.valueOf(propValue)), new Text(propName));
11.             } else {
12.               System.out.println("Unprocessed Line " + value);
13.             }
14.         }
15.     }
16.public static class SimpleKeyReducer extends Reducer<IntWritable, Text, Text,
  IntWritable> {
17.     public void reduce(IntWritable key, Iterable<Text> values, Context context)
  throws
18.             IOException, InterruptedException {
19.                 for (Text valtemp : values) {
20.                     IntWritable recordFrequency = new IntWritable(-1*key.get());
21.                     context.write(valtemp, recordFrequency);
22.                     System.out.println(valtemp + "=" + recordFrequency);
23.         }
24.     }
25. }
```

Pattern 类创建一个正则表达式，第 3～13 行为 map( )函数，Matcher 类通过正则表达式提取会员 ID、购买商品的数量，map( )函数输出的 key 为商品数量（propValue），value 为会员 ID（propName）；第 17～24 行中 reduce( )函数接收 map( )函数输出的 key，通过降序排列后将其作为 value 输出。

习　题

1. 一个.gzip 文件的大小为 139MB，客户端设置 Block 大小为 128MB，请问其占用几个 Block？（　）

   A. 1　　B. 2　　C. 3　　D. 4

2. HDFS 有一个.gzip 文件，大小为 75MB，客户端设置 Block 大小为 64MB。当运行 MapReduce 任务读取该文件时，输入分片大小为（　　）。

   A. 64MB　　B. 75MB　　C. 一个 Map 读取 64MB，另外一个 Map 读取 11MB

3. HDFS 有一个 LZO（with index）文件，大小为 75MB，客户端设置 Block 大小为 64MB。当运行 MapReduce 任务读取该文件时，输入分片大小为（　　）。

   A. 64MB　　B. 75MB　　C. 一个 Map 读取 64MB，另外一个 Map 读取 11MB

4. 下面哪个是 MapReduce 适合做的？（　　）

   A. 迭代计算　　B. 离线计算
   C. 实时交互计算　　D. 流式计算

5. MapReduce Shuffle 过程中，Reduce 是如何知道要从哪个节点获取 Map 输出的？

6. 资源调度框架 YARN 由多个组件构成，请陈述核心组件及其作用。

第 6 章 Hive 大数据仓库

Hive 让精通 SQL 技能的数据分析师能够对存放在 HDFS 中的大规模数据集执行查询，且不要求数据分析师精通 Java 编程。如今 Hive 已经是一个成熟的 Apache 项目，是一个通用的、可伸缩的数据处理平台。本章介绍如何使用 Hive，并假设读者使用过 SQL 和具有常见的数据库体系结构的基础知识。

6.1 Hive 简介

Hive 是基于 Hadoop 文件系统的数据仓库架构，为数据仓库的管理提供了数据抽取、转换、加载（Extract Transformation Load，ETL）工具，增强了数据存储管理和大型数据集的查询和分析能力。为了使用户能够使用与 SQL 相似的操作，使开发人员能够使用 Mapper 和 Reducer 操作，Hive 定义了类 SQL 语言——HiveQL。

Hive 在处理不变的大规模数据集上的批量任务时有很大优势，但在处理小规模数据集时会出现延迟现象，此时，Hive 的性能就不能与传统的关系数据库进行比较了。Hive 不提供数据排序和查询缓存（Cache）功能，不提供在线事务处理，也不提供实时的查询和记录级的更新。

Hive 建立在 Hadoop 的体系结构之上，可将命令转换为 MapReduce 任务，然后交给 Hadoop 集群进行处理。Hive 的体系结构如图 6-1 所示。

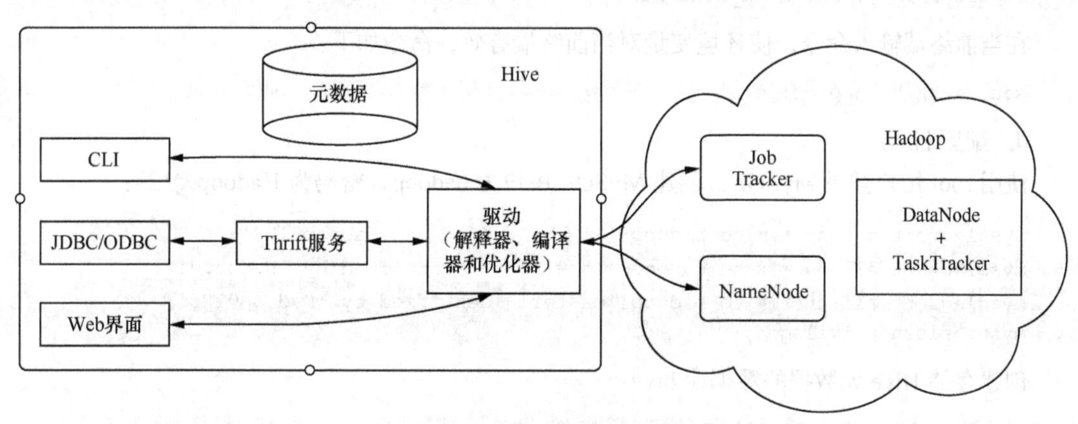

图 6-1 Hive 的体系结构

CLI（Command-Line Interface）即 Hive 命令行界面，是使用 Hive 的常用方式。CLI 可用于创

建表、检查模式以及查询表等，也可用于提供交互式的界面供输入语句或者供用户执行含有HiveQL语句的脚本。Thrift服务允许客户端使用Java、C++、Ruby和其他多种语言通过编程来远程访问Hive，也提供使用Java数据库互连（Java Database Connectivity，JDBC）和开放式数据库互连（Open Database Connectivity，ODBC）访问Hive的功能。所有的Hive客户端都需要一个元数据服务（Metastore Service），Hive使用这个服务存储表模式信息和其他元数据信息，通常会使用关系数据库中的表存储这些信息。解释器、编译器和优化器分别完成HiveQL查询语句从词法分析、语法分析、编译、优化到查询计划的生成各个阶段的任务。生成的查询计划存储在HDFS中，随后由MapReduce调用执行。最后，Hive还提供了一个简单的Web界面，也就是Hive Web界面（Hive Web Interface，HWI），提供了远程访问Hive的服务。

6.2 Hive安装及配置

1. 先决条件

Hive的安装需要在Hadoop已经成功安装的基础上，并且要求Hadoop的各项服务已经正常启动。

2. 下载Hive安装包

Hive官方网站上提供了Hive的文档、各个版本的安装包。读者可以根据自己的网络状态选择对应的镜像下载站点下载Hive 2.3.3的二进制安装包，并将其放置在/home/hadoop/software目录下进行解压，命令如下：

```
tar -zxvf apache-hive-2.3.3-bin.tar.gz
```

3. 配置环境变量

配置环境变量文件.bash_profile，命令如下：

```
vim ~/.bash_profile
```

在文件末尾添加以下内容：

```
export HIVE_HOME=/home/hadoop/software/apache-hive-2.3.3-bin
export PATH=$PATH:$HIVE_HOME/bin
```

在当前终端输入命令，使环境变量对当前终端有效，命令如下：

```
source ~/.bash_profile
```

4. 配置Hive

使用root用户登录MySQL，创建MySQL用户为hadoop，密码为Hadoop@123：

```
mysql>grant all on *.* to hadoop@'%' identified by 'Hadoop@123';
mysql>grant all on *.* to hadoop@'localhost' identified by 'Hadoop@123';
mysql>grant all on *.* to hadoop@'master' identified by 'Hadoop@123';
mysql>flush privileges;
```

创建存储Hive元数据的数据库hive：

```
mysql> create database hive;
Query OK, 1 row affected (0.80 sec)
```

安装MySQL Connector：

```
$ wget
https://dev.mysql.com/get/Downloads/Connector-J/mysql-connector-java-5.1.46.tar.gz
$ tar -zxvf mysql-connector-java-5.1.46.tar.gz
$ cp mysql-connector-java-5.1.46/mysql-connector-java-5.1.46-bin.jar
$HIVE_HOME/lib/
```

Hive 的配置文件名称为 hive-site.xml，这个文件默认是不存在的，需要手动在 Hive 根目录下的 conf 目录中创建。创建 hive-site.xml 的命令如下：

```
vim $HIVE_HOME/conf/hive-site.xml
```

将如下内容添加到 hive-site.xml 文件中：

```
<?xml version="1.0"?>
<?xml-stylesheet type="text/xsl" href="configuration.xsl"?>
<configuration>
    <property>
        <name>hive.metastore.local</name>
        <value>true</value>
    </property>
    <property>
        <name>javax.jdo.option.ConnectionURL</name>
        <value>
    jdbc:mysql://master:3306/hive?characterEncoding=UTF-8&amp;useSSL=false </value>
    </property>
    <property>
        <name>javax.jdo.option.ConnectionDriverName</name>
        <value>com.mysql.jdbc.Driver</value>
    </property>
    <property>
        <name>javax.jdo.option.ConnectionUserName</name>
        <value>hadoop</value>
    </property>
    <property>
        <name>javax.jdo.option.ConnectionPassword</name>
        <value>Hadoop@123</value>
    </property>
</configuration>
```

一切准备完成后，通过 Hive 的 schematool 初始化 Hive 元数据，命令如下：

```
[hadoop@master software]$ schematool -dbType mysql -initSchema
Metastore connection URL:
jdbc:mysql://master:3306/hive?characterEncoding=UTF-8&useSSL=false
Metastore Connection Driver :   com.mysql.jdbc.Driver
Metastore connection User:  hadoop
Starting metastore schema initialization to 2.3.0
Initialization script hive-schema-2.3.0.mysql.sql
Initialization script completed
schemaTool completed
```

5. 运行 Hive

在上述配置完成后，启动 Hive 客户端，命令如下：

```
[hadoop@master software]$ hive
hive>
```

6.3 从创建数据库到创建表

6.3.1 数据类型

Hive 不仅支持关系数据库中大多数的基本数据类型，同时也支持很少出现的 3 种集合数据类型（STRUCT、MAP、ARRAY）。Hive 的数据类型可以分为两大类：基础数据类型和复杂数据类型。表 6-1 中列举了 Hive 的基础数据类型。

表 6-1　基础数据类型

序号	数据类型	描述	示例
1	TINYINT	微整型，长度为 1 BYTE	10
2	SMALLINT	小整型，长度为 2 BYTE	10
3	INT	整型，长度为 4 BYTE	10
4	BIGINT	大整型，长度为 8 BYTE	10
5	BOOLEAN	布尔类型，取值为 FALSE 或 TRUE	TRUE
6	FLOAT	单精度浮点数	1.23456
7	DOUBLE	双精度浮点数	1.23456
8	STRING	字符序列，可指定字符集。可使用单引号和双引号	'hello hive' "hello hadoop"
9	TIMESTAMP	时间戳	'2017-04-07 15：05：56.1231352'
10	BINARY	字节数组	

表 6-2 中列举了 Hive 的复杂数据类型。

表 6-2　复杂数据类型

序号	数据类型	描述	示例
1	STRUCT	STRUCT 用于封装一组有名字的字段，其类型可以是任意的基本数据类型，可以通过"."号来访问元素的内容	names('Zoro', 'Jame')
2	MAP	MAP 用于表示一组键值对元组集合，使用数组表示法可以访问其元素。其中键只能是基本数据类型的，值可以是任意类型的	money('Zoro',1000, 'Jame'，800)
3	ARRAY	ARRAY 用于表示一系列相同数据类型的元素，每个元素都有一个编号，从 0 开始。例如 fruits['apple','orange','mango']，可通过 fruits[1]来访问 orange	fruits('apple','orange','mango')
4	UNION	UNION 类似于 C 语言中的 UNION 结构。在给定的任何一个时间点，UNION 类型可以保存指定数据类型中的任意一种，类型声明语法为 UNIONTYPE<data_type, data_type,…>。每个 UNION 类型的值都通过一个整数来表示其类型，这个整数位声明时的索引从 0 开始	——

6.3.2　创建数据库

Hive 是一种数据仓库技术，可以定义数据库和表来分析结构化数据。如果用户没有显式指定数据库，那么将使用默认的数据库 default。

下面是创建数据库的语句：

```
CREATE DATABASE|SCHEMA [IF NOT EXISTS] <database name>
```

这里方括号中的 IF NOT EXISTS 是可选子句，若创建的数据库已经存在，没有这个子句时，会抛出一个错误信息。创建数据库时，SCHEMA 和 DATABASE 是可以互换的，它们的含义一致。

创建一个数据库，名称为 first：

```
    hive> CREATE DATABASE [IF NOT EXISTS] first;
```

或采用如下命令创建：

```
    hive> CREATE SCHEMA first;
```

使用如下命令查看 Hive 中所有的数据库：

```
hive> SHOW DATABASES;
default
first
```

Hive 会为每一个数据库创建一个目录，数据库中的表会以数据库目录的子目录形式存储。但是 default 数据库没有本身的目录，为了便于区分，default 数据库的表目录是以.db 结尾的。

用户可以使用 DROP DATABASE 删除数据库，命令如下：

```
DROP DATABASE StatementDROP (DATABASE|SCHEMA) [IF EXISTS] database_name
[RESTRICT|CASCADE];
```

IF EXISTS 子句同样是可选的，加上这个子句能避免因数据库不存在而抛出警告信息。

注意，Hive 不允许用户删除含有表的数据库。用户可先删除数据库中的所有表，再删除数据库；或者在删除命令的最后加上关键字 CASCADE，这意味着在删除数据库前删除所有相应的表。使用 RESTRICT 关键字就和默认情况一样，若删除数据库，必须先删除数据库中的所有表。

6.3.3　创建表

CREATE TABLE 是用于在 Hive 中创建表的语句。语法如下：

```
CREATE [TEMPORARY] [EXTERNAL] TABLE [IF NOT EXISTS] [db_name.] table_name
[(col_name data_type [COMMENT col_comment], ...)]
[COMMENT table_comment]
[ROW FORMAT row_format]
[STORED AS file_format]
LIKE table_name1
[LOCATION hdfs_path]
```

关键字说明如下。

- CREATE TABLE，指定名字创建一个表。若数据库中相同名字的表已经存在，则抛出异常；若用户添加了 IF NOT EXISTS，则会忽略这个异常。
- EXTERNAL，创建一个外部表，后面的 LOCATION 语句是外部表的存放路径。
- COMMENT，为表的属性添加注释。

• LIKE，用户将已存在的表复制给新建表，复制定义而不复制数据。

创建普通表：

```
CREATE TABLE employees(
id          int COMMENT 'employee id'
name        string  COMMENT 'employee name'
salary      float COMMENT 'employee salary'
address     struct<city:string,state:string,street:string> COMMENT 'employee home'
)
```

复制表 employees：

```
CREATE TABLE IF NOT EXISTS test like employees;
```

使用关键字 SHOW TABLES 命令列举所有的表。完成以上操作之后，查看已存在的表，进行如下操作：

```
hive>show tables;
employees
test
```

可以使用 DESCRIBE EXTENDED database.tablename 来查看所建表的详细信息。查看 employees 表详细信息的命令如下：

```
Hive> DESCRIBE EXTENDED first.employees;
id        int   employee id
name      string  employee name
salary    float  employee salary
address   struct<city:string,state:string,street:string>  employee home
```

还可以只查看某一些想要的信息，只需要在表名后增加相应字段的名称即可，命令如下：

```
Hive>DESCRIBE first.employees.name;
name  string  employee name
```

本书之前创建的表都是内部表，也可以称为管理表。当我们删除一个管理表时，Hive 会删除这个表中的数据。通过关键字 EXTERNAL 创建的是外部表，创建外部表时需要使用 LOCATION 语句告诉 Hive 数据所位于的路径。删除外部表时并不会删除数据，只会删除描述表的数据信息。用户可以通过语句 DESCRIBE EXTENDED tablename 查看表是外部表还是管理表。若为管理表，可以看到如下信息：

```
… tableType:MANAGED_TABLE)
```

若为外部表，可以看到如下信息：

```
… tableType:EXTERNAL_TABLE)
```

管理表数据由 Hive 自身管理，外部表数据由 HDFS 管理，它们之间的区别可以归纳为以下几点。

（1）管理表数据存储的位置是 hive.metastore.warehouse.dir（默认位置是/user/hive/warehouse），外部表数据的存储位置由用户指定（如果用户没有指定位置，Hive 将在 HDFS 上的/user/hive/warehouse 文件夹下以外部表的表名创建一个文件夹，并将属于这个表的数据存放在这里）。

（2）删除管理表会直接删除元数据（metadata）及存储数据；删除外部表仅仅会删除元数据，HDFS 上的文件并不会被删除。

（3）对管理表的修改会将修改直接同步给元数据，而对外部表的表结构和分区进行修改，则需要使用修复命令（命令格式为 MSCK REPAIR TABLE table_name;）。

6.3.4　删除表

Hive 中删除表的命令操作为：

```
DROP TABLE [IF EXISTS] employees;
```

关键字 IF EXISTS 可以选择使用，如果表不存在会抛出错误信息；若加上 IF EXISTS 关键字，则不会抛出错误信息。

在 6.3.3 小节中已经提到，若删除管理表，表中的元数据信息和表内的数据都会被删除。若删除外部表，只会删除外部表的元数据信息，不会删除表内的数据。

6.3.5　修改表

通过 ALTER TABLE 语句修改大多数表属性时，只会修改表的元数据，不会修改表内的数据。

1. 表重命名

使用如下语句可以对表进行重命名：

```
ALTER TABLE first_table RENAME TO second_table;
```

2. 增加、修改和删除表分区

可通过 ALTER TABLE table ADD PARTITION 语句为表 table 增加新的分区：

```
ALTER TABLE message ADD [IF NOT EXISTS] PARTITION (year = 2017,month = 4,day = 11)LOCATION
'logs/2017/4/11'
```

同时，用户还能通过以下语句删除某个分区：

```
ALTER TABLE message DROP [IF EXISTS] PARTITION(year = 2017, month = 4,day = 11);
```

上面语句中的 IF EXISTS 为可选语句。对于管理表，即使使用 ALTER TABLE … ADD PARTITION 语句增加分区，分区内的数据也是会同时和元数据一起被删除的；对于外部表，分区内数据不会被删除。

3. 修改信息

用户可以对表中的某个字段进行重命名，同时修改其位置、类型、注释等信息：

```
ALTER TABLE employees
CHANGE COLUMN name employeename string
COMMENT 'change name to employeename'
```

即使字段名或者字段类型没有改变，也需要显式指定原来的字段名，并给出新的字段名及新的字段类型。关键字 COLUMN 和 COMMENT 子句都是可选的。

4. 增加字段

用户可以在分区字段之前增加新的字段到已有的字段之后：

```
ALTER TABLE message ADD COLUMNS(
App_name string COMMENT 'application name',
Session_id LONG COMMENT 'The current session id');
```

其中，COMMENT 可选，为属性注释。如果新增的字段中有某个或者多个字段位置是错误的，那么需要使用“ALTER COLUMN 表名 CHANGE COLUMN”语句逐一将字段调整到正确的位置。

5. 修改表属性

用户可以修改已经存在的属性，但是不能删除表属性：

```
ALTER TABLE message SET TBLPROPERTIES('note'='this column is always NULL');
```

6. 数据操作

当数据被加载至表中时，Hive 不会对数据进行任何转换。LOAD DATA 操作只将数据复制并移动至 Hive 表对应的位置：

```
LOAD DATA [LOCAL] INPATH 'filepath'
[OVERWRITE] INTO TABLE tablename
[PARTITION (partcol1=val1, partcol2=val2 ...)]
```

若分区目录不存在，Hive 会先创建分区目录，然后将数据复制到该目录下。若目标表不是分区表，那么 PARTITION 子句应该省去。一般指定的路径是一个目录，而不是单个独立的文件。Hive 会将所有文件都复制到目录中，这样方便用户组织数据到多个文件中。

如果使用了 LOCAL 关键字，Hive 会复制本地文件系统路径下的数据到目标位置。如果不使用 LOCAL，那么路径应该是 HDFS 的路径。如果使用了 OVERWRITE 关键字，目标文件夹中原有的数据将会被删除，否则 Hive 只会把新增的数据增加到目标文件夹中而不会删除之前的数据。但是如果存在同名的文件，旧的同名文件会被覆盖。

INSERT 语句允许用户向目标表中插入数据，用户可以把一个 Hive 表导入另一个已建的表：

```
INSERT OVERWRITE TABLE tablename
[PARTITION (partcol1=val1, partcol2=val2 ...)]
select_statement FROM from_statement
```

这里使用 OVERWRITE 关键字，因此之前分区中的内容会被覆盖。如果没有使用 OVERWRITE 关键字或者使用 INTO 关键字替换，Hive 会以追加的方式写入数据而不会覆盖之前已存在的数据。

Hive 还提供了一个动态分区功能，可以基于查询参数推断出需要创建的分区名称：

```
INSERT OVERWRITE TABLE employees
PARTTION(country,state)
SELECT …,se.cnty,se.st
FROM staged_employees se;
```

Hive 根据 SELECT 语句中的最后两列来确定分区字段 country 和 state 的值。值得注意的是，字段值和输出分区值之间的关系是根据位置而不是根据命名来匹配的。如果 staged_employees 中共有 100 个国家，执行完上面的查询后，表 employees 将会有 100 个分区。

用户可以在一个语句中完成创建表并将查询结果载入这个表：

```
CREATE TABLE ca_employees
AS SELECT name,salary
FROM employees
WHERE se.state = 'HB'
```

这张表只含有 employees 表中来自 HB（湖北）的雇员的 name 和 salary 信息，新表的模式是根据 SELECT 语句来生成的。

6.4　数据查询及自定义函数运算

HiveQL 是一种查询语言，本章主要介绍如何使用 HiveQL 语句以及 Hive 自定义函数的运算。

6.4.1　HiveQL 操作

HiveQL 的基础查询语句是 SELECT 语句和 WHERE 子句，SELECT 语句从表中查询数据，WHERE 子句类似于一个条件。下面是 SELECT 查询的语法：

```
SELECT [ALL | DISTINCT] select_expr, select_expr, ...
FROM table_reference
[WHERE where_condition]
[GROUP BY col_list]
[HAVING having_condition]
[SORT BY col_list | [DISTRIBUTE BY col_list] [CLUSTER BY col_list]]
[LIMIT number];
```

下面对重要的关键字进行介绍。

1. ALL 和 DISTINCT

ALL 和 DISTINCT 可以定义重复的记录是否要返回。若没有定义，则默认为 ALL，即输出所有匹配的记录而不去掉重复的记录。例如：

```
hive>SELECT col1,col2 FROM t1
    1 2
    1 2
    2 4
    2 4
Hive>SELECT DISTINCT col1,col2 FROM t1
    1 2
    2 4
```

2. table_reference

table_reference 指明查询的信息，可以是一个表、一个视图或一个子查询。

3. WHERE

WHERE 子句用于过滤条件，与 SELECT 结合使用可以查询符合过滤条件的记录。WHERE 语句使用谓词表达式，一些谓词表达式可以使用 AND 和 OR 相连接。where_condition 是一个布尔表达式，当结果为 TRUE 时，匹配到的记录被保留并输出。

4. GROUP BY

GROUP BY 子句将数据按照一个或者多个列的结果进行分组，然后对每组执行聚合函数进行聚合操作。

5. HAVING

HAVING 子句允许用户通过简单的语法，完成原本需要通过子查询才能对 GROUP BY 子句产生的分组进行条件过滤的任务。例如，以下查询语句增加了一个 HAVING 子句来限制输出结果中年平均收入高于 50 元：

```
hive>SELECT year(ymd),avg(price) FROM stocks
    >WHERE exchange = 'NASDAQ' and symbol = 'HUAWEI'
```

```
        >GROUP BY year(ymd)
        >HAVING avg(price)>50.0;
    2001  51.2323423423
    2003  53.2342356756
    2004  55.2323456897
    2005  53.2234001203
```

6. ORDER BY 和 SORT BY

HiveQL 中的 ORDER BY 子句和其他 SQL 中的定义一样，同样是对查询结果进行全局排序，即所有的数据都会通过一个 reduce( )函数进行处理，对于大量数据这会消耗很长的时间。

HiveQL 中增加了一个可选的方式——SORT BY。使用 SORT BY 子句只会在每个 Reduce 中对数据进行排序，保证局部有序，这样提高了全局排序的效率。

7. DISTRIBUTE BY

DISTRIBUTE BY 控制 Map 的输出在 Reduce 中是如何划分的，按照指定的字段将数据划分到不同的 Reduce 输出文件中：

```
hive>INSERT overwrite local directory '/home/hadoop/out' SELECT * FROM test
    >ORDER BY name DISTRIBUTE BY length(name);
```

此方法会根据 name 的长度将数据划分到不同的 Reduce 中，最终输出到不同的文件中。

8. CLUSTER BY

CLUSTER BY 除了具有 DISTRIBUTE BY 的功能外，还兼具 SORT BY 的功能，但是排序只能是倒序排序，不能指定排序规则。ASC 关键字（默认的）表示按升序排序，DSEC 关键字表示降序排序。

9. LIMIT

LIMIT 子句用于限制查询返回的记录行数，并随机选择检索结果中的对应数量输出。

10. 使用正则表达式和基于分区的查询

SELECT 声明可以匹配使用一个正则表达式的列。通常 SELECT 查询需要扫描全部的表，但如果使用了 PARTITIONED BY 语句，那么可以对输入进行“剪枝”，只对表的相关部分进行扫描。

6.4.2 JOIN 语句

Hive 支持通常的 SQL 的 JOIN 语句，但是只支持等值连接。

1. 内连接

内连接（Inner Join）中，只有进行连接的两个表中都存在与连接标准相匹配的数据才会被保留下来。这和 SQL 的内连接相似。

执行以下语句查询每个学生的编号和教师名：

```
hive>SELECT a.stuNo,b.teacherName FROM student a
    >JOIN teacherb ON a.classNo = b.classNo;
```

如果需要查看 Hive 的执行计划，可以在语句前加上 EXPLAIN，比如：

```
    EXPLAIN SELECT a.stuNo,b.teacherName FROM student a JOIN teacher b ON a.classNo =
b.classNo;
```

2. 左外连接

左外连接（Left Outer Join）通过关键字 LEFT OUTER 进行识别。在这种连接操作中，JOIN 操作符左边表中符合 WHERE 子句的所有记录将会被返回。JOIN 操作符右边表中如果没有符合 ON 后面连接条件的记录，那么从右边表选择的列的值将会是 NULL。

```
hive>SELECT s.ymd,s.symbol,s.price,d.dividend
    >FROM stocks s LEFT OUTER JOIN dividends d ON s.ymd = d.ymd AND s.symbol = d.symbol
WHERE s.symbol = 'HUAWEI';
```

3. 右外连接

右外连接（Right Outer Join）会返回右边表所有符合 WHERE 子句的记录，左边表中匹配不上的字段值用 NULL 代替。

4. 完全外连接

完全外连接（Full Outer Join）将会返回左、右边表中所有符合 WHERE 子句的记录。如果任一表的指定字段没有符合条件的值，就使用 NULL 值替代。

5. 半连接

Hive 不提供 IN 子查询。此时可以用左半开连接（Left Semi Join）实现同样的功能。左半开连接会返回左边表的记录，前提是其记录对右边表满足 ON 语句中的判定条件。对常见的内连接来说，这是特殊的优化的情况。

半连接（Semi Join）比通常的内连接要更高效。对于左边表中一条指定的记录，在右边表中一旦找到匹配的记录，Hive 就会停止扫描，所以左边表中选择的列是可以预测的。

```
SELECT * FROM teacher LEFT SEMI JOIN student ON student.classNo = teacher.classNo;
```

注意：右边表（student）中的字段只能出现在 ON 子句中，不能出现在其他地方，比如不能出现在 SELECT 子句中。

6. Map 连接

当一个表非常小，足以直接装载到内存中时，可以使用 Map 连接（Map Join）以提高效率，比如：

```
SELECT /*+ MAPJOIN(teacher) */ a.stuNo, b.teacherName FROM student a JOIN teacher b
ON a.classNo = b.classNo;
```

当连接用到不等值判断时，也比较适合 Map 连接。具体原因需要深入了解 Hive 和 MapReduce 的工作原理。因为篇幅，在此就不展开讲述了。

6.4.3　内置操作符和函数

1. 内置操作符

表 6-3 描述了关系操作符，这些操作符同样可以用于 JOIN … ON 和 HAVING 语句中。

表 6-3　关系操作符

操作符	类型	描述
A = B	数值类型	若 A 与 B 相等则返回 TRUE，否则返回 FALSE
A <=> B	数值类型	对于非 NULL 操作数，结果与=相同。若 A 与 B 都是 NULL，则返回 TRUE；若其中之一是 NULL，则返回 FALSE
A <> B	数值类型	若 A 或 B 为 NULL，返回 NULL。若 A 与 B 不相等，返回 TRUE，否则返回 FALSE
A != B	数值类型	同<> 操作符
A < B	数值类型	若 A 或 B 为 NULL，返回 NULL。若 A 小于 B 返回 TRUE，否则返回 FALSE
A <= B	数值类型	若 A 或 B 为 NULL，返回 NULL。A 小于等于 B 返回 TRUE，否则返回 FALSE
A > B	数值类型	若 A 或 B 为 NULL，返回 NULL。A 大于 B 返回 TRUE，否则返回 FALSE

续表

操作符	类型	描述
A >= B	数值类型	若 A 或 B 为 NULL,返回 NULL。A 大于等于 B 返回 TRUE,否则返回 FALSE
A [NOT] BETWEEN B AND C	数值类型	若 A、B 或 C 为 NULL，返回 NULL。如果 A 大于等于 B 且小于等于 C，返回 TRUE，否则返回 FALSE
A IS NULL	数值类型	若 A 为 NULL，返回 TRUE，否则返回 FALSE
A IS NOT NULL	数值类型	若 A 为 NULL，返回 FALSE，否则返回 TRUE
A [NOT] LIKE B	字符串	若 A 或 B 为 NULL，返回 NULL。若字符串 A 匹配 SQL 正则表达式 B，返回 TRUE，否则返回 FALSE。B 中的_字符匹配 A 中的任何字符，%字符匹配 A 中任意数量的字符，例如'foobar' like 'foo' 得到 FALSE, 'foobar' like 'foo_ _ _'得到 TRUE, 'foobar' like 'foo%'也得到 TRUE
A RLIKE B	字符串	若 A 或 B 为 NULL，返回 NULL。如果 A 中的任何子串（可能是空字符串）匹配 Java 正则表达式 B,返回 TRUE,否则返回 FALSE。例如 'foobar' RLIKE 'foo' 结果为 TRUE， 'foobar' RLIKE '^f.*r$'结果也为 TRUE
A REGEXP B	字符串	同 RLIKE

表 6-4 描述了 Hive 的算术操作符，算术运算的返回值都为数值类型。如果任意的操作数为 NULL，那么结果也是 NULL。算术操作符接受任意的数值类型，如果两种数值类型不同，那么值范围较小的那个数值类型将会向上转换为范围更广的数值类型。例如，INT 类型数据和 BIGINT 类型数据运算，INT 类型数据会将类型向上转换为 BIGINT 类型数据。当进行算术运算时，用户需要注意数据溢出或数据下溢问题。Hive 遵循底层 Java 中数据类型的规则，所以当发生溢出或者下溢时，计算结果不会自动转换为更广泛的数值类型。其中乘法和除法极有可能引发这个问题。

表 6-4 算术操作符

操作符	类型	描述
A + B	数值类型	返回 A 加 B 的结果。结果类型为两个操作数中较高层级的类型，比如一个操作数类型为 INT，另一个为 FLOAT，则结果类型为 FLOAT
A−B	数值类型	返回 A 减 B 的结果
A * B	数值类型	返回 A 乘 B 的结果。如果相乘引起溢出，必须将低层级的类型转换为高层级的类型
A / B	数值类型	返回 A 除以 B 的结果，结果类型为 DOUBLE
A % B	数值类型	返回 A 除以 B 的余数，结果类型为两个操作数中较高层级的类型
A & B	数值类型	返回 A 和 B 按位与的结果，结果类型为两个操作数中较高层级的类型
A | B	数值类型	返回 A 和 B 按位或的结果，结果类型为两个操作数中较高层级的类型
A ^ B	数值类型	返回 A 和 B 按位异或的结果，结果类型为两个操作数中较高层级的类型
~A	数值类型	返回 A 按位非的结果，结果类型与 A 一致

表 6-5 所示的操作符用于逻辑表达式，返回值为 TURE、FALSE 或 NULL。

表 6-5　　逻辑操作符

操作符	类型	描述
A AND B	BOOLEAN	若 A 与 B 都为 TRUE，结果为 TRUE，否则为 FALSE。若 A 或 B 为 NULL，结果为 NULL
A && B	BOOLEAN	同 A AND B
A OR B	BOOLEAN	若 A 或 B 为 TRUE，结果为 TRUE。若 A 或 B 为 FALSE 或 NULL，结果为 FALSE
A || B	BOOLEAN	同 A OR B
NOT A	BOOLEAN	若 A 为 FALSE，结果为 TRUE。若 A 为 NULL，结果为 NULL，否则为 FALSE
! A	BOOLEAN	同 NOT A
A IN (val1, val2, ...)	BOOLEAN	若 A 与任意值相等则返回 TRUE
A NOT IN (val1, val2,...)	BOOLEAN	若 A 与任何值都不相等则返回 TRUE
[NOT] EXISTS (subquery)	BOOLEAN	若子查询返回至少一行数据则返回 TRUE

2. 内置函数

下面介绍在 Hive 中可用的内置函数。这些函数类似于 SQL 的函数，但是使用上有所区别。Hive 支持表 6-6 所示的内置函数。

表 6-6　　内置函数

函数名	返回类型	描述
round(DOUBLE a)	BIGINT	返回 DOUBLE a 最近的 BIGINT 类型的值
floor(DOUBLE a)	BIGINT	返回等于或小于 DOUBLE a 的最大 BIGINT 类型的值
ceil(DOUBLE a)	BIGINT	返回等于或大于 DOUBLE a 最小 BIGINT 类型的值
rand(), rand(INT seed)	DOUBLE	返回一个随机数，接收一个随机数种子
concat(STRING A, STRING B,...)	STRING	返回 A、B 连接产生的字符串
substr(STRING A, INT start)	STRING	返回一个从起始位置 start 直到 A 结束的子字符串
substr(STRING A, INT start, INT length)	STRING	返回给定长度的从起始位置 start 开始的字符串
upper(STRING A)	STRING	返回 A 中所有字符为大写的字符串
ucase(STRING A)	STRING	同 upper( )
lower(STRING A)	STRING	返回 A 所有字符为小写的字符串
lcase(STRING A)	STRING	同 lower( )
trim(STRING A)	STRING	返回 A 两端去除空格的结果
ltrim(STRING A)	STRING	返回 A 从左边去除空格产生的字符串
rtrim(STRING A)	STRING	返回 A 从右边去除空格产生的字符串

续表

函数名	返回类型	描述
regexp_replace(STRING A, STRING B, STRING C)	STRING	返回 A 中的符合 Java 正则表达式 B 的部分替换为 C 的字符串
size(Map<K.V>)	INT	返回 Map 类型元素的数量
size(Array<T>)	INT	返回数组类型元素的数量
cast(<expr> as <type>)	value of <type>	转换数据类型。如果转换不成功，返回 NULL
from_unixtime(INT unixtime)	STRING	转化 UNIX 时间戳(从 UTC 时间 1970-01-01 00:00:00 到指定时间的秒数)为前时区的时间格式
to_date(STRING timestamp)	STRING	返回一个时间戳字符串的日期部分
year(STRING date)	INT	返回日期或时间戳字符串的年份部分
month(STRING date)	INT	返回日期或时间戳字符串的月份部分
day(STRING date)	INT	返回日期或时间戳字符串的日期部分
get_json_object(STRING json_string, STRING path)	STRING	从指定路径的 JSON 字符串提取 JSON 对象，并返回提取的 JSON 对象。如果输入的 JSON 字符串无效，返回 NULL

6.5 Hive 自定义函数编程

用户自定义函数（User-Defined Function, UDF）接收单行输入，并产生单行输出。自定义函数的类需要继承 UDF 类，且重写其 evaluate( )方法。这个方法接收参数，并返回字符串。evaluate( )方法并没有作为接口提供，因为实际使用时，函数的参数个数及类型是多变的。

创建自定义函数的流程如下。

（1）自定义一个 Java 类。

（2）继承 UDF 类。

（3）重写 evaluate( )方法。

（4）打包成.jar 包。

（5）在 Hive 中执行 add jar 命令。

（6）在 Hive 中执行创建模板函数。

（7）在 HiveQL 中使用。

下面的实例创建了两个自定义函数，并演示了其使用方式。

6.5.1 数据准备

本小节用来测试用户自定义函数的数据，来自国家统计局公开的第六次人口普查数据，其中包含全国各省、自治区、直辖市的人口统计数据。

数据文件名为 population.csv，其中第 1 列为地区，第 2、3 列分别为男性和女性人口的数量，如下所示。

```
全国,682329104,650481765
北京,10126430,9485938
天津,6907091,6031602
河北,36430286,35423924
山西,18338760,17373341
内蒙古,12838243,11868048
…
```

将数据导入 Hive 之前，需要先创建其表定义。

进入 Hive，使用 Hive 的默认数据库：

```
hive> use default;
```

执行以下操作，创建一个 Hive 表 population，表中字段分别为 district、man、woman，其中 district 以 STRING 格式保存，man 和 woman 以 INT 格式保存。ROW FORMAT DELIMITED 关键字用来设置创建的表在加载数据时，支持的列分隔符；FIELDS TERMINATED BY ','关键字指定了分隔符为逗号；STORED AS TEXTFILE 关键字表示加载数据的文件格式为文本文件。

```
hive> CREATE TABLE population
    > (district STRING, man INT, woman INT)
    > ROW FORMAT DELIMITED
    > FIELDS TERMINATED BY ','
    > STORED AS TEXTFILE;
```

创建完成后，从 population.csv 中导入数据：

```
hive> load data local inpath "/home/hadoop/population.csv" into table population;
```

6.5.2 编程实现

使用 Java 编程实现类 AddUDF，继承 UDF 类，代码如下：

```
1.package com.test.hive.udf;
2.
3.import org.apache.hadoop.hive.ql.exec.UDF;
4.
5.public class AddUDF extends UDF{
6.    public Integer evaluate(Integer a, Integer b) {
7.        if (null == a || null == b) {
8.            return null;
9.        }
10.        return a + b;
11.    }
12.    public Double evaluate(Double a, Double b) {
13.        if (a == null || b == null)
14.            return null;
15.        return a + b;
16.    }
17.    public Integer evaluate(Integer... a) {
18.        int total = 0;
19.        for (int i = 0; i < a.length; i++)
```

```
20.             if (a[i] != null)
21.                 total += a[i];
22.         return total;
23.     }
24. }
```

通过对 evaluate( )方法的重写，使 AddUDF 可以将两个整数，或两个浮点数，或多个整数加起来，并返回结果。

使用 Java 编程实现类 Translations，继承 UDF 类，代码如下：

```
1. package com.test.hive.udf;
2.
3. import net.sourceforge.pinyin4j.PinyinHelper;
4. import net.sourceforge.pinyin4j.format.HanyuPinyinCaseType;
5. import net.sourceforge.pinyin4j.format.HanyuPinyinOutputFormat;
6. import net.sourceforge.pinyin4j.format.HanyuPinyinToneType;
7. import net.sourceforge.pinyin4j.format.HanyuPinyinVCharType;
8. import net.sourceforge.pinyin4j.format.exception.BadHanyuPinyinOutputFormatCombination;
9. import org.apache.hadoop.hive.ql.exec.UDF;
10.
11. public class Translations  extends UDF{
12.
13.     //pinyin4j 开源包
14.     public String evaluate(String src) {
15.         if(src==null) {
16.             return null;
17.         }
18.         char[] chars = src.toCharArray();
19.
20.         HanyuPinyinOutputFormat format = new HanyuPinyinOutputFormat();
21.         format.setCaseType(HanyuPinyinCaseType.LOWERCASE);
22.         format.setToneType(HanyuPinyinToneType.WITHOUT_TONE);
23.         format.setVCharType(HanyuPinyinVCharType.WITH_U_UNICODE);
24.
25.         String results = "";
26.         for(int i=0; i<chars.length; i++) {
27.             String[] strs;
28.             try {
29.                 strs = PinyinHelper.toHanyuPinyinStringArray(chars[i], format);
30.                 results += strs[0];
31.             } catch (BadHanyuPinyinOutputFormatCombination e) {
32.                 e.printStackTrace();
33.             }
34.         }
35.         return results;
36.     }
37. }
```

Translations 类通过开源包 pinyin4j，重写 evaluate( )方法，实现了从汉字到拼音的转换。

6.5.3 使用自定义函数

将编写好的代码打包为 myUDF.jar，并通过文件传送协议（File Transfer Protocol，FTP）工具上传至/home/hadoop 目录。

进入 Hive，使用默认数据库：

```
hive> use default;
```

向 Hive 中导入.jar 包：

```
hive> add jar /home/hadoop/myUDF.jar;
```

创建临时函数 myaddUDF( )，指向编写好的 AddUDF 类：

```
hive> create temporary function myaddUDF as "com.test.hive.udf.AddUDF";
```

使用 myaddUDF( )查询 population 表，可以得到各地区男、女人口的总和：

```
hive> select man,woman,myaddUDF(man,woman) from population;
OK
682329104 650481765 1332810869
10126430 9485938 19612368
6907091 6031602 12938693
36430286 35423924 71854210
18338760 17373341 35712101
12838243 11868048 24706291
22147745 21598578 43746323
13907218 13545597 27452815
19426106 18887885 38313991
11854916 11164280 23019196
39626707 39034234 78660941
27965641 26461250 54426891
30245513 29254955 59500468
18981054 17913163 36894217
23003521 21564276 44567797
48446944 47345775 95792719
47493063 46536876 94029939
29391247 27846480 57237727
33776459 31924303 65700762
54400538 49919921 104320459
23924704 22099057 46023761
4592283 4079202 8671485
14608870 14237300 28846170
40827834 39589694 80417528
17905471 16843085 34748556
23856696 22110070 45966766
1542652 1459513 3002165
19287575 18039804 37327379
13064193 12511070 25575263
2913793 2712930 5626723
3227404 3073946 6301350
11270147 10545668 21815815
Time taken: 1.673 seconds, Fetched: 32 row(s)
```

创建临时函数 toPinyin( )：

```
hive> create temporary function toPinyin as "com.test.hive.udf.Translations";
```

使用 toPinyin( )查询 population 表，可以将地区的汉字转换为拼音：

```
hive> select district, toPinyin(district) from population;
OK
全国 quanguo
北京 beijing
```

```
天津 tianjin
河北 hebei
山西 shanxi
内蒙古 neimenggu
辽宁 liaoning
吉林 jilin
黑龙江 heilongjiang
上海 shanghai
______________
Time taken: 0.145 seconds, Fetched: 32 row(s)
```

6.6 Hive 实战

下面用一个实例来演示 Hive 的功能。数据集放置于/home/hadoop/data 路径下，名称为 shot_logs.txt。

数据集包括了 2014—2015 赛季 NBA 30 支球队 904 场常规赛的 281 名球员的投篮数据，数据包括比赛双方、主客场、胜负情况、投篮球员、防守球员、投篮距离、投篮命中次数等 21 个字段。

数据字段如下。

- game_id：比赛 ID。
- matchup：比赛时间及球队。
- loc：主场（H）/客场（A）。
- w：比赛结果。
- final_margin：最终得分差距。
- shot_number：投篮次数。
- period：第几节。
- game_clock：小节比赛时间。
- shot_clock：投篮时间（设置为 24 秒）。
- dribbles：运球次数。
- touch_time：触球时间。
- shot_dist：投篮距离。
- pts_type：得分类型，2 分或 3 分。
- shot_result：投篮结果。
- closest_defender：防守球员。
- closest_defender_player_id：防守球员 ID。
- close_def_dist：防守球员距离。
- fgm：投篮命中次数。
- pts：得分。
- player_name：球员名字。
- player_id：球员编号。

6.6.1 数据准备

首先是建表，语句如下：

```
    hive> create table nbaLogs (game_id INT, matchup STRING, loc STRING, w STRING,
final_margin INT, shot_number INT, period INT, game_clock STRING, shot_clock STRING,
dribbles INT, touch_time STRING, shot_dist STRING, pts_type INT, shot_result STRING,
closest_defender STRING, closest_defender_player_id INT, close_def_dist STRING, fgm INT,
pts INT, player_name STRING, player_id INT)
        > ROW FORMAT DELIMITED
        > FIELDS TERMINATED BY '\t'
        > STORED AS TEXTFILE;
    OK
    Time taken: 5.508 seconds
```

其中，FIELDS TERMINATED BY '\t'指定数据文件的列使用制表符分隔。

从本地文本文件中导入数据：

```
    hive> LOAD DATA LOCAL INPATH '/home/hadoop/data/shot_logs.txt' INTO TABLE nbaLogs;
    Loading data to table default.nbalogs
    OK
    Time taken: 1.434 seconds
```

6.6.2 实战步骤

1. 查询每场比赛投篮的次数

每条数据都表示一次投篮，统计每场比赛的投篮次数，即统计相同 game_id 数据的条数：

```
    hive> select count(game_id), game_id from nbaLogs group by game_id;
    OK
    162    21400001
    137    21400002
    117    21400003
    161    21400004
    111    21400005
    …
```

2. 查询每场比赛命中次数最多的球员

首先把每场比赛中参加比赛的球员姓名、每个球员命中次数的查询结果放到一个中间表 sqlsave 中，shot_result='made'限定只查询投篮命中的次数，查询结果以比赛和球员名进行分组，这样就得到了每场比赛每个球员的总命中次数：

```
    hive> create table sqlsave as select game_id, player_name, COUNT(shot_result) cnt from
nbaLogs where shot_result='made' group by game_id,player_name;
```

接下来对中间表 sqlsave 进行查询，括号中的语句 select game_id,MAX(cnt) cnt from sqlsave group by game_id 查询出每场比赛进球最多的球员，并以比赛编号分组，然后和中间表 sqlsave 进行关联查询，得到最终结果：

```
    hive> select a.game_id, a.player_name, a.cnt from sqlsave a join(select game_id,MAX(cnt)
cnt from sqlsave group by game_id) b on a.game_id=b.game_id and a.cnt=b.cnt;
    …
    21400001    anthony davis    10
    21400002    mnta ellis    11
```

```
    21400003    carlos boozer    7
    21400003    james harden    7
    21400004    kemba walker    9
    21400005    cj miles    6
    21400005    donald sloan    6
    21400005    roy hibbert    6
    21400006    kelly olynyk    8
    …
```

3．查询每场比赛投中 2 分球最多的球员

首先把命中 2 分的记录查询结果放到一个中间表 twoPoints 中，shot_result='made' AND pts_type='2'限定只查询每场比赛中进球且为 2 分球的投篮，并按照球员和比赛编号进行分组：

```
    hive> create table twoPoints as select game_id, player_name, COUNT(shot_result) cnt
from nbaLogs where shot_result='made' AND pts_type='2' group by game_id,player_name;
```

再对中间表 twoPoints 进行查询，括号中的语句 select game_id,MAX(cnt) abc from twoPoints GROUP BY game_id 的结果为每场比赛中进 2 分球最多的次数，与 twoPoints 表关联查询，得到最终结果：

```
    hive>selecta.game_id,a.player_name,a.cntfrom twoPoints a JOIN (select game_id,MAX(cnt)
abc from twoPoints GROUP BY game_id ) b where a.game_id=b.game_id AND a.cnt=b.abc;
    …
    21400001    anthony davis    10
    21400002    mnta ellis    10
    21400003    carlos boozer    7
    21400004    khris middleton    7
    21400005    roy hibbert    6
    21400006    kelly olynyk    7
    21400007    marcin gortat    7
    21400008    amir johnson    7
    21400008    jeff teague    7
    21400009    marc gasol    12
    …
```

4．查询每场比赛投中 3 分球最多的球员

同上，首先把命中 3 分的记录查询结果放到一个中间表 threePoints 中：

```
    hive> create table threePoints as select game_id,player_name,COUNT(shot_result) cnt
from nbaLogs where shot_result='made' AND pts_type='3' group by game_id,player_name;
```

再对中间表 threePoints 进行查询，得到最终结果：

```
    hive> select a.game_id,a.player_name,a.cnt from threePoints a JOIN (selectgame_id,MAX(cnt)abc
from threePoints GROUP BY game_id ) b where a.game_id=b.game_id AND a.cnt=b.abc;
    …
    21400001    ryan anderson    3
    21400002    tony parker    4
    21400003    trevor ariza    5
    21400004    kemba walker    3
    21400004    marvin williams    3
    21400004    oj mayo    3
    21400005    chris copeland    2
    21400005    cj miles    2
    …
```

5．查询每场比赛防守成功率最高的球员

首先查询每场比赛防守队员的防守次数，并保存到 totalblocks 表中：

```
hive> create table totalblocks as select game_id,closest_defender,COUNT(shot_result) cnt from nbaLogs group by game_id,closest_defender;
```

把防守次数不大于 1 的数据过滤掉，将余下数据保存到 eliminate 表中，这样能够提高查询的效率：

```
hive> create table eliminate as select game_id,closest_defender,cnt from totalblocks where cnt>1;
hive> select * from eliminate limit 5;
OK
21400001    "Ajinca, Alexis"    2
21400001    "Anderson, Ryan"    4
21400001    "Asik, Omer"    18
21400001    "Davis, Anthony"    16
21400001    "Dedmon, Dewayne"    7
Time taken: 0.095 seconds, Fetched: 5 row(s)
```

计算每场比赛中防守成功的次数，防守成功意味着投篮失败，即筛选条件为 shot_result='missed'，并将结果保存到 shotsblocked 表中：

```
hive> create table shotsblocked as select game_id,closest_defender,COUNT(shot_result) cnt from nbaLogs where shot_result='missed' group by game_id,closest_defender;
```

shotsblocked 表的前 5 行数据如下：

```
hive> select * from shotsblocked limit 5;
OK
21400001    "Anderson, Ryan"    3
21400001    "Asik, Omer"    10
21400001    "Babbitt, Luke" 1
21400001    "Davis, Anthony"    11
21400001    "Dedmon, Dewayne"    2
Time taken: 0.123 seconds, Fetched: 5 row(s)
```

防守成功率=防守成功次数÷总防守次数。总防守次数可以通过查询表 eliminate 得到，防守成功次数可以通过查询表 shotsblocked 得到，通过内连接只能保留经过筛选的数据，最终计算防守成功率，保存到 third 表中：

```
hive> create table third as select a.game_id,a.closest_defender,(b.cnt/a.cnt)*100 cnt from eliminate a inner join shotsblocked b on a.game_id=b.game_id and a.closest_defender=b.closest_defender;
```

third 表的前 5 行数据如下：

```
hive> select* from third limit 5;
OK
21400001    "Anderson, Ryan"    75.0
21400001    "Asik, Omer"    55.55555555555556
21400001    "Davis, Anthony"    68.75
21400001    "Dedmon, Dewayne"    28.57142857142857
21400001    "Evans, Tyreke" 55.55555555555556
Time taken: 1.827 seconds, Fetched: 5 row(s)
```

从 third 表中获得每场比赛中防守成功率最高的球员以及他们各自的防守成功率：

```
hive> select a.game_id,a.closest_defender,a.cnt from third a join (select game_id,MAX(cnt) cnt from third group by game_id) b on a.game_id=b.game_id and a.cnt=b.cnt;
```

```
…
21400001    "Gordon, Ben"   100.0
21400002    "Harris, Devin"   100.0
21400003    "Henry, Xavier"   100.0
21400004    "Pachulia, Zaza"   100.0
21400005    "Allen, Lavoy"   100.0
21400005    "Hill, Solomon"   100.0
21400006    "Jefferson, Cory"   75.0
21400006    "Plumlee, Mason"   75.0
…
```

6. 获取赛季得分最高的前 10 名球员

获取赛季每个球员的得分，使用 SUM( )函数进行统计，结果保存到 seasonscore 表中：

```
hive> create table seasonscore as select player_name,SUM(pts) cnt from nbaLogs group by player_name;
```

将统计结果进行排序，得到前 10 名球员和他们的总得分：

```
hive> select *from seasonscore order by cnt desc limit 10;
…
Total MapReduce CPU Time Spent: 4 seconds 30 msec
OK
stephen curry   1130
james harden   1103
klay thompson   1075
lebron james   1041
mnta ellis  1018
kyrie irving   998
damian lillard  995
lamarcus aldridge   971
nikola vucevic  962
chris paul  947
Time taken: 21.639 seconds, Fetched: 10 row(s)
```

习　　题

1. Hive 是由哪家公司开源的大数据处理组件？（　　）

   A. Google　　B. Apache　　C. Facebook

2. Hive 的计算引擎是什么？（　　）

   A. Spark　　B. MapReduce　　C. HDFS

3. Hive 加载数据文件到数据表中的关键语法是什么？（　　）

   A. LOAD DATA [LOCAL] INPATH filepath [OVERWRITE] INTO TABLE tablename

   B. INSERTDATA [LOCAL] INPATH filepath [OVERWRITE] INTO TABLE tablename

   C. LOAD DATA INFILE d:\car.csv APPEND INTO TABLE t_car_temp FIELDS TERMINATED BY

4. Hive 定义一个自定义函数类时，需要继承以下哪个类？（　　）

   A. FunctionRegistry　　B. UDF　　C. MapReduce

5. 命令 hadoop fs -du -h /user/hive/warehouse 的作用是什么？（　　）

A. 查看 Hive 中各个数据库存储使用情况

B. 显示/user/hive/warehouse 中的文件列表

C. 查看 Hive 中/user/hive/warehouse 下的文件个数

6. 当发现 Hive 脚本执行时报错信息中包含如下内容：

FAILED:ClassCastExceptionorg.apache.hadoop.hive.serde2.typeinfo.PrimitiveTypeInfocannotbecasttoorg.apache.hadoop.hive.serde2.typeinfo.DecimalTypeInfo

则此脚本最可能存在的问题是什么？（　　）

A. 网络问题

B. GROUP BY 中包含详单的字段（字段重复）

C. 字符串和数值类型转换错误

7. Hive 建表时，数值列的字段类型选取 decimal(x,y)与 FLOAT、DOUBLE 的区别，下列说法正确的是哪个？（　　）

A. decimal(x,y)表示整数，FLOAT、DOUBLE 表示是小数

B. FLOAT、DOUBLE 在进行 sum 等聚合运算时，会出现 Java 精度问题

C. decimal(x,y)是数值截取函数，FLOAT、DOUBLE 是数据类型

8. Hive 数据表插入数据时，insert( ) table…括号中可使用哪些关键字？（　　）（多选题）

A. into　　B. append　　C. overwrite

9. 使用以下哪些关键字可以通过 Shell 连接 Hive 客户端进行数据操作？（　　）（多选题）

A. hive−e"HQL"　　B. hive beeline−e"HQL"　　C. hive

10. Hive 自定义函数 jar 发布有哪几种方法？（　　）（多选题）

A. 使用 add jar…

B. 修改配置文件 hive-site.xml 中的 hive.aux.jars.path

C. 在${HIVE_HOME}中创建目录 auxlib，将自定义函数.jar 包上传至该目录

第7章 HBase数据库部署与操作

HBase是一个面向列的分布式数据库，能够承担实时随机访问超大规模数据集的任务。HBase采用自底向上进行构建的模式，能够简单地通过增加节点达到线性扩展的目的。HBase与传统关系数据库不同，不支持SQL，但HBase有自己的优势，在特定的问题空间里，能够承担关系数据库管理系统（Relational Database Management System，RDBMS）不能做的事，如对硬件设备依赖度低，可在普通硬件构成的集群上管理规模巨大的稀疏表。

7.1 HBase简介

HBase技术来源于 Google公司论文“Bigtable：一个结构化数据的分布式存储系统”，经过不断发展，如今HBase已成为一个面向列的开源分布式数据库。就像Bigtable利用了Google文件系统(File System)所提供的分布式数据存储技术一样，HBase在Hadoop之上提供了类似于Bigtable的能力。HBase是Apache的Hadoop项目的子项目，是一个适用于非结构化数据存储的数据库，是一个基于列存储而不是基于行存储的数据库。

7.1.1 HBase表

HBase以表的形式存储数据，每张表包含Row Key、Timestamp、Column Family等基本信息。

- Row Key：行键，是表的主键，表中的记录默认按照Row Key升序排序。
- Timestamp：时间戳，是每次数据操作对应的时间节点，可以把Timestamp看作数据的版本号。
- Column Family：列族。表在水平方向由一个或者多个Column Family组成。一个Column Family可以由任意多个列组成，即Column Family支持动态扩展，无须预先定义列的数量以及类型，所有列均以二进制格式存储，用户需要自行进行类型转换。

列是HBase存储数据的基本单位。列的数量没有限制，一个列族里可以有数百万个列，列中的数据都是以二进制形式存在的，没有数据类型和长度限制。

7.1.2 HBase基本知识

HBase采用的是典型的主从服务器架构，由HMaster服务器和HRegionServer服务器构成，所有服务器通过ZooKeeper协调并处理。

• HMaster：HMaster 可避免单点故障问题。HBase 中可以启动多个 HMaster，通过 ZooKeeper 的 Master Election 机制保证总有一个 HMaster 运行。HMaster 在功能上主要负责表和分区（Region）的管理工作。

（1）管理用户对表的增、删、改、查操作。

（2）管理 HRegionServer 的负载均衡，调整 Region 分布。

（3）在 Region Split 后，负责新 Region 的分配。

（4）在 HRegionServer 停机后，负责失效 HRegionServer 上的 Regions 迁移。

• HRegionServer：HRegionServer 主要负责响应用户 I/O 请求，向 HDFS 中读/写数据，是 HBase 中最核心的模块之一。HRegionServer 内部管理了一系列 HRegion 对象，每个 HRegion 对应了表中的一个 Region，HRegion 由多个 HStore 组成。每个 HStore 对应了表中的一个 Column Family，可以看出每个 Column Family 其实就是一个集中的存储单元，因此最好将具备共同 I/O 特性的 Column 放在一个 Column Family 中，这样最高效。

• ZooKeeper：ZooKeeper Quorum 中除了存储了-ROOT-表的地址和 HMaster 的地址，HRegionServer 也会把自己以短时（Ephemeral）方式注册到 ZooKeeper 中，使得 HMaster 可以随时感知到各个 HRegionServer 的健康状态。此外，ZooKeeper 也避免了 HMaster 的单点问题。

7.2　HBase 的安装

本节主要介绍如何安装 HBase 以及对 HBase 的初始化配置。

7.2.1　必要条件

1. Java

需要安装 Java 1.8 以上版本，只有使用 Java 才能运行 HBase。安装软件可在 Sun 官网下载。安装完成后在命令行输入命令 java -version 可以获取安装的 Java 版本信息，具体的安装过程可参考前文，本小节不再详细介绍。

2. Hadoop

由于 HBase 与 Hadoop 之间的远程通信过程使用的 RPC 协议是版本化的，需要调用方和被调用方相互匹配，若有差异就可能造成通信失败，影响 HBase 系统的稳定性，因此要求 Hadoop 的.jar 包必须部署在 HBase 的 lib 目录下。

3. SSH

SSH 是必须安装的，要保证用户可以通过 SSH 跳转到系统的其他节点，用户必须能够通过密码登录系统。SSH 需要一个公钥，让脚本可以没有阻碍地访问任意一台服务器。

4. 同步时间

集群中的节点时间必须是同步的，若有偏差可能会产生奇怪的现象，所以用户需要在集群中运行 NTP 或同等功能的应用来同步集群的时间。

5. 设置 DataNode 处理线程数

设置 DataNode 可处理线程数上限的参数为 xcievers。在加载前，用户必须确保 Hadoop 配置文件 conf-site.xml 中的 xcievers 参数至少为以下的数值：

```
<property>
    <name>dfs.datanode.max.xcievers</name>
    <value>4096</value>
</property>
```

7.2.2 安装配置 HBase

1. 解压并安装 HBase

下载HBase安装包，复制到master节点，本次安装使用的版本为HBase 1.2.6。可以使用FileZilla软件将安装包从本地上传到服务器，如图 7-1 所示。

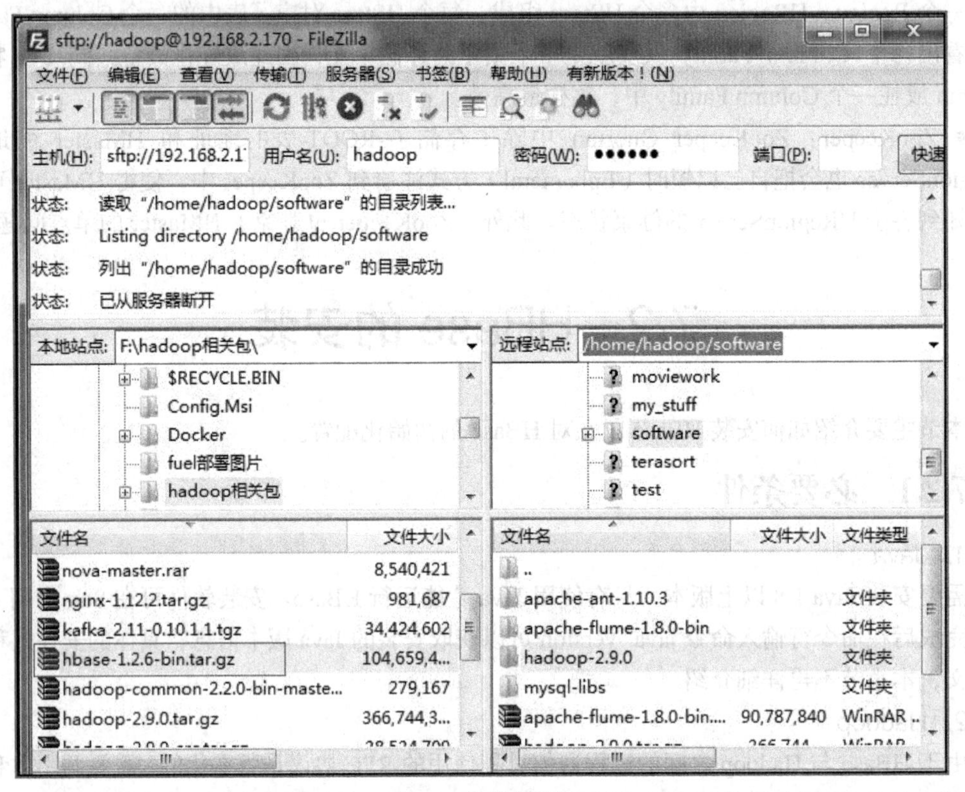

图 7-1 上传安装包

在/home/hadoop/software/目录中可以查看 HBase 安装包，如图 7-2 所示。

```
[hadoop@master ~]$ ls /home/hadoop/software/
apache-ant-1.10.3             hadoop-2.9.0                 mysql-libs
apache-flume-1.8.0-bin        hadoop-2.9.0.tar.gz
apache-flume-1.8.0-bin.tar    hbase-1.2.6-bin.tar.gz
```

图 7-2 查看安装包

解压 HBase 安装包，命令如下：

```
[hadoop@master ~]$ cd /home/hadoop/software/
[hadoop@master software]$ tar -zxf hbase-1.2.6-bin.tar.gz
[hadoop@master software]$ ls
```

解压后如图 7-3 所示。

```
[hadoop@master ~]$ cd /home/hadoop/software/
[hadoop@master software]$ tar -zxf hbase-1.2.6-bin.tar.gz
[hadoop@master software]$ ls
apache-ant-1.10.3              hadoop-2.9.0.tar.gz
apache-flume-1.8.0-bin         hbase-1.2.6
apache-flume-1.8.0-bin.tar     hbase-1.2.6-bin.tar.gz
hadoop-2.9.0                   mysql-libs
```

图 7-3　解压 HBase

2. 配置 HBase

进入 HBase 安装主目录，修改配置文件，命令如下：

```
[hadoop@master hbase-1.2.6]$ cd conf
[hadoop@master conf]$ ls
```

conf 目录包含 HBase 的配置文件，如图 7-4 所示。

```
[hadoop@master hbase-1.2.6]$ cd conf
[hadoop@master conf]$ ls
hadoop-metrics2-hbase.properties  hbase-policy.xml   regionservers
hbase-env.cmd                     hbase-site.xml
hbase-env.sh                      log4j.properties
```

图 7-4　配置 hbase-env.sh

（1）修改环境变量 hbase-env.sh，命令如下：

```
[hadoop@master conf]$ vi hbase-env.sh
```

在文件前面部分有如下内容：

```
# export JAVA_HOME=/usr/java/jdk1.6.0/
```

将其修改：

```
export JAVA_HOME=/home/hadoop/java/jdk1.8.0_161
```

注意：去掉首部“#”。

修改后的 hbase-env.sh 如图 7-5 所示。

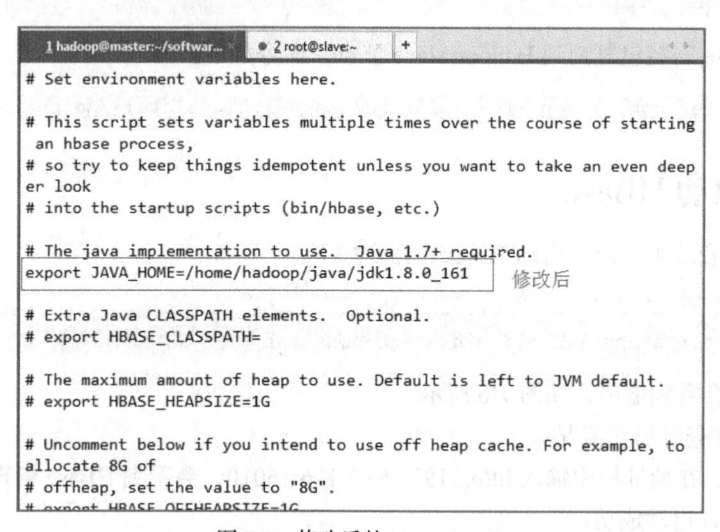

图 7-5　修改后的 hbase-env.sh

（2）修改配置文件 hbase-site.xml，用如下内容替换原 hbase-site.xml 中的内容：

```
   <?xml version="1.0"?>
   <?xml-stylesheet type="text/xsl" href="configuration.xsl"?>
<configuration>
   <property>
       <name>hbase.cluster.distributed</name>
       <value>true</value>
   </property>
   <property>
       <name>hbase.rootdir</name>
       <value>hdfs://master:9000/hbase</value>
   </property>
   <property>
       <name>hbase.zookeeper.quorum</name>
       <value>master</value>
   </property>
</configuration>
```

（3）配置 HRegionServers。

将 HRegionServers 中的 localhost 修改为下面的内容：

```
slave
```

（4）配置环境变量。

编辑系统配置文件，命令如下：

```
[hadoop@master conf]$ vi ~/.bash_profile
```

将下面的代码添加到文件尾部：

```
export HBASE_HOME=/home/hadoop/software/hbase-1.2.6
export PATH=$HBASE_HOME/bin:$PATH
export HADOOP_CLASSPATH=$HBASE_HOME/lib/*
```

然后执行：

```
source ~/.bash_profile
```

（5）将 HBase 安装包复制到 Hadoop slave 节点，命令如下：

```
[hadoop@master conf]$ scp -r ~/software/hbase-1.2.6 slave:~/software/
```

7.2.3 启动 HBase

进入 HBase 安装主目录，启动 HBase，命令如下：

```
[hadoop@master conf]$ cd ~/software/hbase-1.2.6
[hadoop@master hbase-1.2.6]$ bin/start-hbase.sh
```

执行命令后会看到输出，如图 7-6 所示。

使用 Web UI 查看启动情况。

打开浏览器，在地址栏中输入 http://192.168.2.176:16010，会看到 HBase 管理界面，如图 7-7 所示，表明 HBase 启动成功。

```
[hadoop@master conf]$ cd ~/software/hbase-1.2.6
[hadoop@master hbase-1.2.6]$ bin/start-hbase.sh
hadoop@master's password:
master: starting zookeeper, logging to /home/hadoop/software/hbase-1.2.
6/bin/../logs/hbase-hadoop-zookeeper-master.out
starting master, logging to /home/hadoop/software/hbase-1.2.6/logs/hbas
e-hadoop-master-master.out
Java HotSpot(TM) 64-Bit Server VM warning: ignoring option PermSize=128
m; support was removed in 8.0
Java HotSpot(TM) 64-Bit Server VM warning: ignoring option MaxPermSize=
128m; support was removed in 8.0
slave: starting regionserver, logging to /home/hadoop/software/hbase-1.
2.6/bin/../logs/hbase-hadoop-regionserver-slave.out
slave: Java HotSpot(TM) 64-Bit Server VM warning: ignoring option PermS
ize=128m; support was removed in 8.0
slave: Java HotSpot(TM) 64-Bit Server VM warning: ignoring option MaxPe
rmSize=128m; support was removed in 8.0
```

图 7-6　启动 HBase

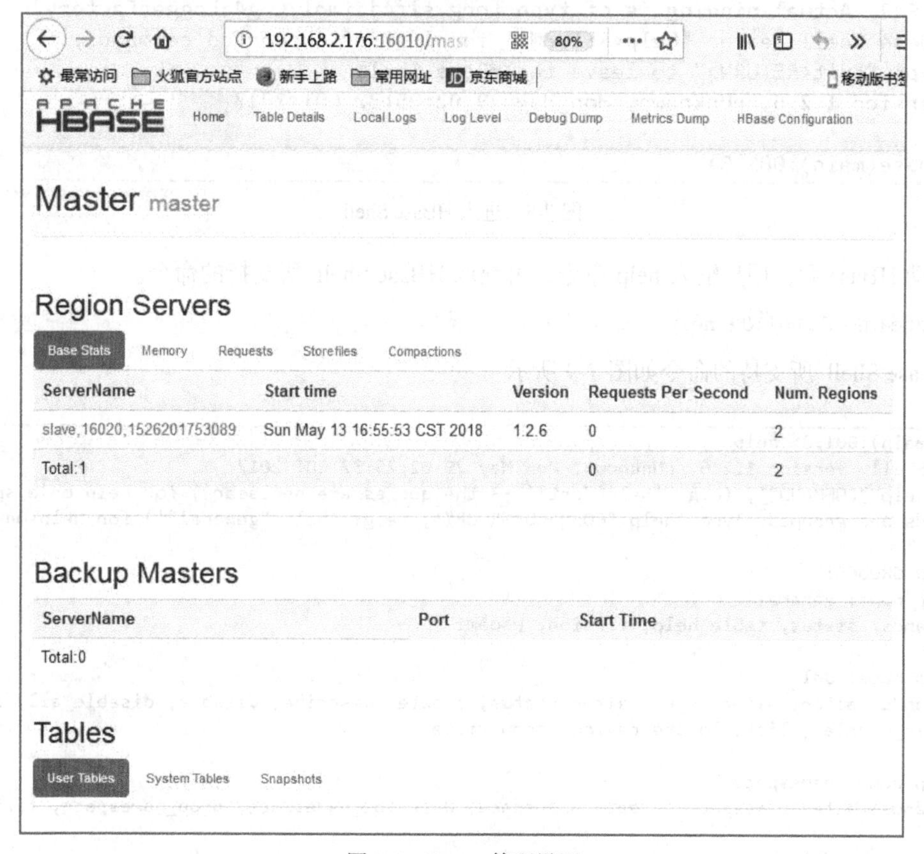

图 7-7　HBase 管理界面

7.3　HBase Shell 操作

HBase 为用户提供了非常方便的使用方式，我们称为 HBase Shell。 HBase Shell 是 HBase 集群的命令行接口，用户可以使用 Shell 访问本地和远程服务器并与其交互。Shell 同时提供了客户端管理功能。

进入 Shell，命令如下：

```
[hadoop@master hbase-1.2.6]$ hbase shell
```

运行结果如图 7-8 所示。

```
[hadoop@master hbase-1.2.6]$ hbase shell
SLF4J: Class path contains multiple SLF4J bindings.
SLF4J: Found binding in [jar:file:/home/hadoop/software/hbase-1.2.6/lib
/slf4j-log4j12-1.7.5.jar!/org/slf4j/impl/StaticLoggerBinder.class]
SLF4J: Found binding in [jar:file:/home/hadoop/software/hadoop-2.9.0/sh
are/hadoop/common/lib/slf4j-log4j12-1.7.25.jar!/org/slf4j/impl/StaticLo
ggerBinder.class]
SLF4J: See http://www.slf4j.org/codes.html#multiple_bindings for an exp
lanation.
SLF4J: Actual binding is of type [org.slf4j.impl.Log4jLoggerFactory]
HBase Shell; enter 'help<RETURN>' for list of supported commands.
Type "exit<RETURN>" to leave the HBase Shell
Version 1.2.6, rUnknown, Mon May 29 02:25:32 CDT 2017

hbase(main):001:0>
```

图 7-8　进入 HBase Shell

启动 HBase 后，用户输入 help 命令，可获取 HBase Shell 所支持的命令。

```
hbase(main):001:0> help
```

HBase Shell 所支持的命令如图 7-9 所示。

```
hbase(main):001:0> help
HBase Shell, version 1.2.6, rUnknown, Mon May 29 02:25:32 CDT 2017
Type 'help "COMMAND"', (e.g. 'help "get"' -- the quotes are necessary) for help on a specific
Commands are grouped. Type 'help "COMMAND_GROUP"', (e.g. 'help "general"') for help on a comm

COMMAND GROUPS:
  Group name: general
  Commands: status, table_help, version, whoami

  Group name: ddl
  Commands: alter, alter_async, alter_status, create, describe, disable, disable_all, drop, d
bled, is_enabled, list, locate_region, show_filters

  Group name: namespace
  Commands: alter_namespace, create_namespace, describe_namespace, drop_namespace, list_names

  Group name: dml
  Commands: append, count, delete, deleteall, get, get_counter, get_splits, incr, put, scan,

  Group name: tools
  Commands: assign, balance_switch, balancer, balancer_enabled, catalogjanitor_enabled, catal
pact, compact_rs, flush, major_compact, merge_region, move, normalize, normalizer_enabled, no
ump

  Group name: replication
  Commands: add_peer, append_peer_tableCFs, disable_peer, disable_table_replication, enable_p
```

图 7-9　HBase Shell 所支持的命令

HBase Shell 中的常用命令如表 7-1 所示。

表 7-1　　HBase Shell 中的常用命令

命令	描述
alter	修改列族模式
count	统计表中行的数量
create	创建表
describe	显示表相关的详细信息
delete	删除指定对象的值（可以为表、行、列对应的值，也可以为指定时间戳的值）
deleteall	删除指定行的所有元素值
disable	使表无效
drop	删除表
enable	使表有效
exists	测试表是否存在
exit	退出 HBase Shell
get	获取行或单元（Cell）的值
incr	增加指定表中行或列的值
list	列出 HBase 中存在的所有表
put	向指向的表单元添加值
tools	列出 HBase 所支持的工具
scan	通过对表扫描来获取对应的值
status	返回 HBase 集群的状态信息
shutdown	关闭 HBase 集群（与 exit 不同）
truncate	重新创建指定表
version	返回 HBase 版本信息

可以对 HBase Shell 支持的命令进行分组，如表 7-2 所示。

表 7-2　　命令分组

分组	命令
general	status，version
ddl	alter，create，describe，disable，drop，enable，exists，is_disabled，is_enabled，list
dml	count，delete，deleteall，get，get_counter，incr，put，scan，truncate
tools	assign，balance_switch，balancer，close_region，compact，flush，major_compact，move，split，unassign，zk_dump
replication	add_peer，disable_peer，enable_peer，remove_peer，start_replication，stop_replication

7.3.1　普通命令

由表 7-2 可知，普通（general）命令有 status 和 version 两个，使用 status 可查询服务器状态；使用 version 可查询 Hive 版本，如图 7-10 所示。

```
hbase(main):003:0> version
1.2.6, rUnknown, Mon May 29 02:25:32 CDT 2017

hbase(main):004:0> status
1 active master, 0 backup masters, 1 servers, 0 dead, 2.0000 average load
```

图 7-10　status 和 version 命令

7.3.2　DDL 操作

1. create

create 命令通过表名或逗号分隔开的列族信息来创建表，具体操作可分为如下 3 种：

```
hbase>create 't1',{NAME => 'f1',VERSIONS => 5}
hbase>create 't1',{NAME => 'f1'},{NAME => 'f2'}
hbase>create 't1',{NAME => 'f1',VERSIONS => 1, TTL => 2592000,BLOCKCACHE => true}
```

下面我们创建一个名为 student 的表：

```
hbase(main):005:0> create 'student' ,'name','score'
```

创建 student 表如图 7-11 所示。

```
hbase(main):005:0> create 'student' ,'name','score'
0 row(s) in 1.3220 seconds

=> Hbase::Table - student
```

图 7-11　创建 student 表

创建表时要加上列族，如果只写 create 'student'就会报错，如图 7-12 所示。

```
hbase(main):006:0> create 'student'

ERROR: Table must have at least one column family

Here is some help for this command:
Creates a table. Pass a table name, and a set of column family
specifications (at least one), and, optionally, table configuration.
Column specification can be a simple string (name), or a dictionary
(dictionaries are described below in main help output), necessarily
including NAME attribute.
Examples:
```

图 7-12　报错提示

检查表是否存在，命令如下：

```
hbase(main):007:0> exists 'student'
```

检查结果如图 7-13 所示。

```
hbase(main):007:0> exists 'student'
Table student does exist
0 row(s) in 0.0840 seconds
```

图 7-13　检查表是否存在

可以使用 drop 命令删除表。删除表之前要先使用 disable 禁用表才能删除成功：

```
hbase(main):008:0> disable 'student'
hbase(main):009:0> drop 'student'
```

删除表如图 7-14 所示。

```
hbase(main):008:0> disable 'student'
0 row(s) in 2.2910 seconds

hbase(main):009:0> drop 'student'
0 row(s) in 1.2490 seconds
```

图 7-14　删除表

再次检查表是否存在，如图 7-15 所示。

```
hbase(main):010:0> exists 'student'
Table student does not exist
0 row(s) in 0.0120 seconds
```

图 7-15　确认表被删除

2. list

list 命令可列出 HBase 中包含的所有表的名称，命令如下：

```
hbase(main):011:0>list
hbase_tb
test
2 row(s) in 0.160 secondshbase>list
```

3. put

put 命令用于向指定的 HBase 表单元添加值，例如，向 t1 表中的行 r1、列 c1:1 单元添加值 v1，并指定时间戳为 ts 的命令如下：

```
hbase>put 't1', 'r1', 'c1:1','v1',ts
```

4. get

get 命令用于获取行或者单元的值，可以指定表名、行值，以及可选的列值和时间戳。获取表 test 行 r1 的值的命令如下：

```
hbase(main):002:0>get 'test', 'r1'
COLUMN              CELL
c1:1                timestamp=1295692753859,value=value1-1/1
c1:2                timestamp=1295692753860,value=value1-1/2
c1:3                timestamp=1297445678442,value=value1-1/3
c2:1                timestamp=1295623234356,value=value1-2/1
4 row(s) in 0.0450 seconds
```

获取表 r1 行、c1:1 列单元值的命令如下：

```
hbase(main):005:0>get 'test', 'r1',{COLUMN =>'c1:1'}
COLUMN                CELL
c1:1                 timestamp=1295692753859,value=value1-1/1
1 row(s) in 0.0040 seconds
```

7.3.3 DML 操作

（1）添加记录，命令如下：

```
put 'users','xiaoming','info:age','24';
put 'users','xiaoming','info:birthday','1987-06-17';
put 'users','xiaoming','info:company','alibaba';
put 'users','xiaoming','address:contry','china';
put 'users','xiaoming','address:province','anhui';
put 'users','xiaoming','address:city','hefei';
```

（2）获取“xiaoming”这个用户的所有信息，命令如下：

```
get 'users','xiaoming'
```

（3）获取一个列族的所有数据，命令如下：

```
get 'users','xiaoming','info'
```

（4）获取一个列族中指定的一个列的所有数据，命令如下：

```
get 'users','xiaoming','info:age'
```

（5）更新表中数据记录，命令如下：

```
put 'users','xiaoming','info:age' ,'27'
get 'users','xiaoming','info:age'
```

（6）获取单元格数据的版本信息，命令如下：

```
get 'users','xiaoming',{COLUMN=>'info:age',VERSIONS=>1}
get 'users','xiaoming',{COLUMN=>'info:age',VERSIONS=>2}
get 'users','xiaoming',{COLUMN=>'info:age',VERSIONS=>3}
```

（7）获取单元格中某个版本数据，命令如下：

```
get 'users','xiaoming',{COLUMN=>'info:age',TIMESTAMP=>1364874937056}
```

（8）对表进行全表扫描，命令如下：

```
scan 'users'
```

（9）删除表中的'info:age'字段，命令如下：

```
delete 'users','xiaoming','info:age'
get 'users','xiaoming'
```

（10）删除整行，命令如下：

```
deleteall 'users', 'xiaoming'
```

（11）对表的行数进行统计，命令如下：

```
count 'users'
```

（12）将表清空，命令如下：

```
truncate 'users'
```

（13）创建一个测试表 test，命令如下：

```
hbase(main):011:0> create 'test','cf1','cf2'
```

（14）使用 put 命令存储一个单元格数据。

注意：test 表中 row-key-0001 行、cf1 列族的 c1 列位置存储的数据为 value1。命令如下：

```
hbase(main):012:0> put 'test','row-key-0001','cf1:c1','value1'
hbase(main):013:0> put 'test','row-key-0001','cf1:c2','value2'
```

（15）创建 test 表，如图 7-16 所示。

```
hbase(main):011:0> create 'test','cf1','cf2'
0 row(s) in 2.2330 seconds

=> Hbase::Table - test
hbase(main):012:0> put 'test','row-key-0001','cf1:c1','value1'
0 row(s) in 0.1080 seconds

hbase(main):013:0> put 'test','row-key-0001','cf1:c2','value2'
0 row(s) in 0.0080 seconds
```

图 7-16　创建 test 表

（16）使用 get 命令获取单元格数据，命令如下：

```
hbase(main):001:0> get 'test','row-key-0001','cf1:c1'
hbase(main):002:0> get 'test','row-key-0001'
```

结果如图 7-17 所示。

```
hbase(main):001:0> get 'test','row-key-0001','cf1:c1'
COLUMN                    CELL
 cf1:c1                   timestamp=1526389865277, value=value1
1 row(s) in 0.3040 seconds

hbase(main):002:0> get 'test','row-key-0001'
COLUMN                    CELL
 cf1:c1                   timestamp=1526389865277, value=value1
 cf1:c2                   timestamp=1526389899147, value=value2
2 row(s) in 0.0260 seconds
```

图 7-17　获取单元格数据

（17）使用 scan 扫描一个范围的数据，命令如下：

```
hbase(main):004:0> scan 'test'
```

结果如图 7-18 所示。

```
hbase(main):004:0> scan 'test'
ROW                        COLUMN+CELL
 row-key-0001              column=cf1:c1, timestamp=1526389865277, value=value1
 row-key-0001              column=cf1:c2, timestamp=1526389899147, value=value2
1 row(s) in 0.0110 seconds
```

图 7-18　扫描一个范围的数据

7.3.4 工具命令

Tools 命令组提供了一些工具命令。这些命令多用于 HBase 集群的管理与调优，涵盖合并、分裂、负载均衡、日志回滚、分配和移动以及 ZooKeeper 信息查看等方面。每种命令的使用方法有多种，适用于不同的场景。例如合并命令 compact，可以合并一张表、一个 Region 的某个列族，或一张表的某个列族。工具命令如表 7-3 所示。

表 7-3　工具命令

命令	命令含义	实例
assign	分配 Region	assign 'Region1'
balance_switch	启用或关闭负载均衡器，并将当前负载均衡器状态作为结果进行返回	balance_switch true, balance_switch false
balancer	触发集群负载均衡器，如果运行成功，则返回 TRUE。如果返回 FALSE，说明某些 Region 处在 Region 迁移（Region In Transition，RIT）状态，不会执行该命令	balancer
close_region	关闭某个 Region	close_region 'REGIONNAME', close_region 'REGUIBBANE', 'SERVER_NAME'
compact	合并表或者 Region	compact 't1', compact 'r1', 'c1', compact 't1', 'c1'
flush	刷新（Flush）表或 Region	flush 'TABLENAME', flush 'REGIONNAME'
hlog_roll	HLog 日志回滚	hlog_roll 'regionName'
major_compact	合并表或 Region	major_compact 't1', major_compact 'r1', 'c1', major_compact 't1', 'c1'
move	移动 Region。可指定 HRegionServer，也可以随机选择	move 'ENCODED_REGIONNAME', move 'ENCODED_REGIONNAME', 'SERVER_NAME'
split	分片表，也可以分片 Region	split 'tablename', split 'tablename', split 'tablename', 'splitKey', split 'regionname', 'splitKey'
unassign	解除指定的 Region	unassign 'REGIONNAME', unassign 'REGIONNAME',true
zk_dump	将 ZooKeeper 的信息输出	zk_dump

7.3.5 复制命令

复制命令用于 HBase 高级特性——复制的管理，可以添加、停止、启用和删除复制功能。表 7-4 所示为复制命令。

表 7-4 复制命令

命令	命令含义	实例
add_peer	添加对等集群，需要指定对等集群的 ID、主机名、端口号和 ZooKeeper 的根路径	add_peer '1', "server1.cie.com:2181:/hbase" add_peer '2', "zk1,zk2,zk3:2181:/hbase-prod"
disable_peer	停止到特定集群的复制流，但仍然保持对新改动的跟踪，参数是对等集群的 ID	disable_peer' 1'
enable_peer	启用到对等集群的复制，从上次停止的位置继续复制，参数是对等参数的 ID	enable_peer' 1'
list_peers	罗列所有正在复制的对等集群	list_peers
remove_peer	停止某个复制流，并且删除其对应的元数据信息，参数是对等集群的 ID	remove_peer' 1'
start_replication	重启所有复制流（只用在负载达到临界的情况下）	start_replication
stop_replication	关闭所有复制流	stop_replication

7.4 HBase 客户端 API

HBase 的主要客户端接口是由 org.apache.hadoop.hbase.client 包中的 HTable 类提供的。通过这个类，用户可以完成向 HBase 存储和检索数据，以及删除无效数据之类的工作。注意所有修改数据的操作都保证了行级别的原子性，但这会影响到这一行数据的所有并发读/写操作。

7.4.1 CRUD 操作

数据库的初始基本操作通常被称为 CRUD（Create、Read、Update、Delete），即增、查、改、删。这些方法都是由 HTable 类提供的，本小节主要介绍这个类的方法。

1. put( )方法

（1）单行数据

下面实现如何向 HBase 中存储数据：

```
void put(Put put) throws IOException
```

这个方法以单个 Put 对象或存储在列表中的一组 Put 对象作为输入参数，其中 Put 对象是由以下几个构造函数创建的：

```
Put(byte[] row)
Put(byte[] row,RowLock rowLock)
Put(byte[] row,long ts)
Put(byte[] row,long ts,RowLock rowLock)
```

创建 Put 实例时，用户需要提供一个行键 row。在 HBase 中，每行数据都有唯一的行键作为标识，与 HBase 的大多数数据类型一样，它是一个 Java 的 byte[]数组。用户可以按自己的需求来指定每行的行键。现在假设用户可以随意设置行键，通常情况下，行键的含义与真实场景相关。例如，它的含义可以是一个用户名或者订单号，内容可以是简单的数字，也可以是较复杂的通用唯一识别码（Universally Unique Identifier，UUID）等。Put 类还提供了很多其他方法，如表 7-5 所示。

表 7-5　Put 类提供的方法

方法	描述
getRow( )	返回创建 Put 实例时所指定的行键
getRowLock( )	返回当前 Put 实例的 RowLock 实例
getLockId( )	将 RowLock 参数传递给构造函数的可选的锁 ID，当该参数未被指定时返回-1L
setWriteToWAL( )	允许关闭默认启用的服务端预写日志（Write-Ahead Log，WAL）功能
getWrietToWAL( )	返回是否启用了 WAL 功能
getTimeStamp( )	返回相应 Put 实例的时间戳，该值可在构造函数中由 ts 参数传入，当未被设定时返回 Long.MAX_VALUE
heapSize( )	计算当前 Put 实例所需的堆大小，既包含其中的数据大小，也包含内部数据结构所需的空间大小
isEmpty( )	检查 FamilyMap 是否含有任何 KeyValue 实例
numFamilies( )	查询 FamilyMap 的大小，即所有的 KeyValue 实例中列族的个数
size( )	返回 Put 添加的 KeyValue 实例

（2）KeyValue 实例

在代码中有时需要直接处理 KeyValue 实例。这些实例都含有一个特定单元格的数据以及坐标。坐标包括行键、列族名、列限定符以及时间戳。KeyValue 类提供了特别多的构造器，允许以各种方式组合这些参数。下面展示了包括所有参数的构造器：

```
    KeyValue(byte[] row, int roffset, int rlength, byte[] family, int foffset, int flength,
byte [] qualifier, int qlength, long timestamp, Type type, byte[] value, int voffset, int
vlength)
```

坐标中任意一个成员都有一个 getter 方法，可以获得字节数组以及它们的参数 offset 和 length；也可以在最顶层访问它们，即直接读取底层字节缓冲区：

```
byte [] getBuffer()
int getOffset()
int getLength()
```

KeyValue 类还提供了一系列实现了 Comparator 接口的内部类，可以在代码里使用它们来实现与 HBase 内部一样的比较器。当需要使用 API 获取 KeyValue 实例并进一步排序或按照顺序处理时，就要用到这些比较器。KeyValue 类提供的比较器如表 7-6 所示。

表 7-6　KeyValue 类提供的比较器

比较器	描述
KeyComparator	比较两个 KeyValue 实例的字节数组格式的行键，即 getKey( )方法的返回值
KVComparator	是 KeyComparator 的封装，基于两个给定的 KeyValue 实例，提供与 KeyComparator 一样的功能
RowComparator	比较两个 KeyValue 实例的行键（getRow( )的返回值）
MetaKeyComparator	比较两个以字节数组格式表示的.META.条目的行键
MetaComparator	KVComparator 类的一个特别版本，用于比较.META.条目，是 MetaKey Comparator 的封装
RootKeyComparator	比较两个字节数组格式表示的-ROOT-条目的行键
RootComparator	KVComparator 类的一个特别版本，用于比较.ROOT.条目，是 RootKey Comparator 的封装

KeyValue 实例还有一个变量（一个额外的属性），代表该实例的唯一坐标，表 7-7 列出了 KeyValue 实例所有可能的类型值。

表 7-7　KeyValue 实例所有可能的类型值

类型	描述
Put	KeyValue 实例代表一个普通的 Put 操作
Delete	KeyValue 实例代表一个 Delete 操作，也称为“墓碑”标记
DeleteColumn	与 Delete 相同，但是会删除一整列
DeleteFamily	与 Delete 相同，但是会删除整个列族，包括该列族的所有列

（3）Put 列表

客户端的 API 可以插入单个 Put 实例，同时也有批量处理操作的高级特性。调用方式如下：

```
void put(List<Put> puts) throws IOException
```

由于用户提交的修改行数据的列表可能涉及多行，所以可能会有部分修改失败。造成修改失败的原因有很多，比如，一个远程的 Region 服务器出现问题，导致客户端的重试次数超过了配置的上限，因此不得不放弃当前操作。如果远程服务器的 put( )调用出现问题，错误会通过随后的一个 IOException 异常反馈给客户端。

当使用基于列表的 put( )调用时，需要特别注意：用户无法控制服务器执行 put( )的顺序，这意味着服务器被调用的顺序也不受用户控制。如果要保证写入的顺序，需要小心使用这个操作。最坏的情况是，减少每次批量处理的操作数，并显式地刷新客户端缓冲区，强制把操作发送到远程服务器。

2. get( )方法

HTable 类提供了 get( )方法和 Get 类，下面简单介绍其中常见的方法和类。

（1）单行数据

可从 HBase 中获取一个特定的值，命令如下：

```
Result get(Get get) throws IOException
```

get( )方法有对应的 Get 类，在使用下面的方法构造 Get 实例时，需要设置行键：

```
Get(byte [] row)
Get(byte [] row,RowLock rowLock)
```

与 Put 操作一样，用户有许多方法可用，可以通过多种标准筛选目标数据，也可以指定精确的坐标获取某个单元格的数据：

```
Get addFamily(byte [] family)
Get addColumn(byte [] family,byte [] qualifier)
Get setTimeRange(long minStamp,long maxStamp) throws IOException
Get setTimeStamp(long timestamp)
Get setMaxVersions()
Get setMaxVersions(int maxVersions)throws IOException
```

表 7-8 列出了 Get 类方法。

表 7-8　　Get 类方法

方法	描述
getRow( )	返回创建 Get 实例时指定的行键
getRowLock( )	返回当前 Get 实例的 RowLock 实例
getLockId( )	返回创建时指定 RowLock 的锁 ID，如果没有指定返回-1L
getTimeRange( )	返回指定的 Get 实例的时间戳范围。注意，Get 类中已经没有 getTimeStamp( )方法了，因为 API 会在内部将 setTimeStamp( )赋的值转换成 TimeRange 实例，设定给定时间戳的最大值和最小值
setFilter( )/getFilter( )	用户可以使用一个特定的过滤器实例，通过多种规则和条件来筛选列和单元格。使用这些方法，用户可以设定或查看 Get 实例的过滤器成员
setCacheBlocks( )/getCacheBlocks( )	每个 HBase 的 Region 服务器都有一个块缓存来有效地保存最近存取过的数据，并以此来加速之后的相邻信息的读取。不过在某些情况下，例如完全随机读取时，最好能避免这种机制带来的扰动。这些方法能够控制当次读取的块缓存机制是否有效
numFamilies( )	快捷地获取 FamilyMap 大小的方法，包括用 addFamily( )方法和 addColumn( )方法添加的列族
hasFamilies( )	检查列族或列是否存在于当前的 Get 实例中
familySet( )/getFamilyMap( )	这些方法能让用户直接访问 addFamily( )和 addColumn( )添加的列族和列。FamilyMap 列族中，键是列族的名称，键对应的值是指定列族的限定符列表。familySet( )方法返回一个所有已存储列族的 Set，即一个只包含列族名的集合

（2）Result 实例

当用户使用 get( )方法获取数据时，HBase 返回的结果包含所有匹配的单元格数据，这些数据将被封装在一个 Result 实例中返回给用户。用 Result 提供的方法，可以从服务器获取匹配指定行的特定返回值，这些值包括列族、列限定符和时间戳等。Result 提供的方法如下：

```
byte[] getValue(byte[] family,byte[] qualifier)
byte [] value()
byte [] getRow()
int size()
boolean isEmpty()
KeyValue[] raw()
List<KeyValue> list()
```

getValue( )方法允许用户取得一个 HBase 中存储的特定单元格的值。因为该方法不能指定时间戳，所以用户只能获取数据最新的版本。value( )方法的使用方法更简单，它会返回第一列对应的最新单元格的值。因为列在服务器中是按字典存储的，所以会返回名称（包括列族和列限定符）

排在首位的那一列的值。

getRow( )方法返回创建 Get 类当前实例使用的行键。size( )方法返回服务器返回值中键值对（KeyValue 实例）的数目。用户可以使用 size( )方法或者 isEmpty( )方法查看键值对的数目是否大于 0，这样可以检查服务器是否找到了与查询对应的结果。

raw( )方法返回原始的底层 KeyValue 的数据结构，具体来说，是基于当前的 Result 实例返回 KeyValue 实例的数组。list( )调用则把 raw( )中返回的数组转化为一个 List 实例，并返回给用户。创建的 List 实例由原始返回结果中的 KeyValue 数组成员组成，用户可以方便地使用数据。

（3）Get 列表

使用列表参数的 get( )方法与使用列表参数的 put( )方法对应，用户可以用一次请求获取多行数据。它允许用户快速高效地从远程服务器获取相关的或完全随机的多行数据。

API 提供的方法签名如下：

```
Result[] get(List<Get> gets)throws IOException
```

这个方法和之前介绍的一样：用户需要创建一个列表，并把之前准备好的 Get 实例添加到其中，然后将这个列表传给 get( )，返回一个与列表大小相同的 Result 数组。

7.4.2　批量处理

7.4.1 小节介绍了查询、添加表中的数据等基本操作，这些操作都是基于单个实例或者基于列表的操作。本小节将介绍一些可以批量处理多行数据的 API 调用。

事实上，许多基于列表的操作是基于 batch( )方法实现的，如 delete(List < Delete > deletes)或者 get(List < Get > gets)，这些方法存在的目的是方便用户使用。如果对 HBase 不熟悉，推荐使用 batch( )方法进行所有操作。下面的客户端 API 提供了批量处理操作，用户可能注意到这里引入了一个新的名为 Row 的类，它是 Put、Get 和 Delete 的“祖先”，或者父类。

```
void batch(List<Row> actions,Object[] results)
throws IOException,InterruptedException
Object [] batch(List<Row> actions)
throws IOException,InterruptedException
```

Get 操作的结果需要观察输出结果的中间部分，即示例代码产生的输出。那些以 Result[n]（n 的范围为 0～3）开头的输出是 actions 参数中对应操作的结果。示例中第一个操作是 Put，对应的结果是一个空的 Result 实例，其中没有 KeyValue 实例。这是批量处理调用返回值的通常规则，它们给每个输入操作返回一个最佳匹配的结果，batch( )返回结果如表 7-9 所示。

表 7-9　batch( )返回结果

结果	描述
null	操作与远程服务器的通信失败
EmptyResult	Put 与 Delete 操作成功后的返回结果
Result	Get 操作成功后的返回结果。如果没有匹配的行或列，会返回空的 Result
Throwable	当服务器返回一个异常时，这个异常会原样返回给客户端。用户可以使用这个异常检查哪里出了错，也许可以在自己的代码中自动处理异常

当用户使用 batch( )功能时，Put 实例并不会被客户端写入缓冲区。batch( )请求是同步的，会把操作直接发送到服务器，这个过程没有延迟或其他中间操作，这与 put( )调用明显不同，所以请

慎重挑选需要的方法。

有两种不同的批量处理操作看起来非常相似。不同之处在于，一个需要用户输入包含返回结果的 Object 数组，而另一个由函数帮助创建这个数组。

```
void batch(List<Row> actions,Object[] results)
 throws IOException,InterruptedException
```

利用上面的方法，用户可以访问部分结果。

```
Object batch(List<Row> actions)
 throws IOException,InterruptedException
```

这个方法如果抛出异常，不会有任何返回结果，因为新结果数组返回之前，控制流就中断了。

两种方法的共同点：Get、Put 和 Delete 都支持。如果执行时出现问题，客户端将抛出异常并报告问题。它们都不使用客户端写缓冲区。

使用下面的命令能够访问成功操作的结果，同时也可以获取远程失败时的异常。

```
void batch(actions,results)
```

使用下面的命令只返回客户端异常，不能访问程序执行中的部分结果。

```
Object[] batch(actions)
```

7.4.3 行锁

我们知道，数据库中存在事务。事务是单个逻辑工作单元执行的一系列操作，要么完全执行，要么完全不执行。事务具有四大特点，即原子性、一致性、分离性和持久性。其中，原子性至关重要，那么在 HBase 内部实现其原子性的重要保证是什么呢？答案就是行锁。行锁就是加在行上的一把“锁”。在它未释放该行前，其他访问者是无法对该行进行修改的，如果要修改，必须获得该行锁才能拥有修改该行数据的权限，这就是行锁的含义。用户应该尽可能地避免使用行锁，就像在 RDBMS 中，两个客户端很可能在拥有对方要请求的锁时，又同时请求对方已拥有的锁，这样便形成了一个死锁。

当使用 put( )访问服务器时，Put 实例可以通过以下构造函数生成：

```
Put(byte[] row)
```

这个构造函数就没有 RowLock 实例参数，所以服务器会在调用期间创建一个锁。实际上，通过客户端的 API，得不到这个生存期短暂的服务端的锁实例。

除了服务器隐式加锁之外，客户端也可以显式地对单行数据的多次操作进行加锁，通过以下调用便可以做到：

```
RowLock lockRow(byte[] row) throws IOException
void unlockRow(RowLock rl) throws IOException
```

调用 lockRow( )需要一个行键作为参数，返回一个 RowLock 的实例，这个实例可以供后续的 Put 或者 Delete 的构造函数使用。一旦不再需要锁时，必须通过 unlockRow( )调用来释放它。

每个排他锁，无论是由服务器提供的，还是通过客户端 API 传入的，都能保护这一行不被其他锁锁定。换句话说，锁必须针对整行，并且指定其行键，一旦它获得锁定权就能防止其他并发修改。

当一个锁被服务器或客户端显式获取之后，其他所有想要对这行数据加锁的客户端将会等待，直到当前锁被释放，或者锁的“租期”超时。后者是为了确保进程的锁不会被占用太长时间或无限期占用。

7.4.4　扫描

扫描操作的使用方法和 get( )方法非常相似。和其他函数类似，这里也提供了 Scan 类。由于扫描的工作方式类似于迭代器，因此用户无须调用 scan( )方法创建实例，只需调用 HTable 的 getScanner( )方法。此方法在返回真正的扫描器（Scanner）实例的同时，用户也可以使用它迭代获取数据。可用方法如下：

```
ResultScanner getScanner(Scan scan) throws IOException
ResultScanner getScanner(byte[] family) throws IOException
ResultScanner getScanner(byte[] family,byte[] qualifier) throws IOException
```

后两行为了方便用户使用，隐式地帮用户创建了一个 Scan 实例，逻辑中最后调用 getScanner (Scan scan)方法。

Scan 类拥有以下构造器：

```
Scan()
Scan(byte[] startRow,Filter filter)
Scan(byte[] startRow)
Scan(byte[] startRow,byte[] stopRow)
```

这与 Get 类的不同点是显而易见的：用户可以选择性地提供 startRow 参数，来定义扫描读取 HBase 表的起始行键，即行键不是必须指定的。同时可选 stopRow 参数限定读取到何处停止。

扫描操作有一个特点：用户提供的参数不必精确匹配起始行，扫描会匹配相等或大于给定的起始行的行键。如果没有显式地指定起始行，它会从表的起始位置开始获取数据。

当遇到了与设置的终止行键相同或大于终止行键的行键时，扫描也会停止。如果没有指定终止行键，会扫描到表尾。

另一个可选参数叫作过滤器（Filter），可直接指向 Filter 实例。尽管 Scan 实例通常由空白构造器构造，但其所有可选参数都有对应的 getter 方法和 setter 方法。

表 7-10 所示为 Scan 类方法。

表 7-10　　Scan 类方法

方法	描述
getStartRow( )/getStopRow( )	查询当前设定的值
getTimeRange( )	检索 Get 实例指定的时间范围或相关时间戳，注意，当需要指定单个时间戳时，API 会在内部通过 setTimeStamp( )将 TimeRange 实例的起止时间戳设为传入值，所以 Get 类中此时已经没有 getTimeStamp( )方法了
getMaxVersions( )	返回当前配置下应该从表中获取的每列的版本
getFilter( )	可以使用特定的过滤器实例，通过多种规则来筛选列和单元格。使用这个方法，用户可以设定或查看 Scan 实例的过滤器成员
setCacheBlocks( )/getCacheBlocks( )	每个 HBase 的 Region 服务器都有一个块缓存，可以有效地保存最近访问过的数据，并以此来加速之后相邻信息的读取。不过在某些情况下，例如全表扫描，最好能避免这种机制带来的扰动。这个方法能够控制版本读取的块缓存机制是否有效
numFamilies( )	快捷地获取 FamilyMap 大小的方法，包括用 addFamily( )和 addColumn( )方法添加的列族和列
hasFamilies( )	检查是否添加过列族和列
getFamilies( )/setFamilyMap( )/getFamilyMap( )	这些方法能够让用户直接访问 addFamily( )和 addColumn( )添加的列族和列。在 FamilyMap 中键是列族的名称，键对应的值是特定列族下列限定符的列表。getFamilies( )方法返回一个只包含列族名的数组

7.4.5 数据过滤

HBase 过滤器的作用是过滤数据，能够为客户端减轻一些数据处理压力，减少数据传输到客户端的网络消耗。在 HBase 中主要的过滤器分为比较过滤器、专用过滤器、附加过滤器。

1. 比较过滤器

比较过滤器基于某些运算做出过滤判定。过滤器有两个基本要素：操作符和比较器。操作符包括 >、<、=、≠、≤、≥、NO_OP 等。比较器包括 BinaryComparator、BinaryPrefixComparator、NullComparator、BitComparator、SubStringComparator、RegexStringComparator 等。

- RowFilter：基于行键比较过滤数据。
- FamilyFilter：基于列族过滤数据。
- QualifierFilter：比较列的名字来过滤数据。
- ValueFilter：比较某一列的值来过滤数据。结合 SubStringComparator 或 RegexStringComparator 可以实现强大的过滤功能。
- DependentColumnFilter：参考列过滤器，可以看作 ValueFilter 和时间戳过滤器的结合。构造函数不传递过滤器时，只会提取时间戳与参考列时间戳相等的列。如果构造函数参数加入操作符和比较器，则组合时间戳和 ValueFilter 的结果进行过滤。

2. 专用过滤器

专用过滤器直接继承 FilterBase，用于特定场景，有些只做行筛选，只适合扫描操作。

- SingleColumnValueFilter：用一列的值决定整行数据是否加入结果集。
- SingleColumnValueExcludeFilter：类似 SingleColumnValueFilter，只是结果集排除参考列。
- PrefixFilter：行键前缀过滤器，只要行键满足给定的前缀就会把该行返回给客户端。
- PageFilter：分页过滤器，一般客户端会记住上一次扫描最后的 RowKey，把这个 RowKey 放入扫描中，结合 PageFilter 实现数据过滤。
- KeyOnlyFilter：只返回键不返回值，值可以用值的长度替代。
- FirstKeyOnlyFilter：只返回第一列的键，扫描完一列之后会跳过这一列，直接扫描下一列。可以用在统计行数的应用中，也可以用在按时间戳生成的列名的应用中。
- InclusiveStopFilter：普通扫描的结果中包含起始行，不包含结束行。该过滤器可以让扫描结果包括结束行。
- TimeStampFilter：只返回满足一定时间戳的值。时间戳可以是一个 List，也可以在此基础上设置范围。
- ColumnCountGetFilter：限制每行返回的列数，适合 Get 操作，不太适合 Scan 操作。
- ColumnPaginationFilter：列分页过滤器，构造函数传入列数和列的偏移量，则返回的每行都是在这个区间段的列。
- ColumnPrefixFilter：对列名称进行前缀匹配。
- RandomRowFilter：构造函数传入概率值，符合条件的行有一定概率被过滤掉。

3. 附加过滤器

附加过滤器如下。

- SkipFilter：用来封装一个过滤器。当内部过滤器满足过滤条件时，整行数据都被过滤掉。
- WhileMatchFilter：当一行数据被过滤掉时，会放弃整个扫描。

7.5 HBase 客户端选择及配置优化

HBase 面对不同的编程语言拥有不同的客户端，本节将简单介绍一些可用的客户端。

用户可使用目前比较流行的编程语言访问 HBase，也可以直接使用 HBase 客户端 API，还可以使用一些能够将请求转换成 API 调用的“代理”。这些代理将原生 Java API 包装成其他协议；这样用户可以使用 API 提供的任意外部语言来编写程序访问 HBase。

客户端与网关之间的协议是由当前可用选择以及远程客户端的需求决定的。常用的就是表述性状态转移（Representational State Transfer，REST）协议，其基于现有网络技术。实际的传输协议是典型的超文本传输协议（Hypertext Transfer Protocol，HTTP）——它是 Web 应用的标准传输协议。由于协议层负责传输可互操作格式的数据，这使得 REST 成为异构系统之间传输数据的理想选择。

在拥有大规模集群的公司中，巨额的带宽消费和许多相互隔离的服务让人觉得需要减少开销并实现自己的 RPC 层。Google 公司就是其中之一，其开发了 Protocol Buffer 框架。由于最初的实现没有发布，Facebook 公司就开发了一个类似的版本，叫作 Thrift。随后 Hadoop 项目创建者启动了第 3 个项目，叫作 Apache Avro，提供了另一种可选的实现。

7.6 HBase 与 MapReduce 集成

无论是在伪分布式模式下还是在真正分布式模式下，HBase 均是构建在 HDFS 上的，因此可以将 HBase 和 MapReduce 编程框架结合起来使用。为充分结合 HBase 大型分布式数据库和 MapReduce 并行计算的优点，可以将 HBase 作为底层“存储结构”，MapReduce 调用 HBase 进行特殊的处理。相应的 MapReduce 的 HBase 实现类包括 InputFormat、Mapper、Reducer 和 OutputFormat 这 4 个类。

1. InputFormat 类

TableInputFormatBase 是 HBase 实现的一个类，该类提供了对表数据的大部分操作，其子类 TableInputFormat 则提供了完整的实现，用于处理表数据并生成键值对。TableInputFormat 类将数据表按照 Region 分割成 Split，Split 数量与 Region 数量相同；然后将 Region 按行键分成键值对，其中 key 值与行键对应，value 值对应该行所包含的数据。

2. Mapper 类和 Reducer 类

TableReducer 类和 TableMapper 类是 HBase 实现的两个类。TableMapper 类没有具体的功能，它的作用是将输入的键、值的类型分别限定为 Result 和 ImmutableBytesWritable。IdentityTableReducer 类和 IdentityTableMapper 类则是上述两个类的具体实现，与 Mapper 类和 Reducer 类一样，只是简单地将键值对输出到下一个阶段。

3. OutputFormat 类

HBase 实现的 TableOutputFormat 将输出的键值对写到指定的 HBase 表中，该类不会对 WAL 进行操作，即如果服务器发生故障，将面临丢失数据的风险。可以使用 MultipleTableOutputFormat 类解决这个问题，该类可以对是否写入 WAL 进行设置。

7.7 HBase 集群监控

在对 HBase 进行集群监控时，有众多监控软件可供选择，本节选用开源软件 Ganglia 作为 HBase 监控系统。Ganglia 是由加利福尼亚大学伯克利分校发起的一个开源监控项目，能够对数以千计的节点进行监控。在每个节点上运行一个名为 Gmond 的守护进程，用来收集和发送度量数据，如处理器速度、内存使用量等。它从操作系统和指定节点中收集数据，接收这些度量数据的主机可以显示这些数据并且可以将这些数据的精简表单传递到层次结构中。正是因为 Ganglia 采用层次结构模式，才具有良好的扩展性。Gmond 占用的系统资源非常少，这使得它成为在集群中各节点上运行的一段“代码”，不会影响用户系统的性能。

Ganglia 监控套件包括 Gmond、Gmetad 和 Ganglia-web 这 3 个主要组件，下面对这 3 个组件进行简单介绍。

• Gmond 是一个守护进程，运行在每个需要监测的节点上，扮演收集监测统计指标的角色，发送和接收在同一个组播或单播通道上的统计信息。如果它是一个发送者（mute=no），会收集基本指标，比如系统负载（load_one）、CPU 利用率，同时也会发送用户通过添加 C/Python 模块自定义的指标。 如果它是一个接收者（deaf=no），会聚合所有从别的节点上发来的指标，并把它们都保存在内存缓冲区中。

• Gmetad 也是一个守护进程，能定期检查各节点上的 Gmond，并从那里获取数据，将它们的指标存储在 RRD 存储引擎中。Gmetad 可以查询多个集群并聚合指标，也被用于生成用户界面的 Web 前端。

• Ganglia-web 应该安装在有 Gmetad 运行的服务器上，以便读取 RRD 文件。 集群是主机和度量数据的逻辑分组，比如数据库服务器、网页服务器、生产、测试、质量保证（Quality Assurance，QA）等，它们都是完全分开的，需要为每个集群运行单独的 Gmond 实例。

图 7-19 显示了 Ganglia 工作原理。Gmond 检测节点并收集检测的数据；Gmetad 定期检查各 Gmond，从中获取数据并保存到缓冲区中；Ganglia-web 读取 RRD 文件，并做相应处理。

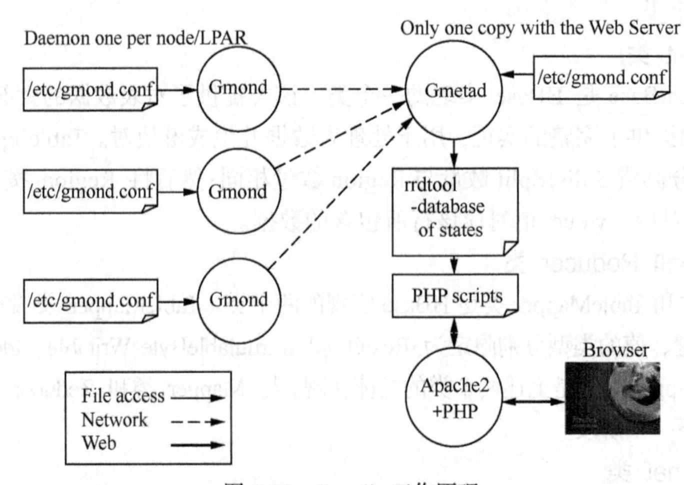

图 7-19　Ganglia 工作原理

安装 Ganglia 监控集群的步骤如下。

（1）在本地配置 EPEL 源，安装软件 ganglia-web、ganglia-gmetad、ganglia-gmond，如图 7-20

所示。命令如下：

```
[root@ master ~]# yum install -y epel-release
[root@master ~]# yum install -y ganglia-web.x86_64 ganglia-gmetad.x86_64 ganglia-gmond.x86_64
```

使用 yum 安装软件之前请确保配置了 EPEL 源，图 7-21 所示界面已经成功安装 3 个软件以及相应的依赖包。

```
[root@master ~]# yum install -y ganglia-gmetad ganglia-devel ganglia-gm
ond rrdtool httpd ganglia-web php
Loaded plugins: fastestmirror
Loading mirror speeds from cached hostfile
Resolving Dependencies
--> Running transaction check
---> Package ganglia-devel.x86_64 0:3.7.2-2.el7 will be installed
---> Package ganglia-gmetad.x86_64 0:3.7.2-2.el7 will be installed
---> Package ganglia-gmond.x86_64 0:3.7.2-2.el7 will be installed
---> Package ganglia-web.x86_64 0:3.7.1-2.el7 will be installed
---> Package httpd.x86_64 0:2.4.6-40.el7.centos will be installed
---> Package php.x86_64 0:5.4.16-36.el7_1 will be installed
---> Package rrdtool.x86_64 0:1.4.8-9.el7 will be installed
--> Finished Dependency Resolution

Dependencies Resolved

======================================================================
 Package            Arch        Version                Repository
                                                                  Size
======================================================================
Installing:
```

图 7-20　使用 yum 安装软件

```
Installed:
  ganglia-gmetad.x86_64 0:3.7.2-2.el7          ganglia-gmond.x86_64 0:3.7.2-2.el7          ganglia-web.x86_64 0:3.7.1-2

Dependency Installed:
  apr.x86_64 0:1.4.8-3.el7                     apr-util.x86_64 0:1.5.2-6.el7               dejavu-sans-mono-fonts.noarch 0:
  ganglia.x86_64 0:3.7.2-2.el7                 httpd.x86_64 0:2.4.6-40.el7.centos          httpd-tools.x86_64 0:2.4.6-40.el
  libconfuse.x86_64 0:2.7-7.el7                libmemcached.x86_64 0:1.0.16-5.el7          libxslt.x86_64 0:1.1.28-5.el7
  libzip.x86_64 0:0.10.1-8.el7                 mailcap.noarch 0:2.1.41-2.el7               php.x86_64 0:5.4.16-36.el7_1
  php-ZendFramework.noarch 0:1.12.20-1.el7     php-bcmath.x86_64 0:5.4.16-36.el7_1         php-cli.x86_64 0:5.4.16-36.el7_1
  php-common.x86_64 0:5.4.16-36.el7_1          php-gd.x86_64 0:5.4.16-36.el7_1             php-process.x86_64 0:5.4.16-36.e
  php-xml.x86_64 0:5.4.16-36.el7_1             rrdtool.x86_64 0:1.4.8-9.el7                t1lib.x86_64 0:5.1.2-14.el7

Complete!
```

图 7-21　软件安装成功

（2）将 Ganglia-web 安装目录链接到/var/www/html 目录，命令如下：

```
[root@master ~]# ln -s /usr/share/ganglia/ /var/www/html
```

（3）修改 http 主站点目录下 ganglia 站点目录的访问权限，并修改 rrds 数据库存放目录权限，命令如下：

```
[root@master ~]# chown -R apache:apache /var/www/html/ganglia
[root@master ~]# chmod -R 755 /var/www/html/ganglia
[root@master ~]# chown -R nobody:nobody /var/lib/ganglia/rrds
```

（4）修改 Ganglia-web 的访问权限，命令如下：

```
[root@master ~]# vi /etc/httpd/conf.d/ganglia.conf
```

配置文件修改为图 7-22 所示的内容。

```
#
# Ganglia monitoring system php web frontend
#

Alias /ganglia /usr/share/ganglia

<Location /ganglia>
  Require all granted
  # Require ip 10.1.2.3
  # Require host example.org
</Location>
~
```

图 7-22 修改 ganglia.conf

（5）配置/etc/ganglia/gmetad.conf，命令如下：

```
[root@master ~]# vi /etc/ganglia/gmetad.conf
```

如图 7-23、图 7-24 所示，分别修改 data_source 和 setuid_username 中的内容。

```
#
# data_source "my cluster" 10 localhost  my.machine.edu:8649  1.2.3.5:8655
# data_source "my grid" 50 1.3.4.7:8655 grid.org:8651 grid-backup.org:8651
# data_source "another source" 1.3.4.7:8655  1.3.4.8

data_source "hbase" 192.168.2.176:8649

#
# Round-Robin Archives
# You can specify custom Round-Robin archives here (defaults are listed below)
```

图 7-23 修改 data_source

```
#-------------------------------------------------------------------------------
# User gmetad will setuid to (defaults to "nobody")
# default: "nobody"
setuid_username nobody
#
#-------------------------------------------------------------------------------
# Umask to apply to created rrd files and grid directory structure
# default: 0 (files are public)
# umask 022
#
#-------------------------------------------------------------------------------
```

图 7-24 修改 setuid_username

（6）配置/etc/ganglia/gmond.conf，命令如下：

```
[root@master ~]# vi /etc/ganglia/gmond.conf
```

修改 3 处内容，如图 7-25 所示。

```
cluster {
  name = "hbase"
  owner = "nobody"
  latlong = "unspecified"
  url = "unspecified"
}
```

```
udp_send_channel {
  #bind_hostname = yes # Highly recommended, soon to be default.
                       # This option tells gmond to use a source address
                       # that resolves to the machine's hostname.  Without
                       # this, the metrics may appear to come from any
                       # interface and the DNS names associated with
                       # those IPs will be used to create the RRDs.
  #mcast_join = 239.2.11.71
  host = 192.168.2.176
  port = 8649
  ttl = 1
}
```

图 7-25 修改 gmond.conf

```
udp_recv_channel {
  #mcast_join = 239.2.11.71
  port = 8649
  #bind = 239.2.11.71
  retry_bind = true
  # Size of the UDP buffer. If you are handling lots of metrics you really
  # should bump it up to e.g. 10MB or even higher.
  # buffer = 10485760
}
```

图 7-25　修改 gmond.conf（续）

（7）启动 Ganglia 服务，命令如下：

```
[root@master ~]# systemctl start httpd.service
[root@master ~]# systemctl start gmetad.service
[root@master ~]# systemctl start gmond.service
```

（8）打开浏览器，输入 http://192.168.2.176/ganglia 可以看到系统监控状态，如图 7-26 所示。

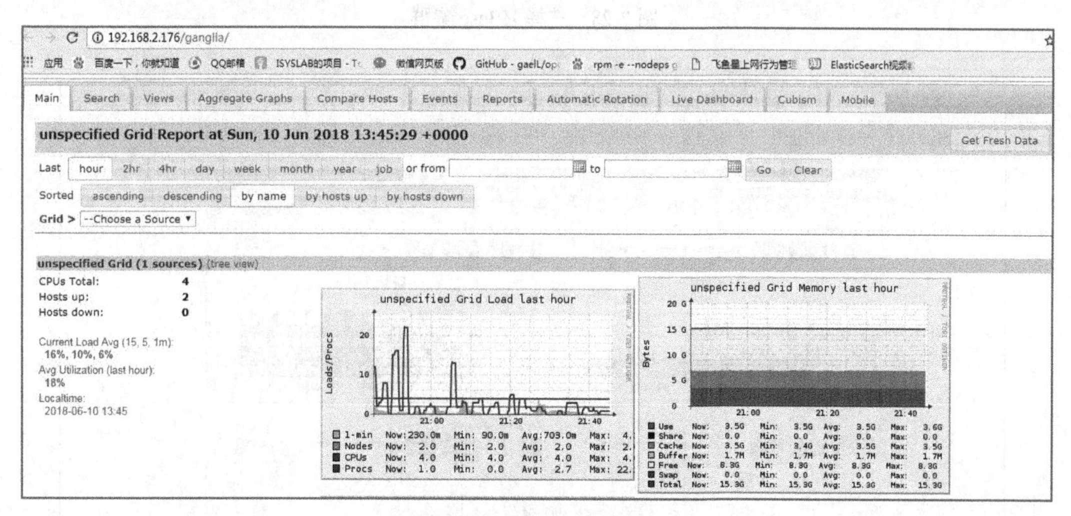

图 7-26　系统监控状态

（9）配置 HBase，让 Ganglia 监控 HBase 集群，命令如下：

```
[root@master ~]#vi /home/hadoop/software/hbase-1.2.6/conf/hadoop-metrics2-hbase.
properties
```

在配置文件末尾添加图 7-27 所示内容。

```
# hbase.sink.file2.class=org.apache.hadoop.metrics2.sink.FileSink
# hbase.sink.file2.context=thrift-two
# hbase.sink.file2.filename=thrift-one.metrics

# hbase.sink.file3.class=org.apache.hadoop.metrics2.sink.FileSink
# hbase.sink.file3.context=rest
# hbase.sink.file3.filename=rest.metrics

*.sink.ganglia.class=org.apache.hadoop.metrics2.sink.ganglia.GangliaSink31
*.sink.ganglia.period=10
hbase.sink.ganglia.period=10
hbase.sink.ganglia.servers=192.168.2.176:8649
```

图 7-27　修改配置文件

（10）切换到 HBase 命令目录，重启 HBase，让配置文件生效：

```
[root@master ~]# cd /home/hadoop/software/hbase-1.2.6/bin/
[root@master bin]# ./stop-hbase.sh
[root@master bin]# ./start-hbase.sh
```

（11）刷新网页，选择 HBase 集群，便可以监控 HBase 集群状态，如图 7-28、图 7-29 所示。

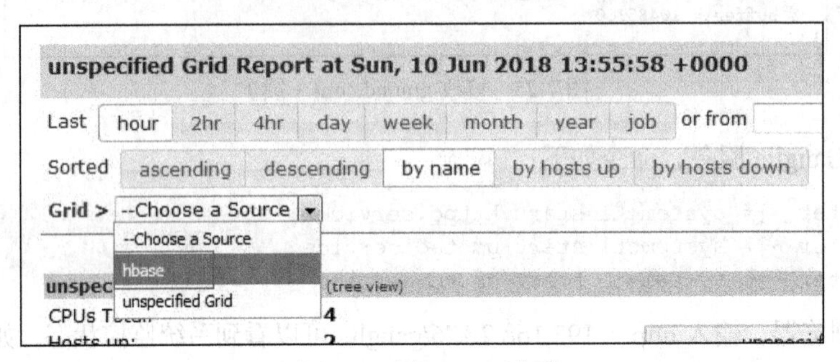

图 7-28　选择 HBase 集群

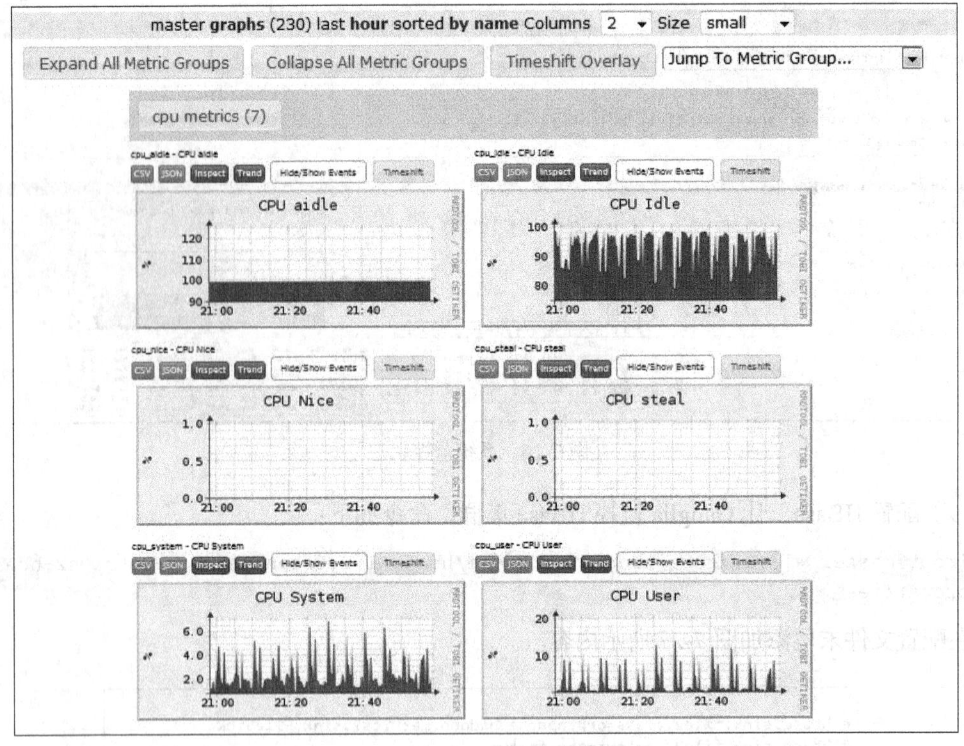

图 7-29　HBase 集群状态

7.8　HBase 实战：公有云网盘系统管理

公有云网盘的应用十分广泛，用户可以将自己的文件上传到公有云网盘，当用户需要使用文件时能够及时下载。这样可以节省用户磁盘空间，同时能够做到随时随地访问文件，十分方便。

本节介绍 HBase 在公有云网盘系统上的应用。

7.8.1　部署公有云网盘

下面讲解如何在大数据主机上部署网盘项目。此网盘项目以 Tomcat 为服务器，用 HBase 存储用户信息，用 HDFS 存储上传的文件。具体步骤如下。

（1）把 apache-tomcat-7.0.42.tar.gz 上传到/home/hadoop/software/目录并解压，命令如下：

```
[hadoop@master ~]$ cd /home/hadoop/software/
[hadoop@master software]$ tar -zxf apache-tomcat-7.0.42.tar.gz
```

解压后的文件如图 7-30 所示。

```
[hadoop@master ~]$ cd /home/hadoop/software/
[hadoop@master software]$ ls
apache-ant-1.10.3            apache-hive-2.3.3-bin.tar.gz  hbase-1.2.6
apache-flume-1.8.0-bin       apache-tomcat-7.0.42.tar.gz   hbase-1.2.6-bin.tar.gz
apache-flume-1.8.0-bin.tar   hadoop-2.9.0                  mysql-libs
apache-hive-2.3.3-bin        hadoop-2.9.0.tar.gz
[hadoop@master software]$ tar -zxf apache-tomcat-7.0.42.tar.gz
[hadoop@master software]$ ls
apache-ant-1.10.3            apache-hive-2.3.3-bin.tar.gz  hadoop-2.9.0.tar.gz
apache-flume-1.8.0-bin       apache-tomcat-7.0.42          hbase-1.2.6
apache-flume-1.8.0-bin.tar   apache-tomcat-7.0.42.tar.gz   hbase-1.2.6-bin.tar.gz
apache-hive-2.3.3-bin        hadoop-2.9.0                  mysql-libs
```

图 7-30　解压后的文件

（2）把网盘项目的.war 文件上传到 apache-tomcat-7.0.42/webapps 子目录，图 7-31 所示的 testnetpan.war 为上传的文件。

```
[hadoop@master webapps]$ pwd
/home/hadoop/software/apache-tomcat-7.0.42/webapps
[hadoop@master webapps]$ ls
docs  examples  host-manager  manager  ROOT  testnetpan.war
```

图 7-31　上传 testnetpan.war

（3）在 Tomcat 解压目录下执行下面的命令，启动 Tomcat，命令如下：

```
[hadoop@master webapps]$ cd ..
[hadoop@master apache-tomcat-7.0.42]$ bin/startup.sh
```

图 7-32 显示了启动 Tomcat 的过程。

```
[hadoop@master webapps]$ cd ..
[hadoop@master apache-tomcat-7.0.42]$ bin/startup.sh
Using CATALINA_BASE:   /home/hadoop/software/apache-tomcat-7.0.42
Using CATALINA_HOME:   /home/hadoop/software/apache-tomcat-7.0.42
Using CATALINA_TMPDIR: /home/hadoop/software/apache-tomcat-7.0.42/temp
Using JRE_HOME:        /home/hadoop/java/jdk1.8.0_161
Using CLASSPATH:       /home/hadoop/software/apache-tomcat-7.0.42/bin/b
ootstrap.jar:/home/hadoop/software/apache-tomcat-7.0.42/bin/tomcat-juli
.jar
```

图 7-32　启动 Tomcat 的过程

（4）使用下面的命令查看启动日志，保证启动过程没有错误，命令如下：

```
[hadoop@master apache-tomcat-7.0.42]$ cat logs/catalina.out
```

从图 7-33 可以看出启动过程正常，没有错误。

```
[hadoop@master apache-tomcat-7.0.42]$ cat logs/catalina.out
May 15, 2018 2:43:35 PM org.apache.catalina.core.AprLifecycleListener init
INFO: The APR based Apache Tomcat Native library which allows optimal performance in p
roduction environments was not found on the java.library.path: /usr/java/packages/lib/
amd64:/usr/lib64:/lib64:/lib:/usr/lib
May 15, 2018 2:43:35 PM org.apache.coyote.AbstractProtocol init
INFO: Initializing ProtocolHandler ["http-bio-8080"]
May 15, 2018 2:43:35 PM org.apache.coyote.AbstractProtocol init
INFO: Initializing ProtocolHandler ["ajp-bio-8009"]
May 15, 2018 2:43:35 PM org.apache.catalina.startup.Catalina load
INFO: Initialization processed in 665 ms
May 15, 2018 2:43:35 PM org.apache.catalina.core.StandardService startInternal
INFO: Starting service Catalina
May 15, 2018 2:43:35 PM org.apache.catalina.core.StandardEngine startInternal
INFO: Starting Servlet Engine: Apache Tomcat/7.0.42
May 15, 2018 2:43:35 PM org.apache.catalina.startup.HostConfig deployWAR
INFO: Deploying web application archive /home/hadoop/software/apache-tomcat-7.0.42/web
apps/testnetpan.war
```

图 7-33　查看启动日志

（5）修改网盘数据库路径，让网盘数据库与本地服务器相连接，使用 Vi 编辑器编辑 db.properties 中的内容，本环境的路径如图 7-34 所示。

```
[hadoop@master ~]$ cd /home/hadoop/software/apache-tomcat-7.0.42/webapps/testnetpan/WEB-INF/classes/
[hadoop@master classes]$ more db.properties
host=192.168.2.176\:9000
hdfs=hdfs\://192.168.2.176\:9000
hbase_dir=hdfs\://192.168.2.176\:9000/hbase

[hadoop@master classes]$
```

图 7-34　让网盘数据库与本地服务器相连接

（6）打开本机的浏览器，在地址栏输入 http://192.168.2.176:8080/testnetpan/。当输入网盘地址后，进入登录界面，如图 7-35 所示。

图 7-35　登录界面

（7）此时的项目还不能使用，因为 HBase 中的表还没有创建，执行 http://192.168.2.176:8080/testnetpan/init 进行项目的初始化。从图 7-36 中可以看出，需要在地址栏增加 init，并刷新浏览器，进行初始化操作。

图 7-36　初始化项目

（8）打开命令行界面，进入 HBase 的 Shell，使用 list 命令查看初始化的表，命令如下：

```
[hadoop@master classes]$ hbase shell
hbase(main):001:0> list
```

从图 7-37 中可以看出，初始化成功后，HBase 中多了 email_user、follow、followed 等数据表。

```
[hadoop@master classes]$ hbase shell
HBase Shell; enter 'help<RETURN>' for list of supported commands.
Type "exit<RETURN>" to leave the HBase Shell
Version 1.2.6, rUnknown, Mon May 29 02:25:32 CDT 2017

hbase(main):001:0> list
TABLE
book
email_user
follow
followed
gid
id_user
share
shareed
user_id
9 row(s) in 0.2360 seconds

=> ["book", "email_user", "follow", "followed", "gid", "id_user", "shar
e", "shareed", "user_id"]
```

图 7-37　列出 HBase 中初始化的数据表

（9）初始化完成后，单击“注册”按钮，使用邮箱注册一个账号，如 test，并设置密码，如图 7-38 所示。

图 7-38　注册新用户

（10）用注册的账号登录系统，单击左侧“我的网盘”链接，然后单击右侧“创建文件夹”按钮新建一个文件夹，如图 7-39 所示。

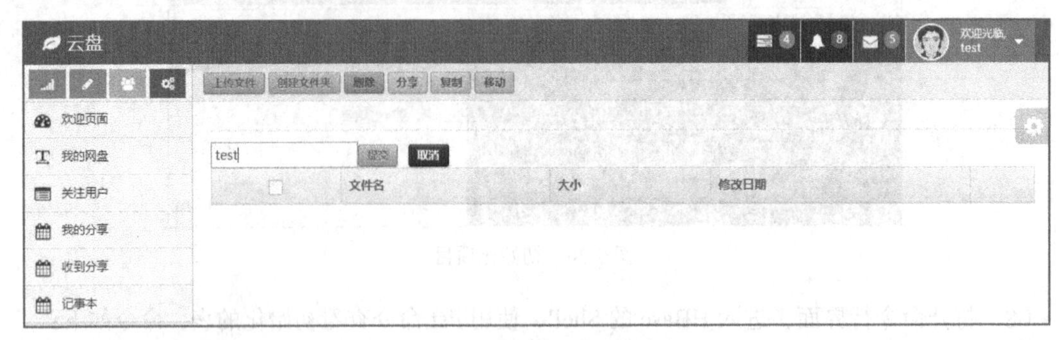

图 7-39　新建 test 文件夹

（11）进入新创建的 test 文件夹，单击“上传文件”按钮，选择一个文件进行上传，上传过程如图 7-40、图 7-41 所示。

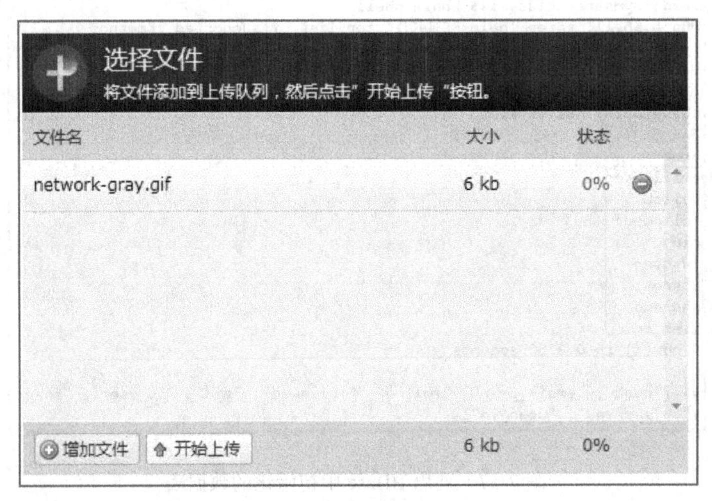

图 7-40　上传文件

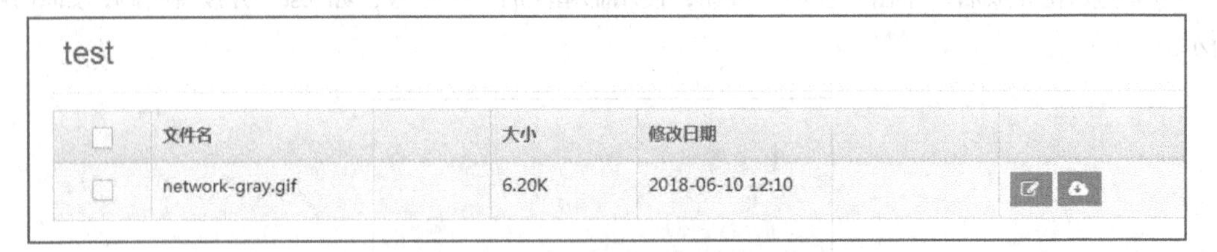

图 7-41　在 test 文件夹中成功上传了文件

7.8.2　网盘核心代码分析

通过 7.8.1 小节的实践，成功部署了公有云网盘，能够在网盘中上传文件，并实现增、删、改、查等功能。本小节分析网盘中与 HBase 相关的核心代码。

网盘的 Web 端采用 Spring MVC 4.0.5，是基于 HDFS 和 HBase 的伪分布式环境进行开发的，网盘功能列表如表 7-11 所示。

表 7-11　　网盘功能列表

功能	说明
登录	用户登录个人账户
注册	用户注册个人账户
网盘管理	上传文件、创建文件夹、修改文件及文件夹、删除文件及文件夹
关注用户管理	关注和取消关注用户
文件分享	分享文件及文件夹、分享列表查看
其他	浏览文档

网盘分为 Web 端和 Hadoop 端。Web 端又分为前端和后端，前端使用 HTML、CSS、JavaScript 等技术，后端使用的是 Spring MVC。Hadoop 端使用 HDFS 作为底层存储系统，数据库则是基于列的 HBase。网盘结构如图 7-42 所示。

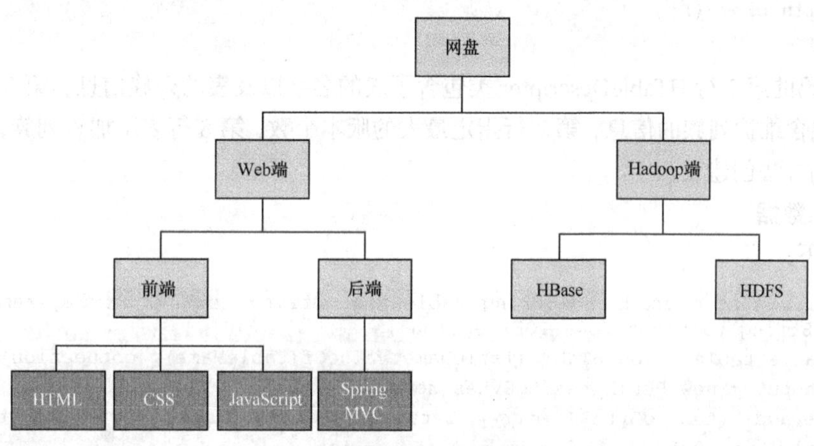

图 7-42　网盘结构

网盘中与 HBase 的交互使用 Java API 进行，包括创建表、插入数据、删除表等。接下来介绍本系统中相关的核心代码。

1. HBase 与项目的基本配置

代码如下：

```
1. private HbaseDB() {
2.   System.setProperty("hadoop.user.name", "hadoop");
3.   System.setProperty("hadoop.home.dir", "D:/hadoop/hadoop-2.6.0");
4.   Configuration conf = HBaseConfiguration.create();
5.   conf.set("hbase.zookeeper.quorum", SiteUrl.readUrl("host"));
6.   conf.set("hbase.rootdir", SiteUrl.readUrl("hbase_dir"));
7.   try {
8.       connection = HConnectionManager.createConnection(conf);
9.   } catch (IOException e) {
10.      e.printStackTrace();
11.  }
12.}
```

上述代码的第 2、3 行设置指定键值对作为系统属性，第一个参数表示系统属性的名称，第二个参数表示系统属性的值。第 4 行创建 HBaseConfiguration 类的实例，该类的主要作用是提供对配置参数的访问路径。HBaseConfiguration.create( )方法返回的是一个能够访问 HBase 配置信息的 Configuration 实例。第 5 行调用 set( )方法设立 ZooKeeper 的地址。第 8 行创建一个连接到集群的实例。

2. 创建表

代码如下：

```
1. public static void createTable(String tableName,String[] fams,int version) throws
   Exception {
2.  HTableDescriptor tableDescriptor = null;
3.  HColumnDescriptor hd = null;
4.  for (int i = 0; i < fams.length; i++) {
5.      tableDescriptor = new HTableDescriptor(TableName.valueOf(tableName));
6.      hd = new HColumnDescriptor(fams[i]);
7.      hd.setMaxVersions(version);
8.      tableDescriptor.addFamily(hd);
9.      admin.createTable(tableDescriptor);
10. }
11. admin.close();
12. }
```

上述代码的第 2 行 HTableDescriptor 类包含了表的名字以及表的列族信息，第 3 行 HColumn Descriptor 包含维护列族的信息，第 7 行指定最大的版本个数，第 8 行表示加入列族，第 9 行使用 createTable( )方法创建表。

3. 插入数据

代码如下：

```
1. public static void add(String tableName, String rowKey, String family, String
qualifier, String value) throws IOException {
2.  HTable table = new HTable(TableName.valueOf(tableName), connection);
3.  Put put = new Put(Bytes.toBytes(rowKey));
4.  put.add(Bytes.toBytes(family), Bytes.toBytes(qualifier), Bytes.toBytes(value));
5.  table.put(put);
6.  table.close();
7. }
```

插入数据时需要先与表建立连接，插入数据之后要及时关闭连接，以保证数据库安全。其中第 2 行表示建立连接，第 3 行用行键实例化 Put，第 4 行指定列族名、列名和值。

4. 删除表

代码如下：

```
1. public static void delTable(String tableName) throws Exception {
2.  HBaseAdmin admin = new HBaseAdmin(connection);
3.  if (admin.tableExists(tableName)) {
4.      admin.disableTable(tableName);
5.      admin.deleteTable(tableName);
6.  }
7.  admin.close();
8. }
```

在删除表的过程中，第 2 行建立表连接，第 4 行禁用表，第 5 行删除表，删除表后关闭连接，确保数据库安全。

5. 删除数据

代码如下：

```
1. public static void deleteRow(String tableName, String[] rowKey) throws Exception {
2.   HTable table = new HTable(TableName.valueOf(tableName), connection);
3.   List<Delete> list = new ArrayList<Delete>();
4.   for (int i = 0; i < rowKey.length; i++) {
5.       Delete delete = new Delete(Bytes.toBytes(Long.valueOf(rowKey[i])));
6.       list.add(delete);
7.   }
8.   table.delete(list);
9.   table.close();
10. }
```

根据 row 可以批量删除数据。先遍历表，将待删除数据存放到 list 中，第 8 行调用 delete( ) 方法删除数据。

习　题

1. HBase 依靠什么存储底层数据？（　　）

   A. HDFS　　B. Hadoop　　C. Memory　　D. MapReduce

2. LSM 更能保证哪种操作的性能？（　　）

   A. 读　　B. 写　　C. 随机读　　D. 合并

3. LSM 结构的数据首先存储在哪儿？（　　）

   A. 硬盘上　　B. 内存中　　C. 磁盘阵列中　　D. 闪存中

4. HFile 数据格式中的 Data 字段用于什么？（　　）

   A. 存储实际的 KeyValue 数据　　B. 存储数据的起点

   C. 指定字段的长度　　D. 存储数据块的起点

5. 下面哪些概念是 HBase 框架中使用的？（　　）（多选题）

   A. HDFS　　B. GridFS　　C. Zookeeper　　D. EXT3

6. HBase 构建二级索引的实现方式有哪些？（　　）（多选题）

   A. MapReduce　　B. Coprocessor　　C. Bloom Filter　　D. Filter

7. HBase 的物理存储单元是什么？（　　）

   A. Region　　B. Column Family　　C. Column　　D. 行

8. Compaction 的目的是什么？（　　）

   A. 减少同一个 Region 中同一个 Column Family 下的文件数量

   B. 提升数据读取性能

   C. 减少同一个 Column Family 的文件数量

   D. 减少同一个 Region 的文件数量

9. 客户端采用批量写入接口写入 10 条数据，某个 HegionServer 节点上包含该表的 2 个 Region，分别为 A 和 B，写入的 10 条数据中有 2 条属于 A，有 4 条属于 B，请问写入这 10 条数据需要向该 HRegionServer 发送几次 RPC 请求？

10. 一张表包含 3 个 Region，即 [10,20)、[20,30)、[30,+∞)，分别编号为①、②、③，那么 11、20、222 分别属于哪个 Region？

11. 阅读 HBase 网盘源码，尝试添加下载功能。

第 8 章 数据获取与 Flume 应用

互联网或计算机日志系统是大数据应用获取数据的重要来源，用户可以针对特定领域的文本、图片、音频、视频等，通过爬虫获取数据，也可以从计算机运行的日志文件中收集数据。通过预处理后，这些数据被上传到 HDFS 供进一步分析使用。本章内容主要包括：

（1）介绍比较常见的公开数据平台；

（2）使用网络爬虫通过一定的策略获取相关网站上的数据；

（3）使用实时日志收集系统 Flume 工具进行数据采集等。

8.1 公开数据资源获取

随着互联网技术的成熟，在过去 20 年间，互联网上的信息量发生了爆炸式的增长。数据作为一种新的“能源”形式，正在源源不断地发挥其巨大的价值，帮助人们改善生活方式。目前国内拥有众多的公开数据平台，可以帮助我们更好地开展大数据研究工作。

首先，在政府类数据平台中，对外公开的数据平台包括国家统计局网站、国家数据网站、中国经济网等。其中国家统计局网站主要提供包括国家经济宏观数据在内的，涉及社会发展与民生相关的重要数据及信息，数据非常全面，且定期发布统计出版物，实用性强；国家数据网站提供包括国计民生各个方面的月度数据、季度数据、年度数据、各地区数据、部门数据以及国际数据，该网站的数据来自国家统计局，排版清晰、简洁；中国经济网收编了 300000 多条按时间顺序排列的数据，数据内容涵盖宏观经济数据、行业经济数据以及地区经济数据。

其次，在各行业机构类数据平台中，对外公开的数据平台包括数据堂、百度预测、友盟+等。其中数据堂平台提供计算机、医疗、人脸、语音、方言、交通、电商等各类的具体数据资源，比较适合语音数据、图像数据等资料的获取；百度预测为百度的大数据产品，百度预测开放平台可以通过上传历史数据或者接入 API 来使用，根据历史数据情况，进行数据结果预测。

最后，国际上的公开数据平台包括 World Bank、Kaggle 等。其中 World Bank 提供世界各国多方面的权威经济数据，包括国际收支平衡表（Balance of Payments，BOP）、开发框架、环境、利率与价格、外债、金融统计、政府财政、国民账户、社会指标、贸易 10 个分类下的 54 个项目，以及可视化的图表，可更清晰地展示数据的动态变化；Kaggle 作为一个数据建模和数据分析竞赛平台，包含了众多企业和研究者上传的医疗、交通等行业数据集，可用于数据分析和建模。

8.2 使用网络爬虫获取数据

通过公开的数据平台下载数据集，为我们获取数据带来了便利。但是数据平台提供的数据集在时效性、数据量等方面存在欠缺。为解决以上问题，本节将介绍网络爬虫技术，通过爬虫实时获取指定网页的内容，从而提高数据获取的速度和质量。

图 8-1 所示为搜索引擎的大致架构。首先，搜索引擎所获得的信息全部来源于互联网上的网页，其网络爬虫程序会根据一定的搜索策略和相关的网络协议在互联网上获取新的网页内容，然后进行网页去重，删掉重复的内容。

得到这些网页后，基于服务器的云存储和云计算平台会对网页内容进行解析，并根据一定的链接关系进行排序，然后将结果存储在一种高效的数据结构中。

对于用户查询，搜索引擎收到用户查询的关键词后，首先进行查询分析，然后根据网页排序算法，将用户关心程度较高的网页优先显示给用户。

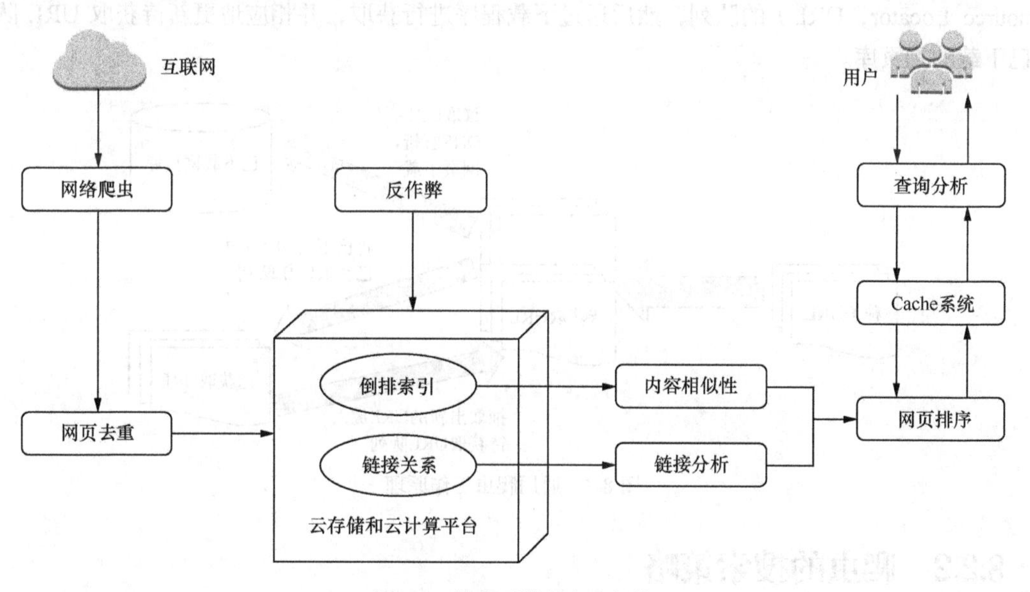

图 8-1 搜索引擎的大致架构

8.2.1 爬虫的工作原理

网络爬虫作为搜索引擎中用于获取信息的重要手段，主要利用 TCP/IP、HTTP 等网络协议，基于超链接和网页文档检索的方法遍历互联网信息空间，是一种功能很强大的、能够自动获取网页信息的程序。它有着以下的特性和功能。

（1）通过输入特定的请求访问某一站点。

（2）自动从一个网站地址移动到下一个网站地址，并且建立索引，将其加入数据库。

（3）如图 8-2 所示，一个爬虫系统，主要由调度器、下载器、网页解析器和传输器四大部分组成。

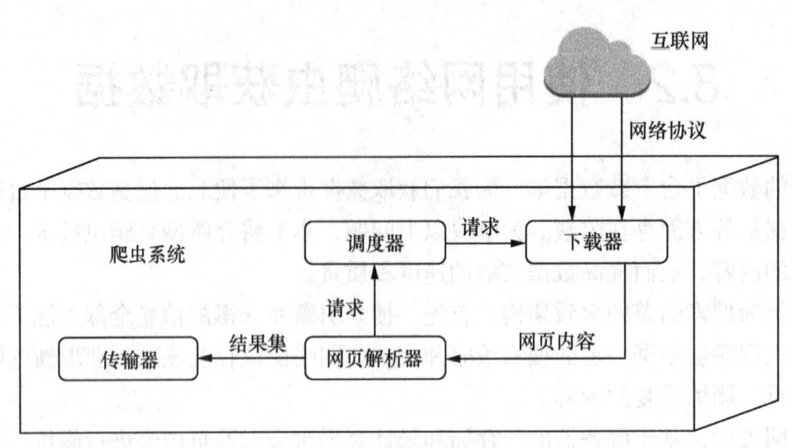

图 8-2　通用爬虫框架

（4）爬虫最重要的功能是，通过 TCP/IP、HTTP 等协议制定相关的获取策略从互联网上下载需要的网页。其工作原理如图 8-3 所示，首先要获得待获取的种子统一资源定位符（Uniform Resource Locator，URL）的队列，然后通过下载程序进行获取，并相应地更新待获取 URL 队列和已下载的网页库。

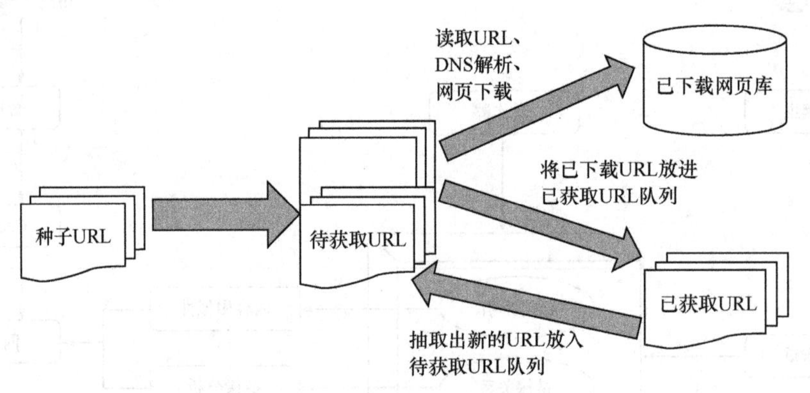

图 8-3　通用爬虫工作原理

8.2.2　爬虫的搜索策略

当网络爬虫遍历某个网页源码时，它利用 HTML 的标记结构来搜索信息并获取指向其他网页的 URL。网络爬虫在搜索时往往采用一定的搜索策略，可以完全不依赖用户干预去实现网络信息的自动获取。传统的爬虫技术一般采用以下 3 种搜索策略。

1. 深度优先搜索策略

深度优先搜索和二叉树的先根遍历很类似。这种策略的方法是，从一个网页链接出发，在解析其网页内容的同时，如果碰到网页中的超链接就进入，一直到超链接的最下层；一条超链接走不通了就退回上一层，选择相邻的另一条超链接继续向下走，直到所有链接访问完毕。

2. 宽度优先搜索策略

宽度优先搜索和二叉树的层次遍历很类似。这种策略的方法是，从一个网页链接出发，在解析其网页内容的同时，如果碰到网页中的超链接就进入；但它不会顺着一条支路一直向下走，而是将一个网页中的所有超链接全部访问完毕后才会向下访问；然后进入子网页，同样将子网页中的超链接访问完毕后再继续向下。

3. 聚焦搜索策略

聚焦搜索策略和前两种搜索策略的注重点不同，因为用户不在乎搜索显示的东西有多么全，只在乎搜索到的内容和自己搜索的关键字的相关度，是否足够匹配。因此聚焦搜索把搜索策略的重点放在主题上面。它在解析一个网页时会分析其包含的超链接中哪些的价值更高，哪些的推广程度更广泛，从而优先去获取这些网站，而不是一味地根据队列中的流程去搜索，这也更能体现如今搜索引擎关注的方向：以用户为中心。

8.2.3　爬虫的简单应用

不论是简单的爬虫程序，还是复杂的爬虫程序，其核心都是对正则表达式的运用。这里使用 Python 编写一个简单的爬虫程序，用于理解爬虫的基本工作原理。该程序主要用于获取特定网页，解析其内容，并将其中包含的超链接提取出来保存到文件中。

（1）安装 Python 环境，Linux 发行版自带 Python 2 开发环境。若读者需要 Python 3 开发环境，可到官网下载适合自己系统的 Python 版本，并配置好环境变量。在命令行中输入 python 验证环境是否安装成功，如图 8-4 所示。

```
[hadoop@master ~]$ python
Python 2.7.5 (default, Nov 20 2015, 02:00:19)
[GCC 4.8.5 20150623 (Red Hat 4.8.5-4)] on linux2
Type "help", "copyright", "credits" or "license" for more information.
>>> 
```

图 8-4　Python 环境安装

（2）程序中需要用到两个 Python 的函数库。

- Requests：采用 Python 语言实现，遵循 Apache License 2.0 开源协议的基于 urllib 的 HTTP 库。比 urllib 更加方便，可以很好地满足 HTTP 测试需求。
- Re：提供了正则表达式功能，正则表达式的模式（Pattern）可以被编译成一系列的字节码，然后用 C 语言编写的引擎执行。

（3）新建 data 目录，存放爬虫文件，命令如下：

```
[hadoop@master ~] $ mkdir /home/hadoop/data
```

（4）进入 data 目录，创建一个名为 crawler.py 的文件，命令如下：

```
[hadoop@master ~] $ cd /home/hadoop/data
[hadoop@master data] $ vim crawler.py
```

在 crawler.py 文件中添加如下代码：

```
1. #!bin/python
2. #-*_coding:utf8-*-
3. import requests
4. import re
5. import sys
6. reload(sys)
7. sys.setdefaultencoding("utf-8")
8. r = requests.get('http://www.hust.edu.cn/')
9. data = r.text
```

```
10. link_list=re.findall(r"(?<=href=\").+?(?=\")|(?<=href=\').+?(?=\')" ,data)
11. fh = open('info.txt', 'w')
12. for url in link_list:
13. fh.write(url+'\n')
14. fh.close()
```

其中第3、4、5行表示导入爬虫所需的模块,第6、7行表示避免存入文件时发生UnicodeEncodeError错误，第8、9行表示获取特定的网页，将网页内容保存到对象中，第10行表示利用正则表达式查找网页中所包含的超链接，第11、12、13行表示创建名为info.txt的文件，将获取的信息存入文件中，最后关闭文件。

（5）在终端运行爬虫程序，命令如下：

```
[hadoop@master data]$ python crawler.py
```

（6）这时在程序所在的同级目录中会生成一个名为info.txt的文件，如图8-5所示。

```
[hadoop@master data]$ ll
total 8
-rw-rw-r-- 1 hadoop hadoop  347 Aug 22 02:34 crawler.py
-rw-rw-r-- 1 hadoop hadoop 3365 Aug 22 02:44 info.txt
[hadoop@master data]$
```

图8-5　爬虫结果文件

（7）查看获取的数据信息，命令如下：

```
[hadoop@master data]$ cat info.txt
```

结果如图8-6所示。

```
http://focus.hustonline.net/
http://focus.hustonline.net/
http://cxcy.hust.edu.cn
http://cxcy.hust.edu.cn
http://dqlh.hust.edu.cn/
http://dqlh.hust.edu.cn/
http://kyzj.fiscal.hust.edu.cn
http://kyzj.fiscal.hust.edu.cn
/740/list.htm
http://www.lib.hust.edu.cn/
http://xsyj.hust.edu.cn/
http://www.tjmu.edu.cn/
http://service.hust.edu.cn/
http://mail.hust.edu.cn
http://xxgk.hust.edu.cn
javascript:;
javascript:;
/750/list.htm
/750/list.htm
javascript:;
javascript:;
http://bbs.whnet.edu.cn/
http://bbs.whnet.edu.cn/
http://www.beian.gov.cn/portal/registerSystemInfo?recordcode=42011102000123
[hadoop@master data]$
```

图8-6　爬虫获取的数据信息

8.3　使用 Flume 获取数据

8.3.1　Flume 简介

在介绍 Flume 之前，先看看使用 Flume 收集数据的 Hadoop 业务流程，如图 8-7 所示。我们可以看出，在大数据业务的处理过程中，数据采集是业务处理的开始，是非常重要且不可避免的一步。

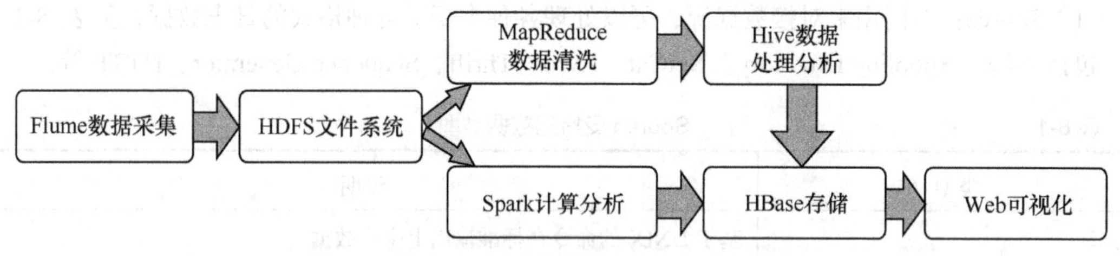

图 8-7　使用 Flume 收集数据的 Hadoop 业务流程

Flume 是一个分布式的日志收集系统，具有高可靠、高可用、事务管理、失败重启等特点。Flume 可有效地从不同的源收集、聚合和传输大量日志数据到集中式数据存储设备。其设计原理是基于数据流的，如图 8-8 所示。可以利用 Flume 将日志数据从各种网站服务器上汇集起来存储到 HDFS、HBase 等分布式存储系统中。

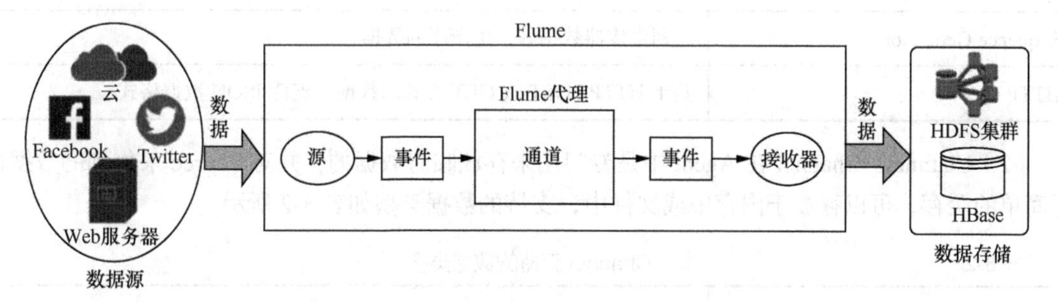

图 8-8　Flume 数据收集结构

Flume 作为业界广泛认可和应用的实时日志收集系统，具有以下几个重要的特点：

（1）Flume 可以高效地将收集的日志信息存入 HDFS、HBase 等分布式存储系统中或者直接移交到 Hadoop 系统中；

（2）除了日志信息，Flume 也可以用来收集规模宏大的社交网络的时间节点数据，比如 Facebook、Twitter 等；

（3）Flume 支持多种接入资源数据的类型以及输出资源数据的类型；

（4）Flume 支持多路径流量、多管道接入和输出、上下文路由等；

（5）可以支持被水平扩展。

8.3.2　Flume 运行机制

Flume 收集数据是通过日志收集节点上运行的 Agent 实现的，Flume 以 Agent 为最小的独立运行单位，一个Agent就是在JVM中运行的一个独立Flume进程，如图8-9所示。Agent包含Source、Channel、Sink三大核心组件。

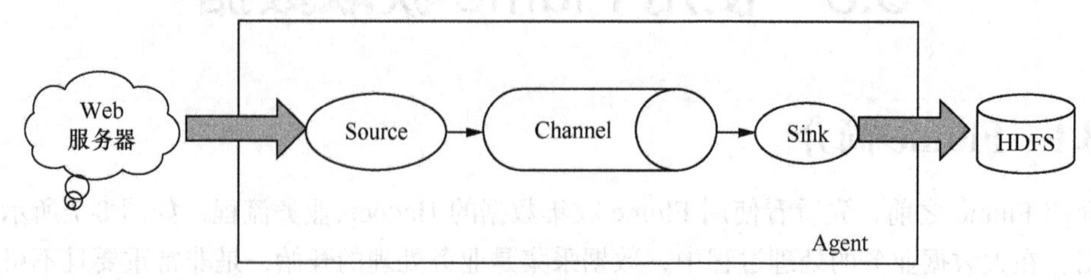

图8-9　Flume架构

（1）Source：专门用来对接数据源，可以处理各种类型、各种格式的日志数据，如表8-1所示，包括Exec、Spooling Directory、NetCat、Avro、Thrift、Sequence Generator、HTTP等。

表8-1　Source支持的数据类型

类型	说明
Exec	基于UNIX的命令在标准输出上生产数据
Spooling Directory	监控指定目录内的数据变更
NetCat	监听指定端口，将流经端口的数据作为输入
Avro	支持Avro协议，内置支持
Thrift	支持Thrift协议，内置支持
Sequence Generator	序列生成器数据源，生产序列数据
HTTP	基于HTTP POST或GET方式的数据，支持JSON数据格式

（2）Channel：Channel在Agent中是专门用来存放临时数据的，并对Source采集到的数据进行简单的缓存，可以存放于内存中或文件中，支持的数据类型如表8-2所示。

表8-2　Channel支持的数据类型

类型	说明
Memory	数据缓存到内存中
File	数据持久化到磁盘文件
Spillable Memory	数据优先缓存到内存，内存不足则持久化到磁盘文件
JDBC Channel	数据存储到一个支持JDBC连接的数据库
Custom Channel	自定义Channel实现

（3）Sink：实现数据的分发功能，支持的数据类型如表8-3所示，分发目的地包括Kafka、Hive、HDFS、Logger、Avro等。

Source接收由外部源（如Web服务器）传递给它的事件，外部源以目标Flume源识别的格式向Flume发送事件。当Source接收事件时，它将其存储到一个或多个Channel。该Channel是一个被动存储器，可以保持事件直到它被 Sink 消耗。Sink 从 Channel 中移除事件并将其写入外部存储库（如 HDFS，通过 HDFS 接收器）或将其转发到流中的下一个 Agent 的 Source。给定 Agent 的 Source、Sink 与 Channel 中暂存的事件可异步运行。

表 8-3　　Sink 支持的数据类型

类型	说明
Kafka	数据写入 Kafka topic
Hive	数据写入 Hive 数据库或分区
HDFS	数据保存到 HDFS
Logger	数据写入日志文件
Avro	数据被转换成 Avro Event，然后发送到配置的 RPC 端口上
Thrift	数据被转换成 Thrift Event，然后发送到配置的 RPC 端口上
HBase	数据写入 HBase 数据库
Custom	自定义 Sink 实现

8.3.3　Flume 安装部署

因为 Flume 的 Agent 是运行在 JVM 上的 Java 进程，所以 Flume 的安装运行需要 Java 1.7 及以上版本的支持，以及对 Agent 目录的读/写权限。本次安装的环境为 CentOS 7，Flume 版本为 Flume 1.8，JDK 版本为 Java 1.8。

安装步骤如下。

（1）下载并解压 Flume。Flume 下载界面如图 8-10 所示，选择二进制版本 Apache Flume binary (tar.gz)中的安装包，将其下载并解压到/home/hadoop/software 目录。

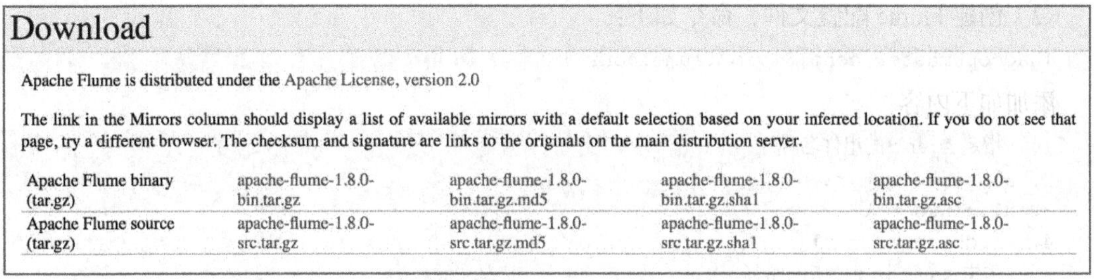

Download

Apache Flume is distributed under the Apache License, version 2.0

The link in the Mirrors column should display a list of available mirrors with a default selection based on your inferred location. If you do not see that page, try a different browser. The checksum and signature are links to the originals on the main distribution server.

Apache Flume binary (tar.gz)	apache-flume-1.8.0-bin.tar.gz	apache-flume-1.8.0-bin.tar.gz.md5	apache-flume-1.8.0-bin.tar.gz.sha1	apache-flume-1.8.0-bin.tar.gz.asc
Apache Flume source (tar.gz)	apache-flume-1.8.0-src.tar.gz	apache-flume-1.8.0-src.tar.gz.md5	apache-flume-1.8.0-src.tar.gz.sha1	apache-flume-1.8.0-src.tar.gz.asc

图 8-10　Flume 下载界面

（2）配置环境变量，切换到 root 用户，编辑/etc/profile 文件，命令如下：

```
[root@master~]# vim /etc/profile
```

在文件末尾添加以下代码：

```
export FLUME_HOME= /home/hadoop/software/apache-flume-1.8.0-bin
export PATH=$PATH:$FLUME_HOME/bin
```

（3）载入环境变量，命令如下：

```
[root@master~]# source /etc/profile
```

（4）切换到 hadoop 用户，修改 flume-env.sh，命令如下：

```
[root@master~]# su - hadoop
[hadoop@master~]$ cd /home/hadoop/software/apache-flume-1.8.0-bin/conf
[hadoop@master conf]$ cp flume-env.sh.template flume-env.sh
[hadoop@master conf]$ vim flume-env.sh
```

修改如下两项内容：

```
export JAVA_HOME=/home/hadoop/java/jdk1.8.0_161     %Java 安装路径
export JAVA_OPTS="-Xms100m -Xmx2000m -Dcom.sun.management.jmxremote"
```

（5）验证配置是否成功，命令如下：

```
[hadoop@master ~]$ flume-ng version
```

出现图 8-11 所示的信息，则安装成功。

```
[hadoop@master ~]$ flume-ng version
Flume 1.8.0
Source code repository: https://git-wip-us.apache.org/repos/asf/flume.git
Revision: 99f591994468633fc6f8701c5fc53e0214b6da4f
Compiled by denes on Fri Sep 15 14:58:00 CEST 2017
From source with checksum fbb44c8c8fb63a49be0a59e27316833d
[hadoop@master ~]$
```

图 8-11 Flume 环境配置验证

8.3.4 Flume 简单应用

安装好 Flume 后，使用 Flume 的步骤分为两步：（1）写一个配置文件，在配置文件当中描述 Source、Channel 与 Sink 的具体实现；（2）运行一个 Agent 实例，在运行 Agent 实例的过程中会读取配置文件的内容，这样 Flume 就会采集数据。下面介绍使用 Flume 监听一条路径，并随时获取写入该路径的信息的步骤。

（1）首先进入/apache-flume-1.8.0-bin/conf 目录，命令如下：

```
[hadoop@master ~]$ cd  /home/hadoop/software/apache-flume-1.8.0-bin/conf
```

（2）创建 Flume 配置文件，命令如下：

```
[hadoop@master conf]$ vim flume.conf
```

添加如下内容：

```
1. #指定 Agent 的组件名称
2. a1.sources = r1
3. a1.sinks = k1
4. a1.channels = c1
5. #指定 Source 要监听的路径
6. a1.sources.r1.type = spooldir
7. a1.sources.r1.spoolDir=/home/hadoop/software/apache-flume-1.8.0-bin/logs
8. #指定 Sink
9. a1.sinks.k1.type = logger
10. #指定 Channel
11. a1.channels.c1.type = memory
12. a1.channels.c1.capacity = 1000
13. a1.channels.c1.transactionCapacity = 100
14. #绑定 Source 和 Sink 到 Channel 上
15. a1.sources.r1.channels = c1
16. a1.sinks.k1.channel = c1
```

（3）启动 Agent，命令如下：

```
[hadoop@master ~]$ cd /home/hadoop/software/apache-flume-1.8.0-bin
[hadoop@master apache-flume-1.8.0-bin]$ bin/flume-ng agent --conf conf
--conf-file conf/flume.conf --name a1 -Dflume.root.logger=INFO,console
```

Flume 启动参数如表 8-4 所示。

表 8-4　Flume 启动参数

参数	作用
-conf 或 -c	指定配置文件夹，包括 flume-env.sh 和 Log4j 的配置文件
-conf-file 或-f	配置文件地址
-name 或-n	Agent 名称
-z	ZooKeeper 连接字符串
-p	ZooKeeper 中的存储路径前缀

（4）另外打开一个命令行界面，在 apache-flume-1.8.0-bin 目录下新建日志目录，命令如下：

```
[hadoop@master ~]$ cd /home/hadoop/software/apache-flume-1.8.0-bin
[hadoop@master apache-flume-1.8.0-bin]$ mkdir logs
```

Flume 目录文件如图 8-12 所示。

```
[hadoop@master apache-flume-1.8.0-bin]$ mkdir logs
[hadoop@master apache-flume-1.8.0-bin]$ ll
total 144
drwxr-xr-x  2 hadoop hadoop    59 Aug 16 17:23 bin
-rw-r--r--  1 hadoop hadoop 81264 Sep 15  2017 CHANGELOG
drwxr-xr-x  2 hadoop hadoop   142 Aug 16 17:28 conf
-rw-r--r--  1 hadoop hadoop  5681 Sep 15  2017 DEVNOTES
-rw-r--r--  1 hadoop hadoop  2873 Sep 15  2017 doap_Flume.rdf
drwxr-xr-x 10 hadoop hadoop  4096 Sep 15  2017 docs
drwxrwxr-x  2 hadoop hadoop  4096 Aug 16 17:23 lib
-rw-r--r--  1 hadoop hadoop 27663 Sep 15  2017 LICENSE
drwxrwxr-x  2 hadoop hadoop     6 Aug 22 03:38 logs
-rw-r--r--  1 hadoop hadoop   249 Sep 15  2017 NOTICE
-rw-r--r--  1 hadoop hadoop  2483 Sep 15  2017 README.md
-rw-r--r--  1 hadoop hadoop  1588 Sep 15  2017 RELEASE-NOTES
drwxrwxr-x  2 hadoop hadoop    67 Aug 16 17:23 tools
[hadoop@master apache-flume-1.8.0-bin]$
```

图 8-12　Flume 目录文件

（5）追加文件到 apache-flume-1.8.0-bin/logs 目录，写入 Hello World 作为测试内容，然后将文件复制到 Flume 的监听路径上，命令如下：

```
[hadoop@master ~]$ echo "Hello World" > /home/hadoop/software/apache-flume-
1.8.0-bin/logs/spool.log
```

（6）当数据写入监听路径后，如图 8-13 所示，控制台就会显示收集到的数据。

```
2018-04-26 16:41:06,985 (SinkRunner-PollingRunner-DefaultSinkProcessor) [INFO -
org.apache.flume.sink.LoggerSink.process(LoggerSink.java:95)] Event: { headers:{
file=/home/hadoop/software/apache-flume-1.8.0-bin/logs/spool.log} body: 48 65 6C
 6C 6F 20 57 6F 72 6C 64                      Hello World }
```

图 8-13　Flume 收集到的 Hello World 信息

8.4 综合案例

本案例通过 Python 爬虫获取豆瓣网最近上映的电影信息，获取的信息通过 Flume 传输到 HDFS 中。Python 的版本是 3.6，Flume 的版本是 1.8，本案例源码在 ch08 目录中。

操作步骤如下。

（1）上传源码 ch08/scrape-douban.py 到集群的/home/hadoop/data/flume-test/目录，可通过 FileZilla 工具实现。

（2）进入/home/hadoop/data/flume-test/目录，查看源码，命令如下：

```
[hadoop@master flume-test]$ cat scrape-douban.py
```

① 编写网络爬虫程序，首先要对网页进行访问，使用 Python 中的 urllib 库，代码如下：

```
1. from urllib import request
2. resp = request.urlopen('https://movie.douban.com/nowplaying/wuhan/')
3. html_data = resp.read().decode('utf-8')
```

其中 https://movie.douban.com/nowplaying/wuhan/ 表示豆瓣网最近上映的电影，可以在浏览器中输入该网址进行查看。html_data 是字符串类型的变量，里面存放了网页的 HTML 代码。

② 对 HTML 代码进行解析，获取我们需要的数据。在 Python 中使用 BeautifulSoup 库进行 HTML 代码的解析（注意：如果没有安装 Beautiful Soup 库，则使用 pip install BeautifulSoup 进行安装即可）。

BeautifulSoup 的使用格式如下：

```
1.BeautifulSoup(html,"html.parser")
```

第一个参数表示需要获取数据的 HTML，第二个参数是指定解析器。

③ 打开已获取网页的 HTML 代码，查看需要的数据所在的 HTML 标签，如图 8-14 所示。

```
<div id="nowplaying">
    <div class="mod-hd">
        <h2>正在上映</h2>
    </div>
    <div class="mod-bd">
        <ul class="lists">
                <li
                    id="26430636"
                    class="list-item"
                    data-title="狂暴巨兽"
                    data-score="6.7"
                    data-star="35"
                    data-release="2018"
                    data-duration="107分钟"
                    data-region="美国"
                    data-director="布拉德·佩顿"
                    data-actors="道恩·强森 / 娜奥米·哈里斯 / 杰弗里·迪恩·摩根"
                    data-category="nowplaying"
                    data-enough="True"
                    data-showed="True"
                    data-votecount="40770"
                    data-subject="26430636"
                >
```

图 8-14　网页 HTML 代码

从图 8-14 中可以看出，从<div id="nowplaying">标签开始是我们想要的数据，里面有电影的名称、评分、主演等信息。所以相应的代码如下：

```
1. from bs4 import BeautifulSoup as bs
2. soup = bs(html_data, 'html.parser')
3. nowplaying_movie = soup.find_all('div', id='nowplaying')
4. nowplaying_movie_list = nowplaying_movie[0].find_all('li', class_='list-item')
5. nowplaying_list = []
6.     for item in nowplaying_movie_list:
7.         nowplaying_dict = {}
8.         nowplaying_dict['id'] = item['data-subject']
9.         nowplaying_dict['title'] = item['data-title']
10.        nowplaying_dict['score'] = item['data-score']
11.        nowplaying_dict['region'] = item['data-region']
12.        nowplaying_dict['director'] = item['data-director']
13.        nowplaying_dict['votecount'] = item['data-votecount']
14.        nowplaying_list.append(nowplaying_dict)
15.     return nowplaying_list
```

④ 存储电影信息到逗号分隔值（Comma-Separated Values，CSV）文件（CSV 是一种编辑方便、可视化效果极佳的数据存储方式），相应的代码如下：

```
1. csv_file = open("/home/hadoop/data/flume-test/movielist.csv","w",newline = '')
2. writer = csv.writer(csv_file)
3. writer.writerow(['id','title','score','region','director','votecount'])
4.    commentList = []
5.    NowPlayingMovie_list = getNowPlayingMovie_list()
6.    for i in range(len(NowPlayingMovie_list)):
7. writer.writerow([NowPlayingMovie_list[i]['id'],NowPlayingMovie_list[i]['title'],
8. NowPlayingMovie_list[i]['score'],NowPlayingMovie_list[i]['region'],
9. NowPlayingMovie_list[i]['director'],NowPlayingMovie_list[i]['votecount']])
10.     csv_file.close()
```

（3）运行爬虫程序，将获取的电影信息放在/home/hadoop/data/flume-test/目录下的 movielist.csv 文件中，命令如下：

```
[hadoop@master flume-test]$ python3 scrape-douban.py
```

（4）查看获取的电影信息，命令如下：

```
[hadoop@master flume-test]$ cat movielist.csv
```

结果如图 8-15 所示。

（5）进入配置文件夹，命令如下：

```
[hadoop@master ~]$ cd /home/hadoop/software/apache-flume-1.8.0-bin/conf
```

（6）创建 flume.conf 文件，命令如下：

```
[hadoop@master conf]$ vim flume.conf
```

```
[hadoop@master flume-test]$ cat movielist.csv
id,title,score,region,director,votecount
26430636,狂暴巨兽,6.7,美国,布拉德·佩顿,40793
26691361,21克拉,5.3,中国,何念,13149
4920389,头号玩家,8.9,美国,史蒂文·斯皮尔伯格,392359
26661189,脱单告急,5.7,中国,柯孟融,7244
26640371,犬之岛,8.4,美国 德国,韦斯·安德森,51832
26942631,起跑线,8.0,印度,萨基特·乔杜里,55674
26384741,湮灭,7.3,美国 英国,亚历克斯·嘉兰,98505
26647117,暴裂无声,8.3,中国,忻钰坤,77423
27077266,米花之味,7.6,中国,鹏飞,3822
26588783,冰雪女王3: 火与冰,4.4,俄罗斯 中国,阿列克谢·特斯蒂斯林,1058
26861685,红海行动,8.5,中国,林超贤,357231
26760161,冰封迷案,4.5,中国,唐宇,229
7056414,通勤营救,6.7,美国 英国,佐米·希尔拉,23796
30152451,厉害了, 我的国,0,中国,卫铁,111
30188193,求求你爱上我,0,中国,董春泽,481
25727544,寻找罗麦,4.6,中国,法国,王超,2484
```

图 8-15　获取的电影信息

往 flume.conf 文件中添加以下内容：

```
1. a1.sources = r1
2. a1.sinks = k1
3. a1.channels = c1
4. #配置 Source
5. a1.sources.r1.type = spooldir
6. a1.sources.r1.spoolDir = /home/hadoop/data/flume-test
7. a1.sources.r1.fileHeader = true
8. #配置 Sink
9. a1.sinks.k1.type = hdfs
10. a1.sinks.k1.hdfs.path = hdfs://master:9000/tmp/flume-movies/%Y%m%d
11. a1.sinks.k1.hdfs.filePrefix = events
12. a1.sinks.k1.hdfs.fileType = DataStream
13. a1.sinks.k1.hdfs.useLocalTimeStamp = true
14. #配置 Channel
15. a1.channels.c1.type = memory
16. a1.channels.c1.capacity = 1000
17. a1.channels.c1.transactionCapacity = 100
18. a1.sources.r1.channels = c1
19. a1.sinks.k1.channel = c1
```

（7）启动 Agent，命令如下：

```
[hadoop@master ~]$ cd /home/hadoop/software/apache-flume-1.8.0-bin
[hadoop@master apache-flume-1.8.0-bin]$ bin/flume-ng agent --conf conf
--conf-file conf/flume.conf --name a1 -Dflume.root.logger=INFO,console
```

启动效果如图 8-16 所示。

```
[hadoop@master apache-flume-1.8.0-bin]$ bin/flume-ng agent --conf conf --conf-fi
le conf/flume.conf --name a1 -Dflume.root.logger=INFO,console
Info: Sourcing environment configuration script /home/hadoop/software/apache-flu
me-1.8.0-bin/conf/flume-env.sh
Info: Including Hadoop libraries found via (/home/hadoop/software/hadoop-2.9.0/b
in/hadoop) for HDFS access
Info: Including Hive libraries found via () for Hive access
+ exec /home/hadoop/java/jdk1.8.0_161/bin/java -Xms100m -Xmx2000m -Dcom.sun.mana
gement.jmxremote -Dflume.root.logger=INFO,console -cp '/home/hadoop/software/apa
che-flume-1.8.0-bin/conf:/home/hadoop/software/apache-flume-1.8.0-bin/lib/*:/hom
e/hadoop/software/hadoop-2.9.0/etc/hadoop:/home/hadoop/software/hadoop-2.9.0/sha
re/hadoop/common/lib/*:/home/hadoop/software/hadoop-2.9.0/share/hadoop/common/*:
/home/hadoop/software/hadoop-2.9.0/share/hadoop/hdfs:/home/hadoop/software/hadoo
p-2.9.0/share/hadoop/hdfs/lib/*:/home/hadoop/software/hadoop-2.9.0/share/hadoop/
hdfs/*:/home/hadoop/software/hadoop-2.9.0/share/hadoop/yarn:/home/hadoop/softwar
e/hadoop-2.9.0/share/hadoop/yarn/lib/*:/home/hadoop/software/hadoop-2.9.0/share/
hadoop/yarn/*:/home/hadoop/software/hadoop-2.9.0/share/hadoop/mapreduce/lib/*:/h
ome/hadoop/software/hadoop-2.9.0/share/hadoop/mapreduce/*:/home/hadoop/software/
hadoop-2.9.0/contrib/capacity-scheduler/*.jar:/lib/*' -Djava.library.path=:/home
/hadoop/software/hadoop-2.9.0/lib/native org.apache.flume.node.Application --con
f-file conf/flume.conf --name a1
SLF4J: Class path contains multiple SLF4J bindings.
SLF4J: Found binding in [jar:file:/home/hadoop/software/apache-flume-1.8.0-bin/l
```

图 8-16　Agent 启动

（8）运行爬虫程序，将获取的电影信息放在/home/hadoop/data/flume-test/目录下的 movielist.csv 文件中，命令如下：

```
[hadoop@master flume-test]$ python3 scrape-douban.py
```

（9）登录 HDFS，查询 Flume 采集的电影信息，命令如下：

```
[hadoop@master ~]$hadoop fs -ls /tmp/flume-movies
```

查询结果如图 8-17 所示。

```
[hadoop@master ~]$ hadoop fs -ls /tmp/flume-movies
Found 1 items
drwxr-xr-x   - hadoop supergroup          0 2018-04-27 16:02 /tmp/flume-movies/2
0180427
[hadoop@master ~]$
```

图 8-17　Flume 采集的电影信息

（10）查询子文件的具体内容，命令如下：

```
[hadoop@master ~]$hadoop fs -cat /tmp/flume-movies/20180427/events.xxxxxx
```

查询结果如图 8-18 所示。

```
[hadoop@master ~]$ hadoop fs -cat /tmp/flume-movies/20180427/events.152481611330
9
id,title,score,region,director,votecount
26430636,狂暴巨兽,6.7,美国,布拉德·佩顿,40793
26691361,21克拉,5.3,中国,何念,13149
4920389,头号玩家,8.9,美国,史蒂文·斯皮尔伯格,392359
26661189,脱单告急,5.7,中国,柯孟融,7244
26640371,犬之岛,8.4,美国 德国,韦斯·安德森,51832
26942631,起跑线,8.0,印度,萨基特·乔杜里,55674
26384741,湮灭,7.3,美国 英国,亚历克斯·嘉兰,98505
26647117,暴裂无声,8.3,中国,忻钰坤,77423
27077266,米花之味,7.6,中国,鹏飞,3822
[hadoop@master ~]$
```

图 8-18　Flume 采集的具体内容

习 题

1. 以下哪项不属于爬虫常用的工具包？（　　）

   A. Requests

   B. BeautifulSoup

   C. Scrapy

   D. Shogun

2. Source 支持的数据类型包括？（　　）

   A. Exec

   B. Avro

   C. Thrift

   D. HTTP

3. 了解 Python 语法，并学习 Python 正则表达式，尝试从网站上下载不同的内容。
4. 配置 Flume 的核心组件，尝试使用不同的数据类型采集数据。
5. 了解 Flume 的过滤机制，尝试在数据采集阶段实现数据的过滤。

第 9 章 基于 Spark 的内存计算

Spark 作为 MapReduce 的同级分布式处理框架，可以集成在原有的 Hadoop 生态系统中，同时也存在着自己的生态系统。本章侧重介绍 Spark 的部署和弹性分布式数据集（Resilient Distributed Dataset，RDD）的计算原理以及 Spark 的工作机制和读/写数据流，最后简单介绍以 Spark 核心衍生的 SQL 查询、流式计算、机器学习等生态系统。

9.1 Spark 简介

Spark 是一个基于内存运算，用来实现高效集群计算的平台。准确地讲，Spark 是一个大数据并行计算框架，是对广泛使用的 MapReduce 计算模型的扩展。Spark 有着自己的生态系统，但是同样兼容 HDFS、Hive 等分布式存储系统，可以完美融入 Hadoop 的生态圈中，代替 MapReduce 去实现更为高效的分布式计算。基于 MapReduce 的计算引擎通常会将中间结果输出到磁盘上，进行存储和容错，这就使得在迭代计算和交互计算的任务上表现得效率低下，而 Spark 可以将中间结果保存在内存中，不需要读/写 HDFS，从而大幅提高计算速度。

Spark 是基于 Scala 语言开发的，但是其本身提供了丰富的接口。除了提供基于 Python、Java、Scala 和 SQL 的接口和丰富的内建函数库，它还为其他一些大数据工具的集成提供了 API，这就允许开发者在自己熟悉的编程环境中进行使用。

Spark 架构如 Hadoop 一样，采用的都是分布式计算中的 Master-Slave 模型。图 9-1 所示为 Spark 官方给出的分布式架构，驱动程序（Driver Program）负责提交应用，触发集群开始处理作业；集群管理器（Cluster Manager）作为主节点控制整个集群，负责集群的正常运行；工作节点（Worker Node）作为计算节点，负责接收主节点命令以及对运行状态的报告，其上的执行器（Executor）组件真正负责任务的执行、启动线程池等运行任务。

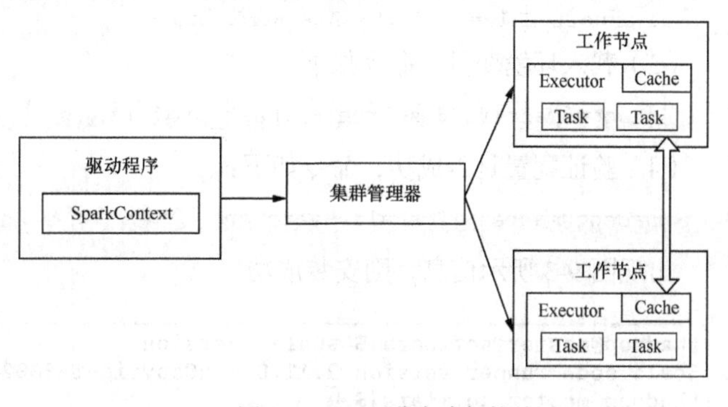

图 9-1　Spark 分布式架构

9.2 Spark 快速部署

9.2.1 Spark 单机模式部署

Spark 是运行在 JVM 上的。要在机器上运行 Spark，则需要保证 JDK 已经正常安装。如果希望使用 Python 接口，则需要安装 Python 解释器。下面使用的是单机模式部署 Spark，计算机操作系统为 CentOS 7。部署步骤如下。

1. 安装配置 JDK

验证 JDK 配置是否成功，命令如下：

```
[hadoop@master~]$ java -version
```

出现图 9-2 所示信息，则安装成功。

```
[hadoop@master ~]$ java -version
java version "1.8.0_161"
Java(TM) SE Runtime Environment (build 1.8.0_161-b12)
Java HotSpot(TM) 64-Bit Server VM (build 25.161-b12, mixed mode)
[hadoop@master ~]$
```

图 9-2 Java 环境配置验证

若没有安装配置 JDK，请参考 2.3.2 小节配置 Java 环境。

2. 安装配置 Scala

（1）这里使用的 Scala 版本为 Scala 2.11.8，下载并解压 Scala，命令如下：

```
[hadoop@master~]$ wget http://www.scala-lang.org/files/archive/scala-2.11.8.tgz
[hadoop@master~]$ tar -zxf scala-2.11.8.tgz
[hadoop@master~]$ mv scala-2.11.8 /home/hadoop/software
```

（2）配置环境变量，编辑 ~/.bash_profile 文件，命令如下：

```
[hadoop@master~]$ vi ~/.bash_profile
```

在文件末尾添加以下配置，保存后退出：

```
1. export SCALA_HOME=/home/hadoop/software/scala-2.11.8
2. export PATH=$PATH:$SCALA_HOME/bin
```

（3）载入环境变量，命令如下：

```
[hadoop@master~]$ source ~/.bash_profile
```

（4）验证配置是否成功，命令如下：

```
[hadoop@master~]$ scala -version
```

出现图 9-3 所示信息，则安装成功。

```
[hadoop@master software]$ scala -version
Scala code runner version 2.11.8 -- Copyright 2002-2016, LAMP/EPFL
[hadoop@master software]$
```

图 9-3 Scala 环境配置验证

3. 安装配置 Spark

（1）本书采用的版本为 Spark 2.2.2，如图 9-4 所示，请到官方网站下载。

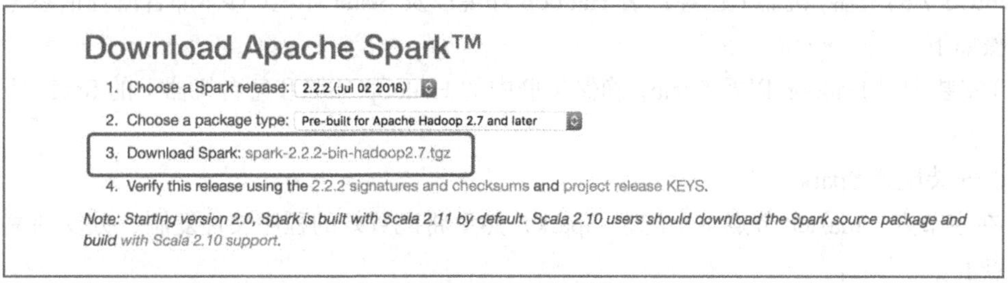

Download Apache Spark™

1. Choose a Spark release: 2.2.2 (Jul 02 2018)
2. Choose a package type: Pre-built for Apache Hadoop 2.7 and later
3. Download Spark: spark-2.2.2-bin-hadoop2.7.tgz
4. Verify this release using the 2.2.2 signatures and checksums and project release KEYS.

Note: Starting version 2.0, Spark is built with Scala 2.11 by default. Scala 2.10 users should download the Spark source package and build with Scala 2.10 support.

图 9-4　Spark 下载界面

解压 Spark 到/home/hadoop 目录，命令如下：

```
[hadoop@master~]$ tar -zxf spark-2.2.2-bin-hadoop2.7.tgz
[hadoop@master~]$ mv spark-2.2.2-bin-hadoop2.7 /home/hadoop/spark-2.2.2
```

（2）配置环境变量，编辑 ~/.bash_profile 文件，命令如下：

```
[hadoop@master~]$ vi ~/.bash_profile
```

在文件末尾添加以下配置，保存后退出。

```
export SPARK_HOME=/home/hadoop/spark-2.2.2
```

（3）载入环境变量，命令如下：

```
[hadoop@master~]$ source ~/.bash_profile
```

（4）验证 Spark 是否安装成功，进入 spark-2.2.2-bin-hadoop2.7 的 bin 目录，打开 Scala 版本的 Shell，命令如下：

```
[hadoop@master bin]$ spark-shell
```

出现图 9-5 所示的信息，则 Spark 安装成功。

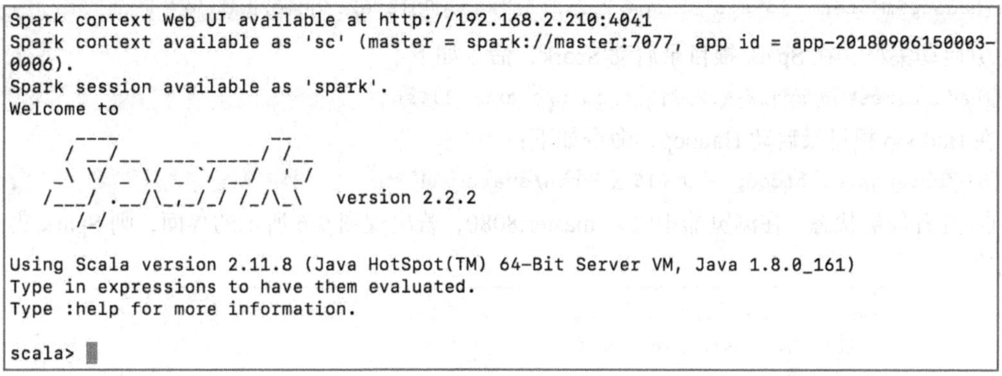

```
Spark context Web UI available at http://192.168.2.210:4041
Spark context available as 'sc' (master = spark://master:7077, app id = app-20180906150003-
0006).
Spark session available as 'spark'.
Welcome to
      ____              __
     / __/__  ___ _____/ /__
    _\ \/ _ \/ _ `/ __/  '_/
   /___/ .__/\_,_/_/ /_/\_\   version 2.2.2
      /_/

Using Scala version 2.11.8 (Java HotSpot(TM) 64-Bit Server VM, Java 1.8.0_161)
Type in expressions to have them evaluated.
Type :help for more information.

scala>
```

图 9-5　Scala Shell 界面

9.2.2　Spark 分布式集群部署

Spark 在生产环境中，主要部署在安装有 Linux 操作系统的集群中。使用 Spark 的一大优势就是可以通过增加机器数量并使用集群模式运行，来扩展程序的计算能力。可以在小数据集上利用本地模式快速开发并验证自己的程序，然后无须修改任何代码就可以在大规模集群上运行。

与单机模式部署一样，安装 Spark 需要在集群机器上预先安装好 JDK、Scala 等环境。但是由于

Spark 是计算框架，所以需要预先在集群内有搭建好的存储数据的持久化层工具，如 HDFS、Hive 等。

该案例选择在预装好 Hadoop 的集群上安装 Spark，使用 HDFS 充当持久化层工具。使用 3 台运行 CentOS 7 的机器，机器上已经部署 Hadoop 环境以及 Scala 环境（具体部署流程请参考前文），部署步骤如下。

（1）安装配置 Hadoop 以及 Scala，确保集群中的 Hadoop 环境和每台机器上的 Scala 环境能正常运行。

（2）安装配置 Spark。

① 在主节点（master 节点）上配置 Spark，然后将配置好的程序文件复制、分发到集群的各个从节点上。

② 下载并安装 Spark（同单机模式部署），配置好环境变量并使之生效。

③ 在 master 节点上配置 Spark，修改 conf/spark-env.sh 配置文件，保存后退出。命令如下：

```
1. export JAVA_HOME=/home/hadoop/java/jdk1.8.0_161
2. export HADOOP_CONF_DIR=/home/hadoop/hadoop-2.9.0
3. export SCALA_HOME=/home/hadoop/software/scala-2.11.8
4. export SPARK_WORKER_MEMORY=2g
5. export SPARK_MASTER_IP=192.168.2.78     #对应 master 主机的地址
6. export SPARK_MASTER_PORT=7077
```

④ 配置 slaves 文件。编辑 conf/slaves 文件，以 2 个工作节点为例，将节点的主机名加入 slaves 文件，保存后退出。命令如下：

```
1. Slave1
2. Slave2
```

⑤ 复制 spark-2.2.2 文件夹到其他的工作节点，以 2 个工作节点为例，命令如下：

```
[hadoop@master ~]$ scp -r spark-2.2.2 hadoop@Slave1:/home/hadoop
[hadoop@master ~]$ scp -r spark-2.2.2 hadoop@Slave2:/home/hadoop
```

⑥ 启动集群，在 Spark 根目录启动 Spark，命令如下：

```
[hadoop@master spark-2.2.2]$ ./sbin/start-all.sh
```

在 Hadoop 根目录启动 Hadoop，命令如下：

```
[hadoop@master hadoop-2.9.0]$ ./sbin/start-all.sh
```

⑦ 查看集群状态，在浏览器中输入 master:8080，若出现图 9-6 所示的界面，则 Spark 集群安装成功。

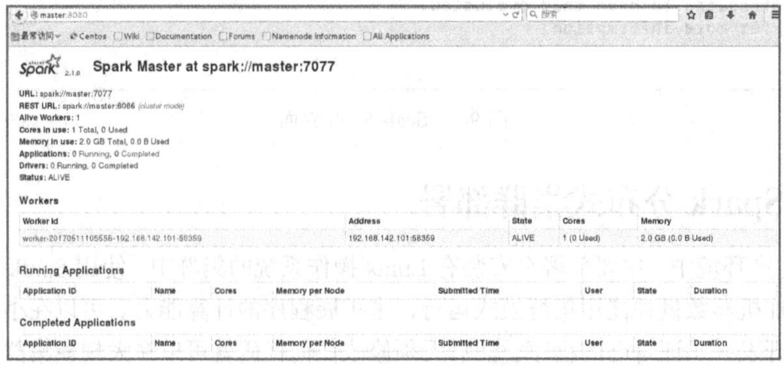

图 9-6　Spark 集群管理界面

⑧ 提交 Spark 自带的应用 SparkPi，命令如下：

```
[hadoop@master spark-2.2.2]$ ./bin/spark-submit \
> --class org.apache.spark.examples.SparkPi \
> --master spark://master:7077 \
> /home/hadoop/spark-2.2.2/examples/jars/spark-examples_2.11-2.2.2.jar 1
```

SparkPi 的运行过程如图 9-7 所示。

```
[hadoop@master spark-2.2.2]$ ./bin/spark-submit \
> --class org.apache.spark.examples.SparkPi \
> --master spark://master:7077 \
> /home/hadoop/spark-2.2.2/examples/jars/spark-examples_2.11-2.2.2.jar 1
18/09/21 14:08:13 INFO spark.SparkContext: Running Spark version 2.2.2
18/09/21 14:08:13 WARN util.NativeCodeLoader: Unable to load native-hadoop library for your plat
form... using builtin-java classes where applicable
18/09/21 14:08:13 INFO spark.SparkContext: Submitted application: Spark Pi
18/09/21 14:08:13 INFO spark.SecurityManager: Changing view acls to: hadoop
18/09/21 14:08:13 INFO spark.SecurityManager: Changing modify acls to: hadoop
18/09/21 14:08:13 INFO spark.SecurityManager: Changing view acls groups to:
18/09/21 14:08:13 INFO spark.SecurityManager: Changing modify acls groups to:
18/09/21 14:08:13 INFO spark.SecurityManager: SecurityManager: authentication disabled; ui acls
disabled; users  with view permissions: Set(hadoop); groups with view permissions: Set(); users
 with modify permissions: Set(hadoop); groups with modify permissions: Set()
18/09/21 14:08:14 INFO util.Utils: Successfully started service 'sparkDriver' on port 48343.
18/09/21 14:08:14 INFO spark.SparkEnv: Registering MapOutputTracker
18/09/21 14:08:14 INFO spark.SparkEnv: Registering BlockManagerMaster
18/09/21 14:08:14 INFO storage.BlockManagerMasterEndpoint: Using org.apache.spark.storage.Defaul
tTopologyMapper for getting topology information
18/09/21 14:08:14 INFO storage.BlockManagerMasterEndpoint: BlockManagerMasterEndpoint up
18/09/21 14:08:14 INFO storage.DiskBlockManager: Created local directory at /tmp/blockmgr-127663
d2-7352-4b6b-8a2d-cb7dc83f51e5
18/09/21 14:08:14 INFO memory.MemoryStore: MemoryStore started with capacity 413.9 MB
18/09/21 14:08:14 INFO spark.SparkEnv: Registering OutputCommitCoordinator
18/09/21 14:08:14 INFO util.log: Logging initialized @2370ms
18/09/21 14:08:14 INFO server.Server: jetty-9.3.z-SNAPSHOT
18/09/21 14:08:14 INFO server.Server: Started @2537ms
18/09/21 14:08:14 INFO server.AbstractConnector: Started ServerConnector@6b9267b{HTTP/1.1,[http/
1.1]}{0.0.0.0:4040}
```

图 9-7　SparkPi 的运行过程

运行完成后，可在集群管理界面查看任务运行的状态，如图 9-8 所示。

Spark 2.2.2 Spark Master at spark://master:7077

URL: spark://master:7077
REST URL: spark://master:6066 (cluster mode)
Alive Workers: 1
Cores in use: 2 Total, 0 Used
Memory in use: 2.0 GB Total, 0.0 B Used
Applications: 0 Running, 1 Completed
Drivers: 0 Running, 0 Completed
Status: ALIVE

Workers

Worker Id	Address	State	Cores	Memory
worker-20180921134841-192.168.2.211-59965	192.168.2.211:59965	ALIVE	2 (0 Used)	2.0 GB (0.0 B Used)

Running Applications

Application ID	Name	Cores	Memory per Executor	Submitted Time	User	State	Duration

Completed Applications

Application ID	Name	Cores	Memory per Executor	Submitted Time	User	State	Duration
app-20180921140815-0000	Spark Pi	2	1024.0 MB	2018/09/21 14:08:15	hadoop	FINISHED	4 s

图 9-8　在 Spark 集群管理界面查看 SparkPi 任务运行的状态

9.3 Spark 程序

通过前文的学习，我们已经掌握了如何在服务器部署 Spark。接下来我们将会介绍如何使用内置的 Spark Shell 工具进行交互式分析，然后使用 Spark 实现经典的大数据案例 WordCount 程序。

9.3.1 Spark Shell

Spark Shell 是 Spark 提供的一种类似于 Shell 的交互式编程环境，能够实时执行用户输入的代码，并输出执行结果。Spark Shell 支持两种语言：Scala 和 Python。由于 Scala 更贴近 Spark 的内部实现，这里仅介绍它。对于 Python 版本（pySpark Shell），读者可以查阅官方文档学习如何使用。操作步骤如下。

（1）首先，在 Spark 的安装目录启动 Spark Shell，命令如下：

```
[hadoop@master spark-2.2.2]$ ./bin/spark-shell --master spark://master:7077
```

启动之后，会输出一些日志，如图 9-9 所示。

```
[hadoop@master spark-2.2.2]$ ./bin/spark-shell --master spark://master:7077
Setting default log level to "WARN".
To adjust logging level use sc.setLogLevel(newLevel). For SparkR, use setLogLevel(newLevel).
18/09/21 19:12:36 WARN util.NativeCodeLoader: Unable to load native-hadoop library for your platform... u
sing builtin-java classes where applicable
Spark context Web UI available at http://192.168.2.210:4040
Spark context available as 'sc' (master = spark://master:7077, app id = app-20180921191237-0006).
Spark session available as 'spark'.
Welcome to
      ____              __
     / __/__  ___ _____/ /__
    _\ \/ _ \/ _ `/ __/  '_/
   /___/ .__/\_,_/_/ /_/\_\   version 2.2.2
      /_/

Using Scala version 2.11.8 (Java HotSpot(TM) 64-Bit Server VM, Java 1.8.0_161)
Type in expressions to have them evaluated.
Type :help for more information.

scala> 
```

图 9-9　Spark Shell 启动界面

（2）上传测试数据集 YoutubeDataSets.txt 到 HDFS 的指定目录，如图 9-10 所示。

```
[hadoop@master ~]$ hadoop fs -ls /user/hadoop/input/
SLF4J: Class path contains multiple SLF4J bindings.
SLF4J: Found binding in [jar:file:/home/hadoop/hadoop-2.7.3/share/hadoop/common/lib/slf4j-log
4j12-1.7.10.jar!/org/slf4j/impl/StaticLoggerBinder.class]
SLF4J: Found binding in [jar:file:/home/hadoop/hbase-1.2.6/lib/slf4j-log4j12-1.7.5.jar!/org/s
lf4j/impl/StaticLoggerBinder.class]
SLF4J: See http://www.slf4j.org/codes.html#multiple_bindings for an explanation.
SLF4J: Actual binding is of type [org.slf4j.impl.Log4jLoggerFactory]
Found 1 items
-rwxrwxrwx   2 hadoop supergroup     969389 2018-08-22 11:08 /user/hadoop/input/YoutubeDataSe
ts.txt
[hadoop@master ~]$ 
```

图 9-10　上传测试数据集

（3）执行以下 Scala 程序，从 HDFS 目录中读取测试数据集，并得到初始 RDD：

```
scala> val rdd1=
sc.textFile("hdfs://master:9000/user/hadoop/input/YoutubeDataSets.txt")
```

执行结果如图 9-11 所示。

```
scala> val rdd1 = sc.textFile("hdfs://master:9000/user/hadoop/input/YoutubeDataSets.txt")
rdd1: org.apache.spark.rdd.RDD[String] = hdfs://master:9000/user/hadoop/input/YoutubeDataSets.txt MapParti
tionsRDD[7] at textFile at <console>:24
```

图 9-11　RDD 初始化

（4）执行以下 Scala 程序，在 rdd1 上进行一些简单的操作，如 count()、take():

```
scala>rdd1.count()      #测试数据集总行数
scala>rdd1.take(3)      #读取测试数据集的前 3 行
```

执行结果如图 9-12 所示。

```
scala> rdd1.count()
res4: Long = 4100

scala> rdd1.take(3)
res5: Array[String] = Array(QuRYeRnAuXM EvilSquirrelPictures     1135    Pets & Animals  252     1075    4.
96      46      86      gFa1YMEJFag     nRcovJn9xHg     3TYqkBJ9YRk     rSJ8QZWBegU     0TZqX5MbXMA     UE
vVksP91kg       ZTopArY7Nbg     0RViGi2Rne8     HT_Ql0JbDpg     YZev1imoxX8     8qQrrfUTmh0     zQ83d_D2MG
s       u6_DQQjLsAw     73Wz9CQFDtE, 3TYqkBJ9YRk        hggh22  1135    Comedy  169     228     5       5Q
uRYeRnAuXM      gFa1YMEJFag     UEvVksP91kg     rSJ8QZWBegU     nRcovJn9xHg     sVkuOk4jmCo     ZTopArY7Nb
g       HT_Ql0JbDpg     0RViGi2Rne8     ShhClb6J-NA     g9e1alirMhc     YZev1imoxX8     I4yKEK9o8gA     zQ
83d_D2MGs       1GKaVzNDbuI     yuZhwV24PmM     DomumdGQSg8     hiSmlmXp-aU     pFUYi7dp1WU     2l6vwAIAqN
U, rSJ8QZWBegU  TimeGem 1135    Entertainment   95      356     4.31    13      1       QuRYeRnAuXM     gF
a1YMEJFag       UEvVksP91kg     3TYqkBJ9YRk     nRcovJn9xHg     sVkuOk4jmCo     ZTopArY7Nbg     gBcu22Vv1n
Y       HT_Ql0JbDpg     0RViGi2Rne8     ShhClb6J-NA     g9e1alirMhc     YZev1imoxX8     I4yKEK9o8gA     zQ
83d_D2MGs       1GKaV...
scala>
```

图 9-12　RDD 简单操作

（5）执行以下 Scala 程序，实现词频统计功能，并输出前 10 行结果：

```
scala>rdd1.flatMap(_.split("\t")).map((_,1)).reduceByKey(_+_).take(10)
```

执行结果如图 9-13 所示。

```
scala> rdd1.flatMap(_.split("\t")).map((_,1)).reduceByKey(_+_).take(10)
res6: Array[(String, Int)] = Array((2828,1), (bNF_P281Uu4,1), (U5Sn0O-4FZw,1), (aA3w2D0U_7U,1), (fEmDWKduL
O0,2), (Uh312eoqwBI,1), (4.24,8), (hMenB9Ywh2Q,5), (Mvz_xzaMvCQ,9), (tu1cs-6bdxA,1))

scala>
```

图 9-13　词频统计的前 10 行结果

9.3.2　在 IDEA 中编写词频统计

在经典的计算框架 MapReduce 中，问题会被拆成两个主要阶段：Map 阶段和 Reduce 阶段。对词频统计来说，MapReduce 程序从 HDFS 中读取一行字符串。在 Map 阶段，将字符串分割成单词，并生成<word,1>这样的键值对；在 Reduce 阶段，将单词对应的计数值（初始为 1）全部累加起来，最后得到单词的总出现次数。

在 Spark 中，并没有 Map、Reduce 这样的划分，而是以 RDD 的转换来呈现程序的逻辑。首先，Spark 程序将从 HDFS 中按行读取的文本作为初始 RDD（即集合的每个元素都是一行字符串）；然后，先通过 split( )操作将每行字符串分割成单词，再通过 flatMap( )操作将数据进行扁平映射成单个个体，并收集起来作为新的单词 RDD；接着，使用 map( )操作将每个单词映射成<word,1>这样的键值对，转换成新的键值对 RDD；最后，通过 reduceByKey( )操作将相同单词的计数值累加起来，得到单词的总出现次数。

WordCount 是大数据领域经典的例子，与 Hadoop 实现的 WordCount 程序相比，Spark 实现的版本要显得更加简洁。操作步骤如下。

（1）打开 IDEA 集成开发工具，新建 Maven 项目，填写 groupId、artifactId，进入 IDEA 编程主界面，如图 9-14 所示。

图 9-14 IDEA 编程主界面

（2）添加 Spark 程序开发基本依赖包到 pom.xml 配置文件：

```
1. <dependency>
2.   <groupId>com.sparkjava</groupId>
3.   <artifactId>spark-core</artifactId>
4.   <version>>${spark.version}</version>
5. </dependency>
```

（3）新建 Scala 类 WordCount，添加以下代码：

```
1. import org.apache.spark.{SparkConf, SparkContext}
2. object WordCount {
3.   def main(args: Array[String]): Unit = {
4.     val conf = new SparkConf().setAppName("wordcount").setMaster("local[1]")
5.     val sc =new SparkContext(conf)
6.     val textFile = sc.textFile(args(0))
7.     Val rdd1 =
  textFile.flatMap(_.split("\t")).map((_,1)).reduceByKey(_+_).sortBy(_._2,false)
8.     try {
9. rdd1.saveAsTextFile(args(1))
10.        print("词频统计已经完成")
11.    }catch {
12.        case e:Exception =>println("结果文件目录已经存在，请先删除！")
13.    }
14.   }
15. }
```

第 4 行创建配置文件 SparkConf，这里设置应用名称、集群名称；第 5 行创建 SparkContext，在程序中主要通过 SparkContext 来访问 Spark 集群；第 6 行使用 textFile( )方法按行读取输入文件；第 7 行创建 rdd1，以 flatMap( )方法将所有行按制表符分割成词，使用 map( )方法将词映射成<word, 1>键值对，使用 reduceByKey( )方法将所有相同的 word 对应的计数累加起来，最后 sortBy( )方法将结果按照值降序输出；第 9 行使用 saveAsTextFile( )方法将结果存储至 HDFS 目录。

（4）选择 Run→Edit Configurations，单击+号，选择 Application，进入图 9-15 所示的界面，添加 Main class、Program arguments 值。

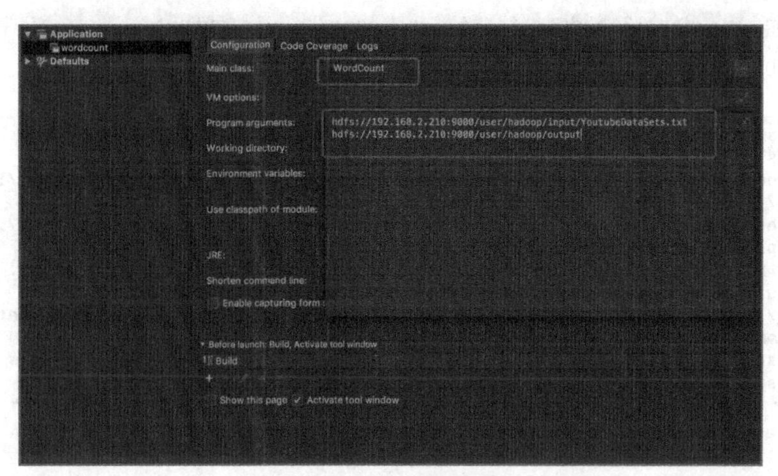

图 9-15 应用配置界面

（5）本地测试运行，单击 run，结果如图 9-16 所示。

图 9-16 IDEA 运行结果

（6）登录集群，查看运行结果文件，命令如下：

```
[hadoop@master spark-2.2.2]$ hadoop fs -cat /user/hadoop/output/part-00000 |
head -n 20
```

输出结果文件前 20 行记录，如图 9-17 所示。

```
[hadoop@master spark-2.2.2]$ hadoop fs -cat /user/hadoop/output/part-00000 | head -n 20
SLF4J: Class path contains multiple SLF4J bindings.
SLF4J: Found binding in [jar:file:/home/hadoop/hadoop-2.7.3/share/hadoop/common/lib/slf4j-
log4j12-1.7.10.jar!/org/slf4j/impl/StaticLoggerBinder.class]
SLF4J: Found binding in [jar:file:/home/hadoop/hbase-1.2.6/lib/slf4j-log4j12-1.7.5.jar!/or
g/slf4j/impl/StaticLoggerBinder.class]
SLF4J: See http://www.slf4j.org/codes.html#multiple_bindings for an explanation.
SLF4J: Actual binding is of type [org.slf4j.impl.Log4jLoggerFactory]
(0,1744)
(5,991)
(Entertainment,908)
(Music,862)
(1,673)
(2,475)
(3,443)
(Comedy,414)
(4,398)
(People & Blogs,398)
(News & Politics,333)
(Film & Animation,260)
(Sports,251)
(6,213)
(1135,206)
(7,198)
(8,171)
(1134,170)
(9,147)
(Howto & Style,137)
cat: Unable to write to output stream.
[hadoop@master spark-2.2.2]$
```

图 9-17 词频统计运行结果

（7）打开本小节源码文件 ch08/wordcount/out/，将打包后的 wordcount.jar 文件上传到集群目录/home/hadoop/data。

（8）删除之前生成的结果文件 output，命令如下：

```
[hadoop@master data]$ hadoop fs -rm -r /user/hadoop/output
```

结果如图 9-18 所示。

```
[hadoop@master data]$ hadoop fs -rm -r /user/hadoop/output
SLF4J: Class path contains multiple SLF4J bindings.
SLF4J: Found binding in [jar:file:/home/hadoop/hadoop-2.7.3/share/hadoop/common/lib/slf4j-log4j12-1.7.10.jar!/org/slf4j/impl/StaticLoggerBinder.class]
SLF4J: Found binding in [jar:file:/home/hadoop/hbase-1.2.6/lib/slf4j-log4j12-1.7.5.jar!/org/slf4j/impl/StaticLoggerBinder.class]
SLF4J: See http://www.slf4j.org/codes.html#multiple_bindings for an explanation.
SLF4J: Actual binding is of type [org.slf4j.impl.Log4jLoggerFactory]
18/09/21 22:16:20 INFO fs.TrashPolicyDefault: Namenode trash configuration: Deletion interval = 0 minutes, Emptier interval = 0 minutes.
Deleted /user/hadoop/output
[hadoop@master data]$
```

图 9-18　删除结果文件 output

（9）提交 Spark 应用，命令如下：

```
[hadoop@master spark-2.2.2]$ ./bin/spark-submit \
> --class Wordcount \
> --master spark://master:7077 \
> /home/hadoop/data/wordcount.jar \
>hdfs://master:9000/user/hadoop/input/YoutubeDataSets.txt \
>hdfs://master:9000/user/hadoop/output
```

结果如图 9-19 所示。

```
[hadoop@master spark-2.2.2]$ ./bin/spark-submit --class WordCount --master spark://master:7077 /home/hadoop/data/wordcount.jar hdfs://master:9000/user/hadoop/input/YoutubeDataSets.txt hdfs://master:9000/user/hadoop/output
18/09/21 22:22:51 INFO spark.SparkContext: Running Spark version 2.2.2
18/09/21 22:22:51 WARN util.NativeCodeLoader: Unable to load native-hadoop library for your platform... using builtin-java classes where applicable
18/09/21 22:22:52 INFO spark.SparkContext: Submitted application: wordcount
18/09/21 22:22:52 INFO spark.SecurityManager: Changing view acls to: hadoop
18/09/21 22:22:52 INFO spark.SecurityManager: Changing modify acls to: hadoop
18/09/21 22:22:52 INFO spark.SecurityManager: Changing view acls groups to:
18/09/21 22:22:52 INFO spark.SecurityManager: Changing modify acls groups to:
18/09/21 22:22:52 INFO spark.SecurityManager: SecurityManager: authentication disabled; ui acls disabled; users  with view permissions: Set(hadoop); groups with view permissions: Set(); users  with modify permissions: Set(hadoop); groups with modify permissions: Set()
18/09/21 22:22:52 INFO util.Utils: Successfully started service 'sparkDriver' on port 34646.
18/09/21 22:22:52 INFO spark.SparkEnv: Registering MapOutputTracker
18/09/21 22:22:52 INFO spark.SparkEnv: Registering BlockManagerMaster
18/09/21 22:22:52 INFO storage.BlockManagerMasterEndpoint: Using org.apache.spark.storage.DefaultTopologyMapper for getting topology information
18/09/21 22:22:52 INFO storage.BlockManagerMasterEndpoint: BlockManagerMasterEndpoint up
18/09/21 22:22:52 INFO storage.DiskBlockManager: Created local directory at /tmp/blockmgr-ca999c02-6fca-4cbb-9b38-6cf170cccce2
18/09/21 22:22:52 INFO memory.MemoryStore: MemoryStore started with capacity 413.9 MB
18/09/21 22:22:52 INFO spark.SparkEnv: Registering OutputCommitCoordinator
18/09/21 22:22:52 INFO util.log: Logging initialized @2428ms
18/09/21 22:22:53 INFO server.Server: jetty-9.3.z-SNAPSHOT
18/09/21 22:22:53 INFO server.Server: Started @2516ms
18/09/21 22:22:53 INFO server.AbstractConnector: Started ServerConnector@70d2e40b{HTTP/1.1,[http/1.1]}{0.0.0.0:4040}
18/09/21 22:22:53 INFO util.Utils: Successfully started service 'SparkUI' on port 4040.
```

图 9-19　spark-submit 运行过程

查看 Spark 运行结果，命令如下：

```
[hadoop@master spark-2.2.2]$ hadoop fs -cat /user/hadoop/output/part-00000 |
head -n 20
```

结果如图 9-20 所示。

```
[hadoop@master spark-2.2.2]$ hadoop fs -cat /user/hadoop/output/part-00000 | head -n 20
SLF4J: Class path contains multiple SLF4J bindings.
SLF4J: Found binding in [jar:file:/home/hadoop/hadoop-2.7.3/share/hadoop/common/lib/slf4j-
log4j12-1.7.10.jar!/org/slf4j/impl/StaticLoggerBinder.class]
SLF4J: Found binding in [jar:file:/home/hadoop/hbase-1.2.6/lib/slf4j-log4j12-1.7.5.jar!/or
g/slf4j/impl/StaticLoggerBinder.class]
SLF4J: See http://www.slf4j.org/codes.html#multiple_bindings for an explanation.
SLF4J: Actual binding is of type [org.slf4j.impl.Log4jLoggerFactory]
(0,1744)
(5,991)
(Entertainment,908)
(Music,862)
(1,673)
(2,475)
(3,443)
(Comedy,414)
(4,398)
(People & Blogs,398)
(News & Politics,333)
(Film & Animation,260)
(Sports,251)
(6,213)
(1135,206)
(7,198)
(8,171)
(1134,170)
(9,147)
(Howto & Style,137)
cat: Unable to write to output stream.
[hadoop@master spark-2.2.2]$
```

图 9-20　spark-submit 运行结果

9.4　Spark RDD 编程

通过前文的 Spark 程序学习，我们基本掌握了执行 Spark 的两种方式。程序中涉及的 RDD 操作我们并没有展开讲解，本节我们将详细地对常用的 RDD 操作进行代码展示，帮助读者读懂日常的 Spark 程序。

9.4.1　RDD 简介

Spark 将数据抽象成 RDD。RDD 实际是分布在集群多个节点上数据的集合，通过操作 RDD 对象来并行化操作集群上的分布式数据。Spark 的主要优势就是来自 RDD 本身的特性。RDD 允许用户在执行多个查询时显式地将工作集缓存在内存中，下一次操作时可直接从内存中读取。与 MapReduce 相比，它省去了大量的磁盘 I/O 操作。另外，持久化的 RDD 能够在错误中自动恢复，如果某部分 RDD 丢失，Spark 会自动重算丢失的部分。

图 9-21 描述了 Spark 的工作原理以及 RDD 计算的工作流，包括对 Spark 系统的输入、运行转换、输出和对 RDD 中数据进行转换和操作。

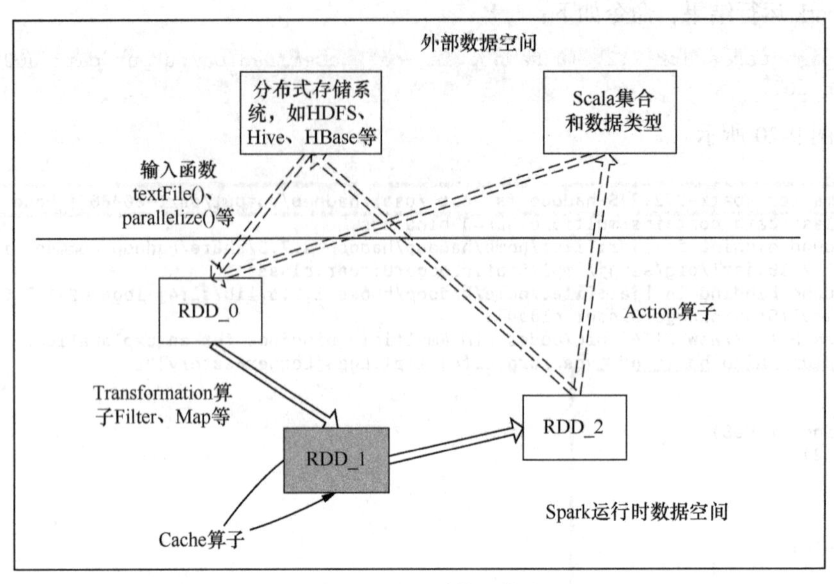

图 9-21　RDD 计算工作流

对 RDD 计算工作流的说明如下。

（1）输入：定义初始 RDD（RDD_0），Spark 程序运行时从外部数据空间读取数据进入系统，转换为 Spark 数据块，形成最初始的 RDD_0。

（2）计算：形成 RDD_0 后，系统根据定义好的 Spark 应用程序对初始的 RDD_0 进行相应的转换操作形成新的 RDD_1，然后再通过行动操作，触发 Spark 驱动器，提交作业。如果数据需要复用，可以通过 Cache 算子对数据进行持久化操作，缓存到内存中。

（3）输出：当 Spark 程序运行结束后，系统会将最终的数据（RDD_2）存储到分布式存储系统中或者存储到 Scala 集合中。

9.4.2　RDD 的操作算子

从相关数据源获取初始数据形成初始 RDD 后，我们就需要根据应用的需求对得到的初始 RDD 进行相关的操作处理，来获取满足需求的数据内容，然后对中间数据进行计算加工，得到最终的数据。

RDD 支持两种操作：转换（Transformation）操作和行动（Action）操作。

（1）转换操作：RDD 执行转换操作的结果，会产生另外一个新的 RDD，转换出来的 RDD 是惰性求值的。这就意味着对 RDD 调用转换操作时，操作并不会立即执行，而是 Spark 在内部记录下所要求执行的操作的相关信息，当在行动操作中需要用到这些转换出来的 RDD 时才会被计算。

（2）行动操作：转换操作返回的是 RDD，而行动操作返回的是其他的数据类型。行动操作会对数据集进行实际的计算，触发 Spark 提交作业，将数据输入 Spark 系统，并将最终求得的结果返回驱动程序，或者写入外部存储系统。

1. RDD 的创建方式

目前 Spark 的 RDD 有以下两种创建方式。

（1）将一个已有的集合传给 SparkContext 的 parallelize( )方法，命令如下：

```
val rdd1 = sc.parallelize(Array(1,2,3,4,5,6,7,8))
```

（2）通过外部存储系统的数据集创建，例如本地的文件系统、Hadoop 支持的数据集（HDFS、Cassandra、HBase 等），命令如下：

```
val rdd2 = sc.textFile("hdfs://localhost:9000/words.txt")
```

2. 基本 RDD 转换操作

（1）map( )转换数据

map( )是一个基础的 RDD 转换操作，通过传入的函数，将每个元素经过函数运算后返回新的 RDD。例如，我们利用一个已有的集合传给 SparkContext 的 parallelize( )方法创建一个 RDD，通过 map( )转换操作的方法将每个值乘以 2，结果生成一个新的 RDD，输出结果如图 9-22 所示。

```
scala> val rdd1 = sc.parallelize(Array(1,2,3,4,5,6,7,8))
rdd1: org.apache.spark.rdd.RDD[Int] = ParallelCollectionRDD[1] at parallelize at <console>:24

scala> val rdd2 = rdd1.map(x=>x*2)
rdd2: org.apache.spark.rdd.RDD[Int] = MapPartitionsRDD[2] at map at <console>:25

scala> rdd2.collect
res0: Array[Int] = Array(2, 4, 6, 8, 10, 12, 14, 16)
```

图 9-22　map( )操作

注意：因为 map( )是一个转换操作，所以不会马上执行。为了方便演示，我们加上 collect 这个行动操作使之立即执行。

（2）flatMap( )转换数据

flatMap( )操作先完成与 map( )一样的操作，为每条输入返回一个数组，然后 flatMap( )的压平操作将所得到的不同级别的数组中的元素，全部转换为同级别的元素输出。

例如，我们利用 map( )和 flatMap( )分割字符串，输出结果如图 9-23 所示。

```
scala> val rdd1 = sc.parallelize(Array("hello world","hadoop spark","welcome to isyslab team"))
rdd1: org.apache.spark.rdd.RDD[String] = ParallelCollectionRDD[12] at parallelize at <console>:24

scala> rdd1.collect
res9: Array[String] = Array(hello world, hadoop spark, welcome to isyslab team)

scala> rdd1.map(x=>x.split(" ")).collect
res10: Array[Array[String]] = Array(Array(hello, world), Array(hadoop, spark), Array(welcome, to, isysl
ab, team))

scala> rdd1.flatMap(x=>x.split(" ")).collect
res11: Array[String] = Array(hello, world, hadoop, spark, welcome, to, isyslab, team)

scala> 
```

图 9-23　flatMap( )操作

由图 9-23 可知，map( )分割后，每个字符串对应返回一个数组。flatMap( )把不同数组的元素扁平化，全部作为同级别的元素输出到一个数组。

（3）filter( )进行过滤

filter( )主要用于过滤 RDD 中的元素，返回一个满足过滤条件的新 RDD。例如，筛选包含"hadoop"的字符串，输出结果如图 9-24 所示。

```
scala> val rdd1 = sc.parallelize(Array("hello world","hadoop spark","welcome to isyslab team"))
rdd1: org.apache.spark.rdd.RDD[String] = ParallelCollectionRDD[15] at parallelize at <console>:24

scala> rdd1.filter(x=>x.contains("hadoop")).collect
res12: Array[String] = Array(hadoop spark)

scala> rdd1.filter(_.contains("hadoop")).collect
res13: Array[String] = Array(hadoop spark)

scala> 
```

图 9-24　filter( )操作

由图 9-24 可知，第二个 filter( )使用匿名参数，下划线“_”与第一个 filter( )中的“x”一样，都表示 RDD 的一个元素。使用匿名参数是 RDD 操作常用的一种写法，这种写法简单，但是不方便代码阅读。我们可以根据自己的代码编写习惯选择不同的写法。

（4）distinct( )进行去重

distinct( )用于 RDD 的数据去重，返回一个只包含不同元素的新 RDD。使用方法如图 9-25 所示。

```
scala> val rdd1 = sc.parallelize(Array(1,2,4,4,5,6,8,8))
rdd1: org.apache.spark.rdd.RDD[Int] = ParallelCollectionRDD[22] at parallelize at <console>:24

scala> rdd1.distinct().collect
res15: Array[Int] = Array(1, 2, 4, 5, 6, 8)

scala> 
```

图 9-25　distinct( )操作

（5）union( )合并多个 RDD

union( )用于将两个 RDD 的元素合并为一个 RDD，不进行去重操作。例如，创建两个存放二元组的 RDD，通过 union( )合并两个 RDD。输出结果如图 9-26 所示。

```
scala> val rdd1 = sc.parallelize(List(("apple",2),("banana",5),("orange",8)))
rdd1: org.apache.spark.rdd.RDD[(String, Int)] = ParallelCollectionRDD[28] at parallelize at <console>:2
4

scala> val rdd2 = sc.parallelize(List(("apple",2),("pear",6),("lemon",8)))
rdd2: org.apache.spark.rdd.RDD[(String, Int)] = ParallelCollectionRDD[29] at parallelize at <console>:2
4

scala> rdd1.union(rdd2).collect
res17: Array[(String, Int)] = Array((apple,2), (banana,5), (orange,8), (apple,2), (pear,6), (lemon,8))

scala> 
```

图 9-26　union( )操作

注意：两个 RDD 中每个元素的值的个数和类型需要保持一致。

（6）join( )连接两个 RDD

join( )是对键值对数据执行的常用操作之一。join( )连接类型如表 9-1 所示。与 union( )的简单合并不同，join( )会将键相同的值进行合并。

表 9-1　join( )连接类型

连接类型	描述
join( )	根据键对两个 RDD 进行内连接，返回两个 RDD 都存在的键的连接结果
leftOuterJoin( )	根据键对两个 RDD 进行左外连接，返回结果保留第一个 RDD 的所有键
rightOuterJoin( )	根据键对两个 RDD 进行右外连接，返回结果保留第二个 RDD 的所有键
fullOuterJoin( )	全外连接，返回结果保留两个连接 RDD 中的所有键

例如，创建存放二元组的 RDD，通过 join( )连接两个 RDD。输出结果如图 9-27 所示。

```
scala> val rdd1 = sc.parallelize(List(("apple",2),("banana",5),("orange",8)))
rdd1: org.apache.spark.rdd.RDD[(String, Int)] = ParallelCollectionRDD[54] at parallelize at <console>:2
4

scala> val rdd2 = sc.parallelize(List(("apple",8),("pear",6),("lemon",7)))
rdd2: org.apache.spark.rdd.RDD[(String, Int)] = ParallelCollectionRDD[55] at parallelize at <console>:2
4

scala> rdd1.leftOuterJoin(rdd2).collect
res26: Array[(String, (Int, Option[Int]))] = Array((apple,(2,Some(8))), (banana,(5,None)), (orange,(8,N
one)))

scala> rdd1.rightOuterJoin(rdd2).collect
res27: Array[(String, (Option[Int], Int))] = Array((apple,(Some(2),8)), (lemon,(None,7)), (pear,(None,6
)))

scala> rdd1.fullOuterJoin(rdd2).collect
res28: Array[(String, (Option[Int], Option[Int]))] = Array((apple,(Some(2),Some(8))), (lemon,(None,Some
(7))), (banana,(Some(5),None)), (pear,(None,Some(6))), (orange,(Some(8),None)))

scala> 
```

图 9-27　join( )操作

（7）sortBy( )排序

sortBy( )对 RDD 进行排序，使用方法如图 9-28 所示。

```
scala> val rdd1 = sc.parallelize(List(("apple",2),("banana",5),("orange",8)))
rdd1: org.apache.spark.rdd.RDD[(String, Int)] = ParallelCollectionRDD[71] at parallelize at <console>:2
4

scala> rdd1.sortBy(x=>x._2,false,1).collect
res31: Array[(String, Int)] = Array((orange,8), (banana,5), (apple,2))

scala> 
```

图 9-28　sortBy( )操作

如图 9-28 所示，sortBy( )包含 3 个可输入参数：①定义元素中要进行排序的值；②排序后 RDD 中的元素是升序还是降序的，默认 true 是升序的；③排序后的新 RDD 的分区个数，默认新 RDD 的分区个数和旧 RDD 的个数保持一致。

（8）sortByKey( )排序

sortByKey( )按照键排序，传入参数默认是 true，即升序。使用方法如图 9-29 所示。

```
scala> val rdd1 = sc.parallelize(List(("banana",5),("apple",2),("orange",8)))
rdd1: org.apache.spark.rdd.RDD[(String, Int)] = ParallelCollectionRDD[81] at parallelize at <console>:2
4

scala> rdd1.sortByKey().collect
res35: Array[(String, Int)] = Array((apple,2), (banana,5), (orange,8))

scala> 
```

图 9-29　sortByKey( )操作

（9）reduceByKey( )数据合并

当数据集以键值对形式存在时，reduceByKey( )会将具有相同键的数据合并。合并的方式是按照传入的匿名函数(x,y)=>x+y 相加，合并后产生一个新 RDD。使用方法如图 9-30 所示。

```
scala> val rdd1 = sc.parallelize(List(("banana",5),("apple",2),("orange",8),("apple",4)))
rdd1: org.apache.spark.rdd.RDD[(String, Int)] = ParallelCollectionRDD[87] at parallelize at <console>:2
4

scala> rdd1.reduceByKey((x,y)=>x+y).collect
res38: Array[(String, Int)] = Array((apple,6), (banana,5), (orange,8))
scala> rdd1.reduceByKey(_+_).collect
res39: Array[(String, Int)] = Array((apple,6), (banana,5), (orange,8))

scala> 
```

图 9-30　reduceByKey( )操作

（10）groupByKey( )分组统计

groupByKey( )是对具有相同键的数据进行分组。通过分组，可以实现对同组数据的统计操作，使用方法如图 9-31 所示。

```
scala> val rdd1 = sc.parallelize(List(("banana",5),("apple",2),("orange",8),("apple",4)))
rdd1: org.apache.spark.rdd.RDD[(String, Int)] = ParallelCollectionRDD[90] at parallelize at <console>:2
4

scala> val rdd2 = rdd1.groupByKey()
rdd2: org.apache.spark.rdd.RDD[(String, Iterable[Int])] = ShuffledRDD[91] at groupByKey at <console>:25

scala> rdd2.collect
res40: Array[(String, Iterable[Int])] = Array((apple,CompactBuffer(2, 4)), (banana,CompactBuffer(5)), (
orange,CompactBuffer(8)))

scala> rdd2.map(x=>(x._1,x._2.size)).collect
res41: Array[(String, Int)] = Array((apple,2), (banana,1), (orange,1))

scala> 
```

图 9-31　groupByKey( )操作

3. 基本 RDD 行动操作

（1）collect 查询

在 driver 驱动程序中，collect 以数组的形式返回数据集的所有元素。因为需要从集群各个节点传输数据到本地，经过网络传输，加载到 driver 内存中，如果数据量比较大，会导致网络负载过重，因此，collect 比较适用于足够小的数据集的过滤器或其他操作之后。collect 使用方法可参考 RDD 转换操作。

（2）first 查询

first 返回数据集的第一个元素。例如，通过 filter( )过滤出集合中包含“hadoop”的元素，first 输出结果集中的第一个元素，如图 9-32 所示。

```
scala> val rdd1 = sc.parallelize(Array("hello world","hadoop spark","hadoop is a distributed technology"))
rdd1: org.apache.spark.rdd.RDD[String] = ParallelCollectionRDD[0] at parallelize at <console>:24

scala> val rdd2 = rdd1.filter(x=>x.contains("hadoop"))
rdd2: org.apache.spark.rdd.RDD[String] = MapPartitionsRDD[1] at filter at <console>:25

scala> rdd2.first
res0: String = hadoop spark

scala> 
```

图 9-32　first 操作

（3）统计运算

RDD 统计运算包括的内容如表 9-2 所示。

表 9-2　　统计运算

命令	说明
min	求最小值
max	求最大值
count	计数
sum	求和
mean	求平均值
stdev	求标准偏差
stats	统计

例如，创建存放二元组的 RDD，根据二元组中第二个元素实现统计运算，结果如图 9-33 所示。

```
scala> val rdd1 = sc.parallelize(List(("banana",5),("apple",2),("orange",8),("apple",4)))
rdd1: org.apache.spark.rdd.RDD[(String, Int)] = ParallelCollectionRDD[27] at parallelize at <console>:24

scala> rdd1.map(x=>x._2).min
res11: Int = 2

scala> rdd1.map(x=>x._2).max
res12: Int = 8
scala> rdd1.map(x=>x._2).count
res13: Long = 4
scala> rdd1.map(x=>x._2).sum
res14: Double = 19.0

scala> rdd1.map(x=>x._2).mean
res15: Double = 4.75

scala> rdd1.map(x=>x._2).stdev
res16: Double = 2.165063509461097

scala> rdd1.map(x=>x._2).stats
res17: org.apache.spark.util.StatCounter = (count: 4, mean: 4.750000, stdev: 2.165064, max: 8.000000, min: 2.000000)

scala> 
```

图 9-33　统计运算

（4）countByKey( )计算

countByKey( )用于计算每个键的条数。例如，创建存放二元组的 RDD，实现 countByKey( ) 计算，运算结果如图 9-34 所示。

```
scala> val rdd1 = sc.parallelize(List((3,5),(4,8),(3,9),(6,9)))
rdd1: org.apache.spark.rdd.RDD[(Int, Int)] = ParallelCollectionRDD[3] at parallelize at <con
sole>:24

scala> rdd1.countByKey()
res2: scala.collection.Map[Int,Long] = Map(3 -> 2, 4 -> 1, 6 -> 1)

scala> 
```

图 9-34　countByKey()操作

（5）saveAsTextFile( )文本保存

saveAsTextFile( )将数据集的元素作为文本文件（或文本文件集）写入本地文件系统、HDFS 或任何其他 Hadoop 支持的文件系统的给定目录中。Spark 将在每个元素上调用 toString( )，将其转换为文件中的一行文本。例如，通过 saveAsTextFile( )将数据集输出到 HDFS 的/tmp/spark-rdd 目录，结果如图 9-35 所示。

```
scala> val rdd1 = sc.parallelize(List((3,5),(4,8),(3,9),(6,9)))
rdd1: org.apache.spark.rdd.RDD[(Int, Int)] = ParallelCollectionRDD[0] at parallelize at <console>:24

scala>  rdd1.partitions.size
res0: Int = 100

scala> rdd1.repartition(1).saveAsTextFile("/tmp/spark-rdd/")

scala> [hadoop@bigdata ~]$
[hadoop@bigdata ~]$ hadoop fs -ls /tmp/spark-rdd
Found 2 items
-rw-r--r--   3 hadoop supergroup          0 2018-12-22 21:29 /tmp/spark-rdd/_SUCCESS
-rw-r--r--   3 hadoop supergroup         24 2018-12-22 21:29 /tmp/spark-rdd/part-00000
[hadoop@bigdata ~]$
```

图 9-35　saveAsTextFile()操作

9.4.3　RDD 的持久化

由于 Spark RDD 是惰性求值的，所以当我们想多次使用同一个转换完的 RDD 时，Spark 会在每次调用行动操作时重新进行 RDD 的转换操作。这样频繁的重算在迭代算法中的开销很大。

为了避免多次计算同一个 RDD，Spark 可以通过 persist( )或 cache( )方法在内存中标记要保存的 RDD，当它首次被行动操作触发计算时，它将会被保留在计算节点的内存中并重用。Spark 数据持久化具有容错机制，如果丢失了 RDD 的任何分区，它将使用最初创建它的转换操作自动重新计算。当需要删除被持久化的 RDD 时，可以用 unpersistRDD( )来完成该工作。

以下示例展示使用 persist( )方法实现 RDD 的持久化：

```
1. import org.apache.spark.storage.StorageLevel
2. val rdd1 = sc.parallelize(List((3,5),(4,8),(3,9),(6,9))).map(x=>x_1*x_2)
3. rdd1.persist(StorageLevel.MEMORY_ONLY)
4. rdd1.max
5. rdd1.min
```

RDD 持久化结果如图 9-36 所示。

```
scala> import org.apache.spark.storage.StorageLevel
import org.apache.spark.storage.StorageLevel

scala> val rdd1 = sc.parallelize(List((3,5),(4,8),(3,9),(6,9))).map(x=>x._1*x._2)
rdd1: org.apache.spark.rdd.RDD[Int] = MapPartitionsRDD[5] at map at <console>:25

scala> rdd1.persist(StorageLevel.MEMORY_ONLY)
res4: rdd1.type = MapPartitionsRDD[5] at map at <console>:25

scala> rdd1.max
res5: Int = 54

scala> rdd1.min
res6: Int = 15

scala> rdd1.collect
res7: Array[Int] = Array(15, 32, 27, 54)

scala>
```

图 9-36　RDD 持久化结果

由图 9-36 可看出，如果不持久化，那么 max 和 min 操作都会对 rdd1 进行计算；而持久化后，max 操作会对 rdd1 进行计算，将二元组中的元素乘积结果保存到内存中。min 操作直接从内存中读取 rdd1 的结果，避免了再次计算同一个 RDD。

此外，每个 RDD 都支持不同的持久化级别进行缓存（如表 9-3 所示），可以选择将持久化数据集保存在硬盘中，或者在内存中作为序列化的 Java 对象缓存，甚至跨节点复制。这些级别选择，是通过将一个 org.apache.spark.storage.StorageLevel 对象传递给 persist( )方法进行确定的，默认情况下会把数据以序列化的形式缓存在 JVM 的堆空间中。

表 9-3　RDD 数据持久化级别

级别	使用的空间	CPU 时间	是否在内存中	是否在磁盘中	备注
MEMORY_ONLY	多	短	是	否	
MEMORY_ONLY_SER	少	长	是	否	
MEMORY_AND_DISK	多	中等	部分是	部分是	内存空间不足，溢写到磁盘
MEMORY_AND_DISK_SER	少	长	部分是	部分是	内存空间不足，溢写到磁盘，在内存中存放序列化后的数据
DISK_ONLY	少	长	否	是	

9.5　Spark 生态系统

Apache Spark 是一个包含着众多工具的大数据计算平台。它提供 Java、Scala、Python 和 R 中的高级 API，以及支持通用执行图的优化引擎。目前，Spark 生态系统已经涉及机器学习、数据挖掘、数据库、信息检索、自然语言处理和语音识别等多个领域。

Spark 生态系统也称为伯克利数据分析栈（Berkeley Data Analytics Stack，BDAS），如图 9-37 所示。以 Spark Core 为核心，从 HDFS、Amazon S3、HBase 等读取数据，以 EC2、MESOS、YARN 等为资源调度器实现 Spark 应用程序的并行计算。不同的应用场景提供了不同的工具，包括用于 SQL 和结构化数据处理的 Spark SQL，用于实时数据处理的 Spark Streaming，用于机器学习的 MLlib，用于图形处理的 GraphX。

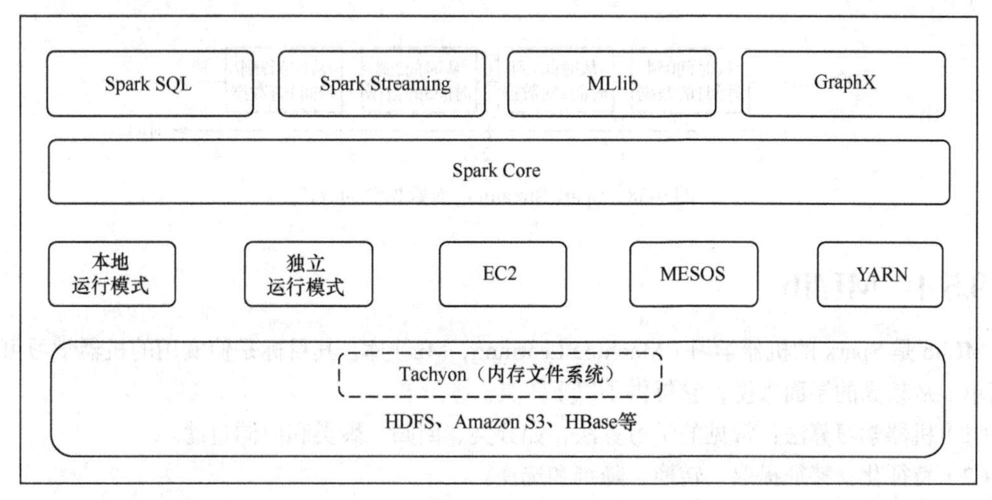

图 9-37　BDAS

9.5.1 Spark Core

Spark Core 是整个 BDAS 的核心组件，实现了 Spark 的基本功能，包括任务调度、内存管理、错误恢复、与各组件间的交互，并且包含着对 RDD 的 API 定义。同时，为运行在其上的上层组件提供 API，为 Python、Java、Scala 等语言提供编程接口。

9.5.2 Spark SQL

Spark SQL 提供在大数据上的 SQL 查询功能，是 Spark 用来操作结构化数据和半结构化数据的模型。结构化数据，是指存放数据的记录或文件带有固定的字段描述，Excel 表格和关系数据库中的数据都属于结构化数据。半结构化数据是不符合严格数据模型结构的数据，但也带有一些数据标记，如 XML 文件和 JSON 文件都是常见的半结构化数据。

相比于基础的 RDD 接口，Spark SQL 提供了数据的结构和计算过程等信息。利用这些信息，Spark 能与传统数据库一样，在具体执行查询和操作前做额外的优化，从而提升系统的整体性能。

Spark SQL 的出现使 Spark 摆脱了 Hive 的限制。Spark SQL 编译时可以包含对 Hive 的支持，也可以不包含。相比于 Hive，Spark SQL 最大的特点就是其表数据在内存中存储不是采用原生态的 JVM 对象存储方式，而是采用内存列存储方式，即任何列都能作为索引，且查询时只有涉及的列才会被读取，这大大提高了查询效率。

9.5.3 Spark Streaming

Spark Streaming 是构建在 Spark 上，用于对实时数据进行流式计算的高通量、高容错处理框架，批处理引擎为 Spark Core。

Spark Streaming 将流式计算分解为一系列连续的小规模批处理作业，从多种数据源（如 Kafka、Flume、TCP 套接字等）中读取数据，然后将数据流以时间片为单位分割，如图 9-38 所示。每块数据都转换为 Spark 中的 RDD，整个 RDD 序列形成一条离散化流（Discretized Stream，DStream）。这样，Spark Streaming 将程序中对 DStream 的操作变为了针对 RDD 的算子操作，然后触发驱动器进行计算，根据需求将计算的中间结果迭代或者存储到外部设备中。

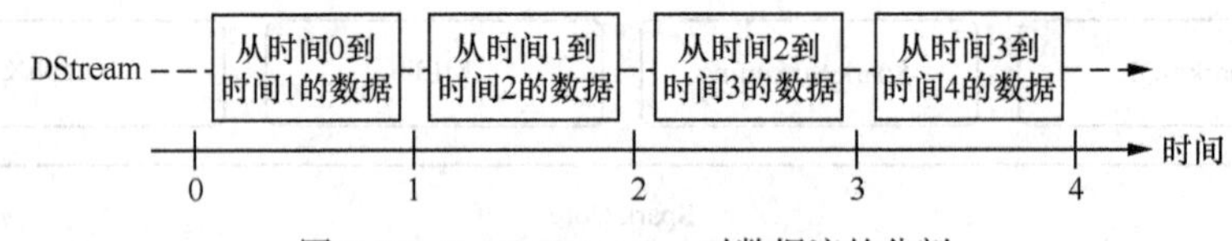

图 9-38　Spark Streaming 对数据流的分割

9.5.4 MLlib

MLlib 是 Spark 的机器学习（Machine Learning，ML）库。其目标是使实用的机器学习可扩展且简单。从较高的层面来说，它提供了以下工具。

（1）机器学习算法：常见的学习算法，如分类、回归、聚类和协同过滤。

（2）特征化：特征提取、转换、降维和选择。

（3）管道：用于构建、评估和调整机器学习管道的工具。

（4）持久性：保存和加载算法、模型和管道。

（5）实用程序：线性代数、统计、数据处理等。

作为 Spark 的机器学习组件，MLlib 继承了 Spark 先进的内存存储模式和作业调度策略，使得其对机器学习问题的处理速度大大高于普通的数据处理引擎，主要特征如下。

1. 速度高

机器学习算法通常需要在多次迭代后获得足够小的误差或者足够收敛才会停止。MapReduce 计算框架进行迭代式计算时，每次计算都要读/写磁盘，会导致非常大的 I/O 和 CPU 开销。而 Spark 是基于内存的运算，将迭代的中间结果缓存在内存中，因此处理速度会有大幅提升。

2. 简单易用

用户不仅可以直接使用 MLlib 中提供的经典机器学习算法 API，还可以基于 RDD 中封装的算法开发自己的机器学习算法。因此，使用和开发基于 Spark 的机器学习算法非常方便和简洁。

3. 集成度高

Spark 中基于 RDD 构建的 MLlib 可以与 Spark 中的 Spark SQL、GraphX、Spark Streaming 等其他组件进行无缝连接。

例如，可以借助 MLlib 对 Spark Streaming 接收到的实时数据流进行训练，这对一些复杂实时流计算场景非常有价值；MLlib 也可以与 GraphX 相结合进行深度的机器学习；MLlib 还可以直接对 Spark SQL 查询到的数据进行分析。

9.5.5　GraphX

在一些复杂的计算场景中，需要使用图对现实世界概念进行抽象，如社交网络、知识图谱等。在社交网络分析中，图的“点”代表人，“边”则代表人与人的关系。

GraphX 是 Spark 中用于图操作和图并行计算的新组件。在较高的层次上，GraphX 通过引入一个新的 Graph 抽象来扩展 Spark RDD：一个定向的多重图，其属性附加到每个顶点和边。为了支持图形计算，GraphX 公开了一组基本操作符（例如 subgraph、joinVertices 和 aggregateMessages）以及 Pregel API 的优化变体。此外，GraphX 包含越来越多的图算法和构建器，可以简化图形分析任务。

9.6　Spark 应用案例

通过前文的学习，我们已经对 Spark 有了一个基本的了解，接下来会使用 Spark 实现微博栏目粉丝统计案例。

9.6.1　案例概述

假设有一个有很多栏目的微博网站，每个栏目都有若干博主，每个博主会有很多粉丝，要求取出各栏目粉丝量排名前三的博主信息。

取一条日志记录，信息格式如下：

美食 user01 user03 user05 user06 user09 user10

其中，美食是专栏名，user01 是博主编号，后面的都是 user01 博主的粉丝。

9.6.2 代码实现

（1）打开 IDEA 集成开发工具，新建 Maven 项目，添加 Spark 基本依赖包（可参考 5.2 节）。

（2）新建 Scala 类 WeiboColumnStatistics，添加如下代码：

```
1. import org.apache.spark.{SparkConf, SparkContext}
2. object WeiboColumnStatistics {
3.   def main(args: Array[String]): Unit = {
4.    valconf = newSparkConf().setAppName("weiboColumnStatistics").setMaster("local")
5.     val sc = new SparkContext(conf)
6.     val text = sc.textFile("./src/files/weibo_log")
7.     val rdd1 = text.map(line => line.split(" "))
8.     valrdd2=rdd1.map(x=>(x(0),x(1),x.length-2))//定义三元组rdd2.collect.foreach(println)
9.     val rdd3 = rdd2.sortBy(x => x._3,false).groupBy(x => x._1)
10.     rdd3.foreach(println)
11.      val rdd4 = rdd3.map( x => {
12.        val xx = x._1
13.        val yy = x._2
14.        (xx,yy.toList.map(x => (x._2,x._3))take(3))
15.      })
16.      rdd4.foreach(println)
17.      sc.stop()
18.    }
19.
20. }
```

代码第 6 行读取 weibo_log 日志文件；第 8 行定义元组，包含 3 个元素：(栏目名,博主编号,粉丝量)；第 9 行 sortBy 实现按照粉丝量倒序排序，groupBy 实现按照栏目名分组；第 11 ~ 15 行 take( )读取排序后的前 3 条信息，即取出各栏目粉丝量排名前三的博主信息。

9.6.3 运行结果

IDEA 本地运行，单击 run，结果如图 9-39 所示。

```
18/12/23 17:59:59 INFO TaskSchedulerImpl: Adding task set 6.0 with 1 tasks
18/12/23 17:59:59 INFO TaskSetManager: Starting task 0.0 in stage 6.0 (TID 4, localhost, executor driver, partition 0, ANY, 7649 bytes)
18/12/23 17:59:59 INFO Executor: Running task 0.0 in stage 6.0 (TID 4)
(美食,List((user03,6), (user01,5), (user02,4)))
(旅游,List((user04,7), (user05,2)))
(公益,List((user10,14), (user11,6), (user23,5)))
18/12/23 17:59:59 INFO ShuffleBlockFetcherIterator: Getting 1 non-empty blocks out of 1 blocks
18/12/23 17:59:59 INFO ShuffleBlockFetcherIterator: Started 0 remote fetches in 0 ms
18/12/23 17:59:59 INFO Executor: Finished task 0.0 in stage 6.0 (TID 4). 1138 bytes result sent to driver
```

图 9-39 运行结果

由运行结果(美食,List((user03,6),(user01,5),(user02,4)))我们可以看到，元组的第一个元素为栏目名，第二个元素为 List 集合，集合内包含了该栏目粉丝量排名前三的博主编号、粉丝数。

习　　题

1. 下列哪个不是 Spark 自带的端口？（　　）
   A. 8080
   B. 4040
   C. 8090
   D. 18080
2. 下列哪个不是 Spark 生态系统的组成部分？（　　）
   A. Spark Streaming
   B. Mlib
   C. Spark R
   D. GraphX
3. Spark 默认的存储级别是（　　）。
   A. MEMORY_ONLY
   B. MEMORY_ONLY_SER
   C. MEMORY_AND_DISK
   D. DISK_ONLY
4. Hadoop 和 Spark 的计算都是并行计算，那么它们有什么相同点和不同点？
5. Spark 有几种部署模式，每种模式的特点是什么？
6. RDD 是 Spark 的“灵魂”，它有几个重要的特征，该如何理解？
7. 如何理解 RDD 的宽窄依赖关系？
8. 如何理解 Spark RDD 的懒加载机制？

第 10 章 利用大数据平台处理图像

本章首先介绍图像的基本概念，包括图像的存储原理、常见格式，然后分析 Hadoop 以及 Spark 等分布式框架处理海量图片所面临的难点以及可能的解决途径，最后介绍 Hadoop 处理海量图像的解决方案 HIPI 及其安装与应用案例。

10.1 图像的基本概念

图像通常以二进制方式存储，而不是 Hadoop 常用的文本格式。因此了解图像与视频存储的基本知识，可以帮助我们理解大数据平台针对处理图像、视频作业的特定设计与实现。从结构上讲，图像可分为栅格图和矢量图两种。

栅格图由许多的屏幕小点（像素）组成，可以用颜色、灰度、明暗对比度等图像属性描述。当栅格图被放大或缩小时，由于像素的数量没有改变，图像的分辨率就会改变。一般来讲，当图像深度高（即存储像素值所需位数多）时，它所能表现的颜色就更多、更丰富，图像的色彩就更艳丽，分辨率就更高。

矢量图用矢量代替栅格图中的“位”。矢量表达方式是先用语句调用调色板描述背景；再用带矢量的数学公式来描述图像景物大小、形状等，仅需要修改公式中的矢量变量，就可以方便地对图形进行放大、缩小和移动等操作。相比栅格图，矢量图的缩放不影响图中景物的清晰度，但在表达复杂景物方面，矢量图不如栅格图简单、直观。

本章以栅格图为例，介绍采用 Hadoop 技术对图像进行操作处理的方法。常见栅格图格式（BMP、TIFF、GIF、JPEG、PNG）介绍如下。

1. BMP 格式

位图（Bit Map，BMP）是一种与硬件设备无关的图像文件格式，使用范围非常广。它采用位映射存储格式，除了图像深度可选以外，不采用其他任何压缩，因此，BMP 文件所占用的空间很大。BMP 文件的图像深度可选 1 位（黑白）、4 位（16 色）、8 位（256 色）及 24 位（16 777 216 色）。BMP 文件存储数据时，图像的扫描方式是按从左到右、从下到上的顺序扫描的。由于 BMP 文件格式是 Windows 环境中交换与图有关的数据的一种标准，因此在 Windows 环境中运行的图形图像软件都支持 BMP 图像格式。

2. TIFF 格式

标签图像文件格式（Tag Image File Format，TIFF）是由 Aldus 公司和 Microsoft 公司为桌面出版系统研制开发的一种较为通用的图像文件格式。TIFF 定义了 4 类不同的格式：TIFF-B 适用于二值图像；TIFF-G 适用于灰度图像；TIFF-P 适用于带调色板的彩色图像；TIFF-R 适用于红绿蓝（Red Green Blue，RGB）真彩图像。TIFF 支持多种编码方法，其中包括 RGB 无压缩、游程编码（Run-Length Encoding，RLE）压缩、JPEG 压缩等。TIFF 是现存图像文件格式中最复杂的，具有扩展性、方便性、可改性。

3. GIF 格式

图形交换格式（Graphics Interchange Format，GIF）是 CompuServe 公司开发的图像文件格式。GIF 格式，是一种基于 LZW 算法的连续色调无损压缩格式，压缩率一般在 50%左右。它不属于任何应用程序，几乎所有相关软件都支持它，公共领域有大量的软件在使用 GIF 图像文件。GIF 图像文件的数据是经过压缩的，而且采用了可变长度等压缩算法。GIF 图像的深度范围为 1～8 位，最多支持 256 种色彩的图像。GIF 格式的另一个特点是在一个 GIF 文件中可以存多幅彩色图像，如果把存于一个文件中的多幅图像数据逐幅读出并显示到屏幕上，就可构成一种最简单的动画（动图）。

4. JPEG 格式

联合图像专家组（Joint Photographic Experts Group，JPEG）也是常见的一种图像格式，文件扩展名为“. jpg”或“. jpeg”。它由一个软件开发联合会组织制定，是一种有损压缩格式，能够将图像压缩在很小的存储空间内。有损压缩会使原始图像数据质量下降，容易造成图像数据的损伤。当编辑和重新保存 JPEG 文件时，可能造成数据的质量下降。JPEG 不适用于所含颜色很少、具有大块颜色相近的区域或亮度差异十分明显的简单图像。

5. PNG 格式

便携式网络图形（Portable Network Graphics，PNG）能够提供大小比 GIF 小 30%的无损压缩图像文件，同时提供 24 位和 48 位真彩色图像支持。PNG 支持高级别无损压缩，支持 alpha 通道透明度，支持 Gamma 校正等。作为网络文件格式，与 JPEG 的有损压缩相比，PNG 提供的压缩量较少。

10.2　Hadoop 处理图像的问题与对策

10.2.1　Hadoop 直接处理图像存在的问题

Hadoop 默认把 HDFS 中的数据作为文本处理，优势是可以处理单个大文件（如大的日志文件），而图像都是小文件，不适合直接用 Hadoop 处理图像数据。主要问题如下。

（1）HDFS 中文件的元数据是直接存储在 NameNode 的内存中的。若直接处理图像，每个图像的目录、块信息等都需要在内存中描述。如果文件数量太大，容易造成 NameNode 节点内存开销过大，导致集群崩溃。

（2）MapReduce 在默认行为下，一个 Map 任务处理一个图像文件，而一个 Map 任务对应一个块。在最新的 Hadoop 中，一个块的大小为 256MB。若每个小文件都占用一个块，文件系统需要极大的存储空间。

（3）由于每个图像文件都需要启动一个任务，在大量小文件的输入下，任务调度需要占用大量资源，极易达到系统瓶颈，导致集群效率低下。

10.2.2 解决途径

要解决上面存在的问题，需要从海量小文件的存储、Map 任务的输入和 Spark 处理图像三个途径进行解决。

1. 海量小文件的存储

对于多个小文件，可以将其合并为一个大文件作为图像容器，用于在 HDFS 上存储，然后自定义切割过程，但必须保证一个图像内部不能被切开。例如，将每个图像转换为二进制记录，然后依次按行存储到文本文件中。

2. Map 任务的输入

将每个图像的二进制文件，作为 Map 任务输入文件的一行。因为 Map 任务的输入是按行读取的，文件中的每一行就是一条二进制记录。例如，将连续 8bit 或 16bit 转换为字符，就可把图像转变成一个字符串表示的文本文件。代表一张需要处理的图像，这样 MapReduce 在对文件进行切割的时候就不会将一行拆开，即不会将一张图像从内部切开；读取以后再将其转换为图像格式进行处理，最后输出。

3. Spark 处理图像的解决途径

首先同样是将每个图像依据像素转换为字符串，并以文本文件并写入 HDFS；文本文件在确定编码方式的情况下可通过读入二进制流的方式恢复为图像。

Spark 对文本文件提供了统一的外部数据源文本接口，将整个文本文件看成一个行集合，即定义了一个基本的 RDD，并在之后进行一系列的 RDD 操作。

将图像对应的二进制文本文件批量读入 Spark 平台。在文件读入后，每个文件对应一个基本的 RDD，即每个图像对应一个 RDD。图像的处理操作即可视为简单的 RDD 转换操作。通过传递函数方式将图像处理算法传递到 Spark 主驱动程序上，实现了图像的并行处理。

10.3 HIPI 安装与部署

Hadoop 图像处理接口（Hadoop Image Processing Interface，HIPI）是一个用于与 MapReduce 并行编程框架一起使用的图像处理库，提供了高效和高吞吐量的图像处理功能。HIPI 不仅可以在 HDFS 中存储大量图像方案，而且可以用于高效的分布式图像处理。此外，HIPI 还提供了 OpenCV 的集成接口。

图 10-1 表示了典型 MapReduce 程序利用 HIPI 处理图像的架构。HIPI 程序的主要输入为 HIB（Hipi Image Bundle）对象，HIB 是在 HDFS 上表示为单个文件的图像集合。HIPI 分发版不仅包含了用于创建 HIB 的工具，而且包含一个从网上下载的图像列表中构建 HIB 的 MapReduce 程序。

HIPI 程序的第一个处理阶段是筛选阶段，允许根据用户定义的各种条件进行筛选，例如利用空间解析或图像元数据来过滤基于 HIB 的图像。该功能是通过 Culler 类实现的；被筛选掉的图像不会占用过多解码时间，从而提高了系统处理速度。

在筛选阶段中，符合条件被保留下来的图像被分配给单独的 Map 任务。该功能是通过 HibInputFormat 类实现的。最后，从 HipiImage 抽象基类派生出的对象和关联的 HipiImageHeader 对象，将单独的图像呈现给 Mapper。例如，ByteImage 类和 FloatImage 类扩展了 HipiImage 基类，并分别提供了作为 Java 字节和浮点数组的图像像素值访问支持。这些类提供了如图像的裁剪、颜色空间转换和缩放等功能。

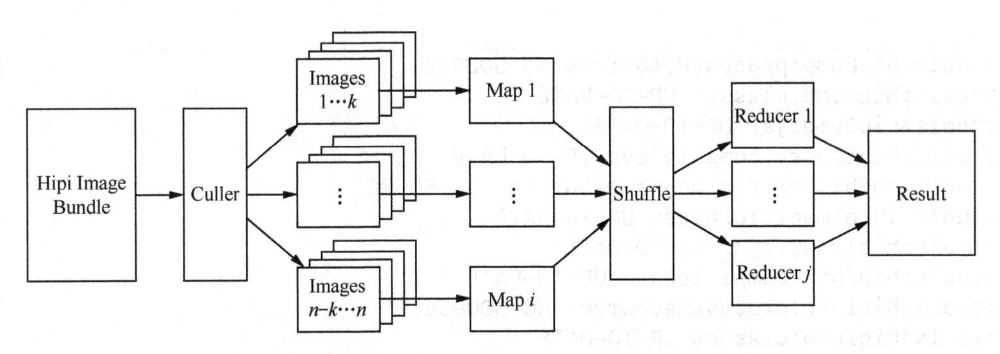

图 10-1　典型 MapReduce 程序利用 HIPI 处理图像的架构

HIPI 还支持开源计算机视觉库 OpenCV。RasterImage（如 ByteImage 和 FloatImage）的子图像类可以使用 OpenCVUtils 类中的例程转换为 OpenCV Java Mat 对象。OpenCVMatWritable 类提供了 OpenCV Java Mat 类的包装器，可以在 MapReduce 程序中使用它作为键或值对象。

然后，根据内置的 MapReduce 洗牌算法，Mapper 的记录被收集并传输到 Reducer 中，以最小化网络流量。最后，用户定义的 Reduce 任务并行执行，它们的输出被聚合并写入 HDFS。

HIPI 程序的运行需要集群中存在 Hadoop 环境，在开始部署 HIPI 之前，需要保证 Java 和 Hadoop 环境已经能正常运行。本次安装的机器操作系统为 CentOS 7，并安装 JDK 1.8、Hadoop 2.7.2。安装步骤如下。

1. 检查 Gradle

HIPI 使用了 Gradle 自动化系统来管理编译和包的组装，首先要确定系统上已经安装了 Gradle，如图 10-2 所示。

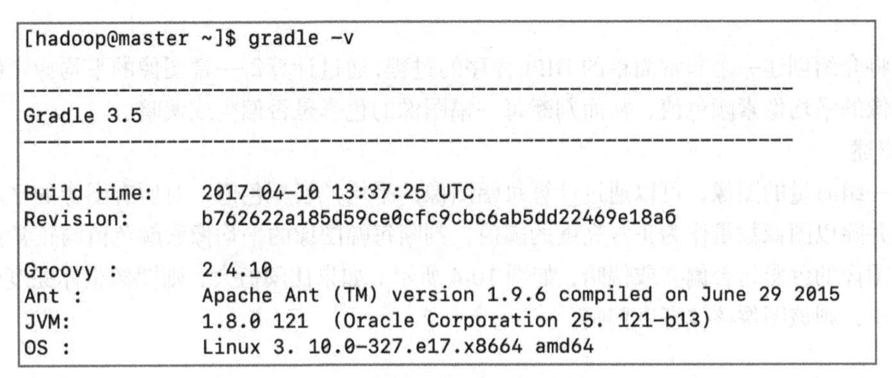

```
[hadoop@master ~]$ gradle -v

------------------------------------------------------------
Gradle 3.5
------------------------------------------------------------

Build time:   2017-04-10 13:37:25 UTC
Revision:     b762622a185d59ce0cfc9cbc6ab5dd22469e18a6

Groovy :      2.4.10
Ant :         Apache Ant (TM) version 1.9.6 compiled on June 29 2015
JVM:          1.8.0 121  (Oracle Corporation 25. 121-b13)
OS :          Linux 3. 10.0-327.e17.x8664 amd64
```

图 10-2　Java 环境配置验证

2. 下载 HIPI

获得新版本的 HIPI 的方式是复制官方的 GitHub 存储库，并将其与所有工具和示例程序一起构建，命令如下：

```
$> git clone git@github.com:uvagfx/hipi.git
```

3. 构建 HIPI 库

从 HIPI 根目录运行 Gradle 来构建 HIPI 库以及所有的工具和示例程序，命令如下：

```
$>cd hipi
$>gradle
```

若输出结果如图 10-3 所示，则说明 HIPI 构建成功。

```
:tools:hibDump:processResources NO-SOURCE
:tools:hibDump:classes UP-TO-DATE
:tools:hibDump:jar UP-TO-DATE
:tools:hibImport:compileJava UP-TO-DATE
:tools :hibImpor t:processResources NO- SOURCE
:tools :hibImport:classes UP-To-DATE
:tools:hibImport:jar UP-TO-DATE
:tools:hibInfo:compi leJava UP-TO-DATE
:tools:hibInfo:processResources NO SOURCE
:tools:hibInfo:classes UP-TO-DATE
:tools:hibInfo:jar UP-TO-DATE
:tools:hibToJpeg:compileJava UP-TO-DATE
:tools:hibToJpeg:processResources NO-SOURCE
:tools:hibToJpeg:classes UP-TO-DATE
:tools:hibToJpeg:jar UP-TO-DATE
:install

Finished building the HIPI library along with all tools and examples.

BUILD SUCCESSFUL
Total time: 12.172 secs
```

图 10-3　HIPI 库构建成功

10.4　使用 HIPI 进行图像处理

本节将介绍创建一个非常简单的 HIPI 程序的过程，通过计算每一幅图像的平均像素颜色值以及全部图像的平均像素颜色值，从而判断每一幅图像的色彩是否偏亮或偏暗。

1. 概述

给定一组海量的图像，可以通过计算每幅图像的平均像素颜色值，对所有图像的平均像素颜色值求和并除以图像数量作为正常亮度的阈值，判断每幅图像的平均像素颜色值与正常阈值的差值来判断图像的色彩是否偏亮或偏暗，如图 10-4 所示。如果比阈值大，则图像整体亮度偏亮；如果比阈值小，则该图像整体亮度偏暗。

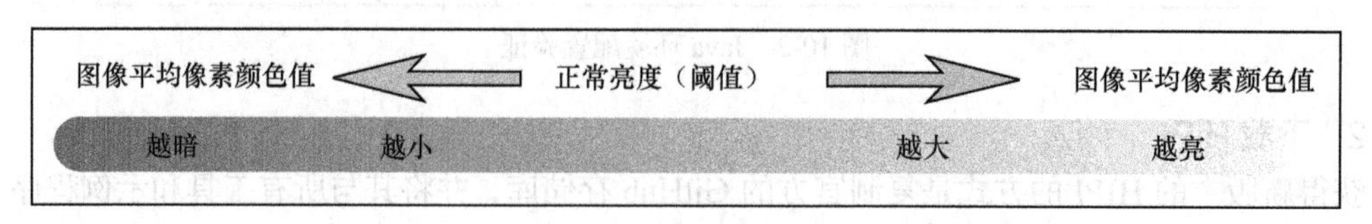

图 10-4　图像明暗判断

对于海量的图像，我们通过 Hadoop 分布式框架并行计算每幅图像的平均像素颜色值，利用 HIPI 将海量小文件集成一个大的二进制 HIB 文件进行处理，文件中的每行代表一幅图像，对应一个 map( )方法。HIPI 计算平均像素颜色值流程如图 10-5 所示。

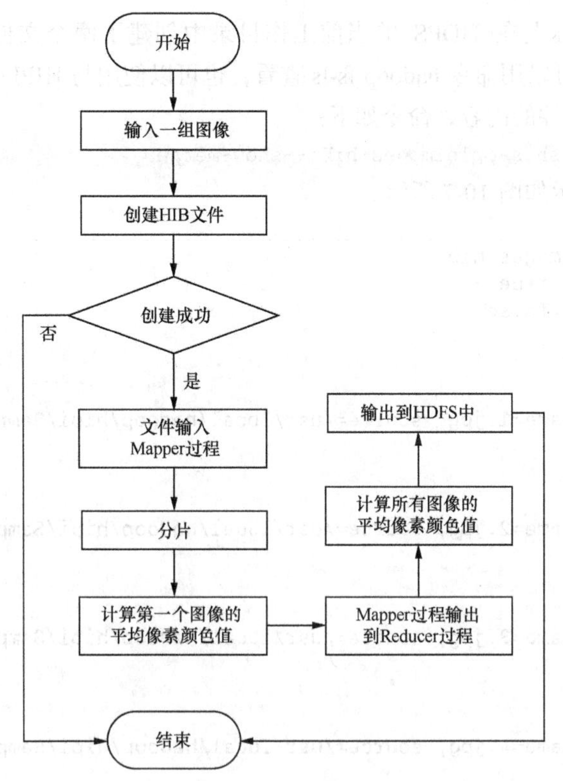

图 10-5　HIPI 计算平均像素颜色值流程

HIPI 程序的主要输入类型是 HIB，它在 HDFS 中存储了一组图像。HIB 文件创建成功后将作为输入文件进入 Mapper 过程，Hadoop 将 HIB 文件进行分片，其中文件的每行对应一个 map( )方法，同时也对应每幅图像，输出结果为该图像的平均像素颜色值。然后将该输出结果作为 Reducer 过程的输入，计算所有图像的平均像素颜色值，将该值作为明暗判断的阈值。比该值越大，说明图像整体越亮；比该值越小，说明图像整体越暗。

下面以 4 幅图像为例，简单介绍 HIPI 计算图像平均像素颜色值的过程。海量图像的操作和代码与此过程相同。

2. 处理步骤

（1）使用 hibImport 工具，根据位于目录/SampleImages 下的图像，创建一个本地文件系统的 HIB 文件（图像集），命令如下：

```
$> tools/hibImport.sh/usr/local/hadoop/hipi/SampleImages/ sampleimages.hib
```

HIB 文件创建完毕后会输出图 10-6 所示的信息。

```
[hadoop@master hipi]$ tools/hibImport.sh /usr/local/hadoop/hipi/SampleImages/ sampleimages.hib
Input image directory: /usr/local/hadoop/hipi/SampleImages/
Input FS : local FS
Output HIB: sample images.hib
Overwrite HIB if it exists: false
17/06/07 11:05:42 WARN util.NativeCodeLoader: Unable to load native-hadoop library for your
platform. . . using builtin-java classes where applicable
 ** added: 1.jpg
 ** added: 2.jpg
 ** added: 3.jpg
 ** added: 4.jpg
Created: sampleimages.hib and sampleimages.hib. dat
```

图 10-6　将 4 幅图像构建成 HIB 文件

（2）hibImport 实际上在 HDFS 的当前工作目录中创建了两个文件：sampleimages.hib 和 sampleimages.hib.dat，可以用命令 hadoop fs-ls 查看，也可以使用与 HIPI 一起的 hibInfo 工具来检查这个新创建的 HIB 文件的内容，命令如下：

```
$> tools/hibInfo.sh sampleimages.hib --show-meta
```

4 幅图像的信息显示如图 10-7 所示。

```
Input HIB: sampleimages.hib
Display meta data: true
Display EXIF data: false
IMAGE INDEX: 0
      1024 x 640
      format: 1
      meta: {filename=1.jpg, source=/usr/local/hadoop/hipi/SampleImages 1.jpg}
IMAGE INDEX: 1
      1024 x 727
      format: 1
      meta: {filename=2.jpg, source=/usr/local/hadoop/hipi/SampleImages 2.jpg}
IMAGE INDEX: 2
      680 x 510
      format: 1
      meta: {filename=3.jpg, source=/usr/local/hadoop/hipi/SampleImages 3.jpg}
IMAGE INDEX: 3
      1024 x 685
      format: 1
      meta: {filename=4.jpg, source=/usr/local/hadoop/hipi/SampleImages 4.jpg}
Found[4] images.
```

图 10-7　HIB 文件中的图像信息

（3）接下来，为程序创建一个源目录层次结构，命令如下：

```
$> mkdir -p examples/helloWorld/src/main/java/org/hipi/examples
```

通过创建 examples/helloWorld/build.gradle 文件为程序添加 Gradle build 任务，在文件中添加以下内容：

```
jar {
manifest {
attributes("Main-Class": "org.hipi.examples.HelloWorld")
    }
}
```

然后更新 HIPI 根目录下的 settings.gradle 文件，告诉 Gradle 产生了新的 build 任务：

```
include ':core', ':tools:hibImport', ... ':examples:covar', ':examples:
helloWorld'
```

（4）创建 MapReduce 程序。

创建一个 examples/helloWorld/src/main/java/org/hipi/examples/HelloWorld.java 的 Java 源文件，包含以下 MapReduce 代码：

```
1. package org.hipi.examples;
2.
3. import org.hipi.image.FloatImage;
4. import org.hipi.image.HipiImageHeader;
5. import org.hipi.imagebundle.mapreduce.HibInputFormat;
6.
7. import org.apache.hadoop.conf.Configured;
```

```
8. import org.apache.hadoop.util.Tool;
9. import org.apache.hadoop.util.ToolRunner;
10. import org.apache.hadoop.fs.Path;
11. import org.apache.hadoop.io.IntWritable;
12. import org.apache.hadoop.io.Text;
13. import org.apache.hadoop.mapreduce.lib.input.FileInputFormat;
14. import org.apache.hadoop.mapreduce.lib.output.FileOutputFormat;
15. import org.apache.hadoop.mapreduce.Job;
16. import org.apache.hadoop.mapreduce.Mapper;
17. import org.apache.hadoop.mapreduce.Reducer;
18. import org.apache.hadoop.mapreduce.lib.input.FileInputFormat;
19. import org.apache.hadoop.mapreduce.lib.output.FileOutputFormat;
20.
21. import java.io.IOException;
22.
23. public class HelloWorld extends Configured implements Tool {
24.
25.   public static class HelloWorldMapper extends Mapper<HipiImageHeader, FloatImage,
  IntWritable, FloatImage> {
26.     public void map(HipiImageHeader key, FloatImage value, Context context)
27.       throws IOException, InterruptedException {
28.     }
29.   }
30.
31.   public static class HelloWorldReducer extends Reducer<IntWritable, FloatImage,
  IntWritable, Text> {
32.       public void reduce(IntWritable key, Iterable<FloatImage> values, Context context)
33.       throws IOException, InterruptedException {
34.     }
35.   }
36.
37.   public int run(String[] args) throws Exception {
38.     //注意输入的参数
39.     if (args.length != 2) {
40.       System.out.println("Usage: helloWorld<input HIB><output directory>");
41.       System.exit(0);
42.     }
43.
44.     // 初始化和配置 MapReduce job
45.     Job job = Job.getInstance();
46.
47.     // 设置输入格式类，该类解析输入的 HIB 和生成的 Map 任务
48.     job.setInputFormatClass(HibInputFormat.class);
49.
```

```
50.     // 设置 driver、 Mapper 和 Reducer 类
51.     job.setJarByClass(HelloWorld.class);
52.     job.setMapperClass(HelloWorldMapper.class);
53.     job.setReducerClass(HelloWorldReducer.class);
54.
55.     // 设置 Map 任务传递的键值对数据类型
56.     job.setMapOutputKeyClass(IntWritable.class);
57.     job.setMapOutputValueClass(FloatImage.class);
58.     job.setOutputKeyClass(IntWritable.class);
59.     job.setOutputValueClass(Text.class);
60.
61.     // 设置 HDFS 的输入和输出路径
62.     FileInputFormat.setInputPaths(job, new Path(args[0]));
63.     FileOutputFormat.setOutputPath(job, new Path(args[1]));
64.
65.     // 执行 MapReduce 任务，直到所有的块执行完毕
66.     boolean success = job.waitForCompletion(true);
67.
68.     //返回成功或失败
69.     return success ? 0 : 1;
70.   }
71.
72.   public static void main(String[] args) throws Exception {
73.   ToolRunner.run(new HelloWorld(), args);
74.     System.exit(0);
75.   }
76.
77. }
```

（5）计算平均像素颜色值。

下面我们在程序中添加一些实际的 HIPI 图像处理代码。在本例中，将计算输入 HIB 图像中的像素的平均像素颜色值。map( )方法将计算单幅图像上的平均像素颜色值，而 reduce( )方法将把这些平均像素颜色值加在一起，然后除以它们的总数来计算总的平均像素颜色值。

首先，修改上面 MapReduce 程序中的 map( )方法：

```
1. public static class HelloWorldMapper extends Mapper<HipiImageHeader, FloatImage,
   IntWritable, FloatImage> {
2.
3.     public void map(HipiImageHeader key, FloatImage value, Context context)
4.         throws IOException, InterruptedException {
5.
6.       // 验证图像是否正确解码，是否足够大，是否有 3 种颜色通道（RGB）
7.       if (value != null && value.getWidth() > 1 && value.getHeight() > 1 && value.
   getNumBands() == 3) {
8.
```

```
9.         // 得到图像规模
10.         int w = value.getWidth();
11.         int h = value.getHeight();
12.
13.         // 得到图像位图数据的指针
14.         float[] valData = value.getData();
15.
16.         // 初始化 3 个元素去保存平均像素颜色值
17.         float[] avgData = {0,0,0};
18.
19.         // 遍历扫描图像中的像素数据，并更新平均像素颜色值
20.         for (int j = 0; j < h; j++) {
21.           for (int i = 0; i < w; i++) {
22.            avgData[0] += valData[(j*w+i)*3+0]; // R
23.            avgData[1] += valData[(j*w+i)*3+1]; // G
24.            avgData[2] += valData[(j*w+i)*3+2]; // B
25.           }
26.         }
27.
28.         // 创建一个 FloatImage 对象去存储平均像素颜色值
29.         FloatImageavg = new FloatImage(1, 1, 3, avgData);
30.
31.         //除以图像中的像素总量
32.         avg.scale(1.0f/(float)(w*h));
33.
34.         // 提交给 Reduce 任务
35.         context.write(new IntWritable(1), avg);
36.
37.       }
38.
39.     }
40.
41.   } // HelloWorldMapper
```

在本例中，map( )方法的两个参数类型分别是 HipiImageHeader（键的类型）和一个 FloatImage（值的类型）。在 HIPI 中，map( )方法的第一个参数类型必须始终是 HipiImageHeader，但是第二个参数类型可以是任何扩展抽象基类 HipiImage 的类型，这让开发人员可以控制图像如何被解码到内存中。

map( )方法为 HIB 中的每幅图像生成一个记录，该记录将使用 context.write( )方法发送到 Reduce 处理阶段。这些记录包括一个 IntWritable 和另一个 HIPI FloatImage 对象，该对象包含图像计算的平均像素颜色值。这些记录是由 MapReduce 框架收集的，并成为 reduce( )方法的输入，作为一个可迭代的 FloatImage 对象列表，这些对象被添加到一起并进行规范化以获得最终结果。

接着修改 reduce( )方法：

```
1. public static class HelloWorldReducer extends Reducer<IntWritable, FloatImage, IntWritable, Text> {
2.
3.    public void reduce(IntWritable key, Iterable<FloatImage> values, Context context) throws IOException, InterruptedException {
4.
5.      // 创建一个 FloatImage 对象去存储最终的结果
6.     FloatImage avg = new FloatImage(1, 1, 3);
7.
8.      // 初始化一个计数器，并且从 Mapper 中迭代 IntWritable/FloatImage 记录
9.      int total = 0;
10.      for (FloatImage val : values) {
11.      avg.add(val);
12.        total++;
13.      }
14.
15.      if (total > 0) {
16.        // 获取平均值
17.        avg.scale(1.0f / total);
18.        // 将最终输入包装成字符串
19.        float[] avgData = avg.getData();
20.        String result = String.format("Average pixel value: %f %f %f", avgData[0], avgData[1], avgData[2]);
21.        // 将结果写到 HDFS 中
22.        context.write(key, new Text(result));
23.      }
24.
25.    }
26.
27.  } // HelloWorldReducer
28.
```

使用 gradle 命令构建 helloWorld.jar，并且用开始创建的 HIB 来运行它，命令如下：

```
$>gradle jar
```

构建成功后，输出如图 10-8 所示。

```
[hadoop@master hipi]$ cd examples/helloWorld/
[hadoop@master helloWorld]$ gradle jar
:examples:helloWorld:compileJava
:examples:helloWorld:processResources NO-SOURCE
:examples:helloWorld:classes
:examples:helloWorld:jar

BUILD SUCCESSFUL

Total time: 8.684 secs
[hadoop@master helloWorld]$ hadoop jar build/libs/helloWorld.jar
Hello HIPI!
```

图 10-8　构建 helloWorld.jar

将构建成功的.jar 包提交到 Hadoop 集群中运行，命令如下：

```
$>hadoop jar build/libs/helloWorld.jar sampleimages.hib sampleimages_average
```

运行结果如图 10-9 所示。

```
Map-Reduce Framework
        Map input records=4
        Map output records=4
        Map output bytes=168
        Map output materialized bytes=182
        Input split bytes=116
        Combine input records=0
        Combine output records=0
        Reduce input groups=1
        Reduce shuffle bytes=182
        Reduce input records=4
        Reduce output records=1
        Spilled Records=8
        Shuffled Maps =1
        Failed Shuffles=0
        Merged Map outputs=1
        GC time elapsed (ms)=236
        CPU time spent (ms )=2290
        Physical memory (bytes) snapshot=360128512
        Virtual memory (bytes) snapshot=4157313024
        Total committed heap usage (bytes) =233242624
Shuffle Errors
        BAD_ID=0
        CONNECTION=0
        IO_ERROR=0
        WRONG_LENGTH=0
        WRONG_MAP=0
        WRONG_REDUCE=0
File Input Format Counters
        Bytes Read=522855
File Output Format Counters
        Bytes Written=50
```

图 10-9　MapReduce 任务运行完成

（6）查看文件结果。

HDFS 的输出目录 sampleimages_average 中包含的文件如图 10-10 所示。

```
[hadoop@master helloWorld]$ hadoop fs -ls sampleimages_average
17/06/07 17:16:33 WARN util.NativeCodeLoader: Unable to load native-hadoop library
for your platform... using builtin-java classes where applicable
Found 2 items
-rw-r--r--   1 hadoop  supergroup       0 2017-06-07 17:12 sampleimages_average/_SUCCESS
-rw-r--r--   1 hadoop  supergroup      50 2017-06-07 17:12 sampleimages_average/part-r-00000
```

图 10-10　HDFS 输出目录中包含的文件

当 MapReduce 程序成功完成时，它就会在输出目录中成功创建文件，并为每个 Reduce 任务创建一个 r-×××××文件，可以使用 cat 命令检索平均像素颜色值，结果如图 10-11 所示。

```
[hadoop@master helloWorld]$ hadoop fs -cat sampleimages_average/part-r-00000
17/06/07 17:17:36 WARN util.NativeCodeLoader: Unable to load native-hadoop library for your platform…
using builtin-java classes where applicable
1     Average pixel value: 0.246038 0.295173 0. 349097
```

图 10-11　程序运行完成后的平均像素颜色值

10.5　HIPI 工具 hibDownload

hibDownload 是一个 MapReduce/HIPI 程序，它从互联网上的一组图像中创建 HIB。例如，可以使用 hibDownload 从热门照片共享网站（如 Flickr、Google Images、Bing 或 Instagram）下载图像。hibDownload 也被设计为能与 Yahoo Flickr Creative Commons 100M 数据集无缝协作。

10.5.1　编译 hibDownload

通过在 HIPI 工具目录中执行以下 Gradle 命令来编译 hibDownload（请参阅有关在系统上设置 HIPI 的一般说明）：

```
$> cd tools
$>gradle hibDownload: jar
```

10.5.2　hibDownload 的使用方法

通过执行 tools 目录中的 hibDownload.sh 脚本运行 hibDownload。若不带任何参数运行，则会显示其使用方法，命令如下：

```
$> ./hibDownload.sh
usage: hibDownload.jar <directory containing source files><output HIB> [-f] [--yfcc100m]
[--num-nodes #count]
 -f,--force             force overwrite if output HIB already exists
 -n,--num-nodes <arg>   number of download nodes (default=1) (ignored if --yfcc100m is
specified)
 -y,--yfcc100m          assume input files are in Yahoo/Flickr CC 100M format
```

hibDownload 需要两个必需的参数：HDFS 上目录的路径（该目录包含具有图像 URL 列表的文件）和目标 HIB 的路径（也在 HDFS 上）。该程序尝试下载整个图像 URL 集（可能使用并行工作的多个下载节点）并将这些下载的图像存储在输出 HIB 中。hibDownload 还接收几个可选参数，这些参数允许指定下载节点的数量并使用 Yahoo Flickr Creative Commons 100M 数据集中的输入文件。

10.5.3　hibDownload 的工作原理

因为单个计算机下载大量图像可能很慢，并且因为典型的 Hadoop 集群提供的带宽可以比单个节点提供的带宽更高，所以 hibDownload 旨在将此任务分布在多个节点上。假设 hibDownload 正在尝试使用 n 个下载节点下载 k 个图像，并且底层 Hadoop 集群由 m 个计算节点组成。需要注意的是用户请求的下载节点的数量 n 不一定等于计算节点数量 m。

首先，让我们学习工具/hibDownload/src/main/java/org/hipi/tools/downloader/DownloaderInputFormat.java。此类扩展了 FileInputFormat 类，负责解析输入文本文件并启动下载图像的 Map 任务。特别是，此

类中的 getSplits()方法导致 *n* 个 Map 任务将每个下载图像分成单独的 HIB。在完成这些 Map 任务之后，单个 Reduce 任务将这些 HIB 合并到位于用户指定的输出路径的单个 HIB 中。图 10-12 说明了该工作原理。

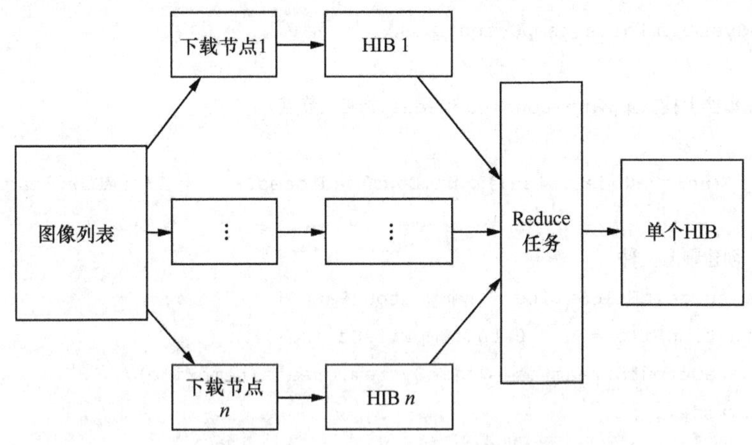

图 10-12　hibDownload 的工作原理

1. Downloader InputFormat 类

在正常操作下，Hadoop 会尝试在集群中的最接近任务的计算节点执行输入数据时的每个 Map 任务。在使用 hibDownload 的情况下，这是有问题的行为，因为输入数据是一个相对较小的文本文件，很可能存储在集群中的单个物理机上。在不采取措施的默认情况下，即使集群包含许多可独立、快速连接到互联网的节点，Hadoop 也会使用某一个节点以串行方式下载所有映像。

为了使 Hadoop 产生 *n* 个 Map 任务，每个任务在集群中不同节点上运行（只有当节点数大于或等于 *n* 时才可以这样做），DownloaderInputFormat 类中的 getSplits( )方法会在 HDFS 上创建 *n* 个临时文件。由于 Hadoop 在分布式文件系统上统一分发文件，因此每个新文件很可能位于一台独立的机器上。根据这些分析，我们在不同机器上生成 *n* 个 Map 任务的过程如下。

（1）在 HDFS 上创建最多 2*n* 个临时文件，目标是识别 *n* 个唯一的计算节点：

```
1.   // 如果未明确设置 downloader.nodes，使用默认值 10
2.   int numDownloadNodes = conf.getInt("downloader.nodes", 10);
3.
4.   // 初始化列表以在集群中存储唯一节点
5.   ArrayList<String>uniqueNodes = new ArrayList<String>(0);
6.
7.   // 初始化列表以存储 InputSplits
8.   List<InputSplit> splits = new ArrayList<InputSplit>();
9.
10.   // 为临时 HIB 文件创建存根
11.   FileSystemfileSystem = FileSystem.get(conf);
12.   String tempOutputPath = conf.get("downloader.outpath") + "_tmp";
13.   Path tempOutputDir = new Path(tempOutputPath);
14.
```

```
15.   // 确保有干净的临时目录
16.   if (fileSystem.exists(tempOutputDir)) {
17.   fileSystem.delete(tempOutputDir, true);
18.   }
19.   fileSystem.mkdirs(tempOutputDir);
20.
21.   // 在集群上搜索最多 numDownloadNodes 个唯一节点
22.   int i = 0;
23.   while (uniqueNodes.size() <numDownloadNodes&& i < 2*numDownloadNodes) {
24.
25.     // 创建临时文件
26.     String tempFileString = tempOutputPath + "/" + i;
27.     Path tempFile = new Path(tempFileString);
28.     FSDataOutputStream os = fileSystem.create(tempFile);
29.     os.write(i);
30.     os.close();
```

（2）确定每个临时文件存储在哪个集群节点上：

```
1.     FileStatus match = fileSystem.getFileStatus(tempFile);
2.     long length = match.getLen();
3.     BlockLocation[] blocks = fileSystem.getFileBlockLocations(match, 0, length);
```

（3）确保此节点尚未位于当前计算节点列表中，否则请重试：

```
1. // 检查用于存储此临时文件的第一个节点是否尚未列在我们的列表中
2.     boolean save = true;
3.     for (int j=0; j<uniqueNodes.size(); j++) {
4.       if (blocks[0].getHosts()[0].compareTo(uniqueNodes.get(j)) == 0) {
5.         save = false;
6.         System.out.println("Repeated host: " + i);
7.         break;
8.       }
9.     }
```

（4）如果节点是新节点，将其保存在唯一节点列表中并重复以上步骤，直到列表包含 n 个节点，或者我们创建了 $2n$ 个临时文件：

```
1. // 如果是新节点，请将其添加到唯一节点列表中
2.     if (save) {
3.       uniqueNodes.add(blocks[0].getHosts()[0]);
4.       System.out.println("Found unique host: " + i);
5.     }
6.     i++;
7.   } // while( hosts.size() < nodes && i < 2*nodes)
```

创建唯一节点列表后，下一步是为每个 Map 任务生成 FileSplit 对象。FileSplit 类在这里以不常见的方式使用，以便为每个 Map 任务提供它将负责下载的 URL 列表的子集。FileSplit 类的构

造函数通常需要 4 个参数：要拆分的文件的路径、拆分的起始字节偏移量、拆分的字节长度以及拆分所在的节点列表。而我们要用行偏移和输入图像 URL 列表中由每个 Map 任务处理的行数替换起始偏移和长度参数：

```
1. // 确定下载时间表（每个节点的图像数）
2.   int span = (int) Math.ceil((float)numImages / (float)uniqueNodes.size());
3.   int last = numImages - span * (uniqueNodes.size() - 1);
4.
5.   if (uniqueNodes.size() > 1) {
6.     System.out.println("First " + (uniqueNodes.size() - 1) + " nodes will each download
  " + span + " images");
7.     System.out.println("Last node will download " + last + " images");
8.   } else {
9.     System.out.println("Single node will download " + last + " images");
10.  }
11.
12.  // 根据下载时间表生成文件拆分
13.  FileSplit[] f = new FileSplit[uniqueNodes.size()];
14.  for (int j = 0; j < f.length; j++) {
15.    String[] node = new String[1];
16.    node[0] = uniqueNodes.get(j);
17.    if (j < f.length - 1) {
18.      splits.add(new FileSplit(path, (j * span), span, node));
19.    } else {
20.      splits.add(new FileSplit(path, (j * span), last, node));
21.    }
22.  }
```

此时，DownloaderInputFormat 类生成了非重叠的 FileSplits，用于划分 Map 任务集中的图像 URL 列表。

2. DownloaderRecordReader 类

Hadoop RecordReader 类负责为从 InputFormat 类接收的每个 InputSplit 对象发出一组键值对（也称为记录）。这些记录由 Map 任务通过重复调用方法 RecordReader :: nextKeyValue( )、RecordReader :: getCurrentKey( )和 RecordReader :: getCurrentValue( )来访问。在使用 hibDownload 的情况下，DownloaderRecordReader 从它接收的输入分割中发出一行到 Map 任务。然后，Map 任务负责下载 URL 块并将其存储在单个 HIB 中，这些 HIB 最终会在 Reduce 阶段附加到单个 HIB 中。如上所述，DownloaderInputFormat 类使用 Hadoop FileSplit 对象与 DownloaderRecordReader 通信，并处理输入文件部分。DownloaderRecordReader 中的 initialize( )方法用于创建输出行块并将其传递给 Mapper 的变量和对象：

```
1. public void initialize(InputSplit split, TaskAttemptContext context) throws IOException {
2.
3.   // 获取输入图像列表和打开输入流的路径
4.   FileSplit fileSplit = (FileSplit)split;
5.   Path path = fileSplit.getPath();
```

```
6.    FileSystem fileSystem = path.getFileSystem(context.getConfiguration());
7.    FSDataInputStream fileIn = fileSystem.open(path);
8.
9.     // 注意 FileSplit 对象中的开始位置和长度字段正在使用
10.    // 在图像 URL 的输入列表中传达一系列行
11.    startLine = fileSplit.getStart();
12.    numLines = fileSplit.getLength();
13.    linesRead = 0;  // 此为特定记录阅读器实例读取的总行数
14.    record reader instance
15.    linesPerRecord = 100; // 可以修改,以更改键值对大小(可以提高效率)
16.
17.     // 如果存在,请获取 FileSplit 对象的相关压缩编解码器
18.    CompressionCodecFactory codecFactory = new Compression CodecFactory (context.get
   Configuration());
19.    CompressionCodec codec = codecFactory.getCodec(path);
20.
21.     // 如果找到编解码器,则使用它来创建解压的输入流
22.     // 否则,假设输入流已经解压
23.     if (codec != null) {
24.       reader=newBufferedReader(newInputStreamReader(codec.createInputStream(fileIn)));
25.     } else {
26.       reader = new BufferedReader(new InputStreamReader(fileIn));
27.     }
28.
29.  }
```

变量 linesPerRecord 确定发送到 Mapper 的块大小，从而确定将生成的中间 HIB 文件的数量，当前设置为 100。

最后 DownloadRecordReader 实现了 Map 任务调用的函数，获取它们的输入键值对。在这种情况下，键为每个块开头的行号，值为一个 Text 对象，其中包含由换行符分隔的整行：

```
1. public LongWritable getCurrentKey() throws IOException, InterruptedException {
2.     return new LongWritable(startLine + linesRead);
3.   }
4.
5.   public Text getCurrentValue() throws IOException, InterruptedException {
6.     return new Text(urls);
7.   }
8.
9.   public boolean nextKeyValue() throws IOException, InterruptedException {
10.
11.     // 如果记录阅读器已到达其分区的末尾，则停止
12.     if (linesRead>= numLines) {
13.       return false;
14.     }
```

```
15.
16.     urls = "";
17.     String line = "";
18.
19.     // linesPerRecord 在上面的 initialize()方法中设置
20.     for (int i = 0; (i <linesPerRecord) && (linesRead<numLines); i++) {
21.
22.       line = reader.readLine();
23.
24.       if (line == null) {
25.         throw new IOException("Unexpected EOF while retrieving next line from input split.");
26.       }
27.
28.       urls += line + "\n";
29.       linesRead++;
30.     }
31.
32.     return !line.isEmpty();
33.
34.   }
```

注意：只要有要下载的其他图像，nextKeyValue()就返回 true。

3. DownloaderMapper 类

DownloaderRecordReader 类将记录发送到 DownloaderMapper 类，其中值由输入文件中的连续行块组成（由换行符分隔的单个 Text 对象）。DownloaderMapper 类中的 map( )方法下载此图像列表并将其存储在 HIB 中。随后，这些 HIB 在 Reduce 任务中连接成单个 HIB。

首先，创建一个临时 HIB 文件来保存 Map 任务下载的图像集：

```
1. public void map(LongWritable key, Text value, Context context)抛出 IOException,
   InterruptedException {
2.     //使用行号和分配给每个 Map 任务的唯一键来生成唯一的文件名
3.     String tempPath = conf.get("downloader.outpath") + key.get()+ uniqueMapperKey
   +"。hd.tmp";
4.
5.     boolean yfcc100m = conf.getBoolean("downloader.yfcc100m", false);
6.
7.     //创建新的临时 HIB 文件
8.     HipiImageBundle hib = new HipiImageBundle(new Path(tempPath), conf);
9.     hib.openForWrite(true);
```

接下来，map( )方法遍历 URL 列表（包含在值中）。要执行实际下载，URLConnection 用于创建可直接写入 HDFS 的流：

```
1.     // value 参数包含由换行符分隔的图像 URL 列表
2.     // \n 设置缓冲读取器以允许处理此字符串
3.     // 逐行
```

```
4.    BufferedReader lineReader = new BufferedReader(new StringReader(value.toString()));
5.    String line;
6.    // 通过 URL 迭代
7.    while ((line = lineReader.readLine()) != null) {
8.      String[] lineFields = null;
9.      String imageUri = null;
10.      if (yfcc100m) {
11.        // 将行拆分为字段
12.        lineFields = line.split("\t"); // 每行中的字段由制表符分隔
13.        if (lineFields[22].equals("1")) { // 0 表示图像，1 表示 YFCC100M 格式的视频
14.          continue;
15.        }
16.        imageUri = lineFields[14];
17.      } else {
18.        imageUri = line; // 否则，假设整行字段是图像 URL
19.      }
20.      long startTime = System.currentTimeMillis();
21.      try {
22.        String type = "";
23.        URLConnection conn;
24.        // 尝试使用 Java.net 在 URL 下载图像
25.        try {
26.          URL link = new URL(imageUri);
27.          numDownloads++;
28.          System.out.println("");
29.          System.out.println("Downloading: " + link.toString());
30.          System.out.println("Number of downloads: " + numDownloads);
31.          conn = link.openConnection();
32.          conn.connect();
33.          type = conn.getContentType();
34.          // 检查是否是支持的图像格式，标头是否是可解析的。如果是，则添加到 HIB
35.          if (type != null && (type.compareTo("image/jpeg") == 0 || type.
compareTo("image/png") == 0)) {
36.            // 获取 URL 连接的输入流
37.            InputStream bis = new BufferedInputStream(conn.getInputStream());
38.            // 在输入流中标记当前位置以便稍后重置
39.            bis.mark(Integer.MAX_VALUE);
40.            // 尝试解码图像标题
41.            HipiImageHeader header = (type.compareTo("image/jpeg") == 0 ?
42.            JpegCodec.getInstance().decodeHeader(bis) :
43.            PngCodec.getInstance().decodeHeader(bis));
44.            if (header == null) {
45.              System.out.println("Failed to parse header, image not added to HIB: " + link.
toString());
```

```
46.          } else {
47.            // 重置为流的开头
48.            bis.reset();
49.            if (yfcc100m) {
50.              // 将字段捕获为后续的图像元数据
51.              for (int i=0; i<lineFields.length; i++) {
52.                header.addMetaData(String.format("col_%03d", i), lineFields[i]);
53.              }
54.              header.addMetaData("source", lineFields[14]);
55.            } else {
56.              // 捕获源 URL 作为后续的图像元数据
57.              header.addMetaData("source",imageUri);
58.            }
59.            // 将图像添加到 HIB
60.            hib.addImage(header, bis);
61.            System.err.println("Added to HIB: " + imageUri);
62.          }
```

注意，addMetaData( )方法用于存储源 URL（如果是 Yahoo Flickr Creative Commons 100M 数据集，则输入文件中的其他字段也一起存储下来）。URL 相关信息与图像像素数据一起保留在 HIB 中。

处理完所有 URL 后，将完成 Mapper 的输出 HIB，并将指示其在文件系统上的位置的记录发送到 Reduce 任务：

```
1.    // 输出键值对以减少由布尔值和 HIB 路径组成的层
2.    context.write(new BooleanWritable(true), new Text(hib.getPath().toString()));
3.    // 清理
4.    lineReader.close();
5.    hib.close();
```

4. DownloaderReducer 类

DownloaderReducer 类（在 DownloaderReducer.java 中定义）非常简单，它将 Map 任务生成的 HIB 连接到用户指定的路径中的单个 HIB：

```
1. public void reduce(BooleanWritable key, Iterable values, Context context) throws
   IOException, InterruptedException {
2.
3.    if (key.get()) {
4.
5.      // 获取输出 HIB 的路径
6.      FileSystem fileSystem = FileSystem.get(conf);
7.      Path outputHibPath = new Path(conf.get("downloader.outfile"));
8.
9.      // 创建 HIB 并打开，用于写数据
10.     HipiImageBundle hib = new HipiImageBundle(outputHibPath, conf);
11.      hib.openForWrite(true);
12.
```

```
13.     // 迭代 Map 任务创建的临时 HIB 文件
14.     for (Text tempString : values) {
15.
16.       // 打开临时 HIB 文件
17.       Path tempPath = new Path(tempString.toString());
18.       HipiImageBundle inputBundle = new HipiImageBundle(tempPath, conf);
19.
20.       // 将临时 HIB 文件附加到输出 HIB（这很快）
21.       hib.append(inputBundle);
22.
23.       // 删除临时 HIB 文件（.hib 文件和.hib.dat 文件）
24.       Path indexPath = inputBundle.getPath();
25.       Path dataPath = new Path(indexPath.toString() + ".dat");
26.       fileSystem.delete(indexPath, false);
27.       fileSystem.delete(dataPath, false);
28.
29.       // 发出输出键值对，表示已处理完临时 HIB 文件
30.       Text outputPath = new Text(inputBundle.getPath().toString());
31.       context.write(new BooleanWritable(true), outputPath);
32.       context.progress();
33.
34.     }
35.
36.     // 完成输出 HIB
37.     hib.close();
38.   }
39. }
```

10.5.4 hibDownload 的使用示例

本小节将介绍一些说明 hibDownload 如何工作的示例。

示例 1　使用 hibDownload 创建 HIB，创建 HIB 的图像来源于从 HIPI 项目网站上提供的一组小样本图像。

第 1 步，列出这些图像 URL 的文本文件并将其复制到 HDFS 的目录上，命令如下：

```
$> cd tools
$>hadoop fs -mkdir download
$>hadoop fs -copyFromLocal ../testdata/downloader-images.txt download / images.txt
```

第 2 步，为了确保命令按预期工作，我们现在可以在 HDFS 上检查文本文件的内容，命令如下：

```
$>hadoop fs -cat download / images.txt
http://hipi.cs.virginia.edu/examples/testimages/01.jpg
http://hipi.cs.virginia.edu/examples/testimages/02.jpg
http://hipi.cs.virginia.edu/examples/testimages/03.jpg
```

```
http://hipi.cs.virginia.edu/examples/testimages/04.jpg
http://hipi.cs.virginia.edu/examples/testimages/05.jpg
http://hipi.cs.virginia.edu/examples/testimages/06.jpg
http://hipi.cs.virginia.edu/examples/testimages/07.jpg
http://hipi.cs.virginia.edu/examples/testimages/08.jpg
http://hipi.cs.virginia.edu/examples/testimages/09.jpg
http://hipi.cs.virginia.edu/examples/testimages/10.jpg
http://hipi.cs.virginia.edu/examples/testimages/11.jpg
http://hipi.cs.virginia.edu/examples/testimages/12.png
```

注意：文本文件 download / images.txt 包含位于 HIPI 项目网站上的 12 个图像的 URL。

第 3 步，使用 tools 目录中提供的脚本运行 hibDownload，命令如下：

```
$> ./hibDownload.sh download/images.txt download.hib --num-nodes 10
...
Downloading: http://hipi.cs.virginia.edu/examples/testimages/01.jpg
> Took 0.467 seconds
Downloading: http://hipi.cs.virginia.edu/examples/testimages/02.jpg
> Took 0.443 seconds
...
```

在这种情况下，我们请求了 10 个下载节点。如果我们的 Hadoop 集群包含至少 10 个节点，那么这将使下载操作分布在 10 个节点上（使用--yfcc100m 标志可禁用此分布式下载功能，若该数据集的作者明确请求不使用分布式下载）。

第 4 步，hibDownload 运行完成后，我们可以使用 hibInfo 工具验证输出的 HIB，命令如下：

```
$> ./hibInfo.sh download.hib --show-meta
Input HIB: download.hib
Display meta data: true
Display EXIF data: false
IMAGE INDEX: 0
  3456×2304
  format: 1
  meta: {source=http://hipi.cs.virginia.edu/examples/testimages/01.jpg}
IMAGE INDEX: 1
  3072×2304
  format: 1
  meta: {source=http://hipi.cs.virginia.edu/examples/testimages/02.jpg}
IMAGE INDEX: 2
  2592×1944
  format: 1
  meta: {source=http://hipi.cs.virginia.edu/examples/testimages/03.jpg}
IMAGE INDEX: 3
  3072×2304
  format: 1
  meta: {source=http://hipi.cs.virginia.edu/examples/testimages/04.jpg}
IMAGE INDEX: 4
  3456×2304
  format: 1
  meta: {source=http://hipi.cs.virginia.edu/examples/testimages/05.jpg}
```

```
IMAGE INDEX: 5
  4320×3240
  format: 1
  meta: {source=http://hipi.cs.virginia.edu/examples/testimages/06.jpg}
IMAGE INDEX: 6
  3456×2304
  format: 1
  meta: {source=http://hipi.cs.virginia.edu/examples/testimages/07.jpg}
IMAGE INDEX: 7
  3456×2304
  format: 1
  meta: {source=http://hipi.cs.virginia.edu/examples/testimages/08.jpg}
IMAGE INDEX: 8
  1600×1065
  format: 1
  meta: {source=http://hipi.cs.virginia.edu/examples/testimages/09.jpg}
IMAGE INDEX: 9
  1600×1065
  format: 1
  meta: {source=http://hipi.cs.virginia.edu/examples/testimages/10.jpg}
IMAGE INDEX: 10
  2048×1318
  format: 1
  meta: {source=http://hipi.cs.virginia.edu/examples/testimages/11.jpg}
IMAGE INDEX: 11
  1024×767
  format: 2
  meta: {source=http://hipi.cs.virginia.edu/examples/testimages/12.png}
Found [12] images.
```

注意，源 URL 保留在图像元数据中。

示例 2　展示 hibDownload 如何与 Yahoo Flickr Creative Commons 100M 数据集一起使用。

第 1 步，将这些源文件复制到 HDFS 的另一个目录中，命令如下：

```
$>hadoop fs -mkdir download-yf
$>hadoop fs -copyFromLocal ../testdata/yfcc100m_dataset-100-temp-0.bz2 download-yf /
yfcc100m_dataset-100-temp-0.bz2
$>hadoop fs -copyFromLocal ../testdata/yfcc100m_dataset-100-temp-1.bz2 download-yf /
yfcc100m_dataset-100-temp-1.bz2


$> ./hibDownload.sh download-yf download-yf.hib --yfcc100m
$>hadoop fs -copyFromLocal ../testdata/yfcc100m_dataset-100-temp-0.bz2 downloader-yf
/ yfcc100m_dataset-100-temp-0.bz2
$>hadoop fs -copyFromLocal ../testdata/yfcc100m_dataset-100-temp-1.bz2 downloader-yf
/ yfcc100m_dataset-100-temp-1.bz2
```

注意：hibDownload 适用于纯文本输入文件和压缩文本文件。Yahoo Flickr Creative Commons 100M 数据集以 BZ2 格式分发，BZ2 格式为一种压缩格式。

此示例中的两个源文件每个都包含 100 个图像 URL。

第 2 步，在 hibDownload 运行完成后，我们可以检查生成的 HIB，命令如下：

```
$> ./hibInfo.sh download-yf.hib --show-meta
Input HIB: download-yf.hib
Display meta data: true
Display EXIF data: false
IMAGE INDEX: 0
  500×332
format: 1
meta:{col_010=,col_011=,col_012=,col_013=http://www.flickr.com/photos/35271748@N00/9135839752/,col_014=http://farm4.staticflickr.com/3812/9135839752_53fb47cee7.jpg,col_015=Attribution-NonCommercial-NoDerivsLicense,col_016=http://creativecommons.org/licenses/by-nc-nd/2.0/,col_017=3812,col_018=4,col_019=53fb47cee7,source=http://farm4.staticflickr.com/3812/9135839752_53fb47cee7.jpg,col_021=jpg,col_000=9135839752,col_022=0,col_001=35271748@N00,col_002=dvdbramhall,col_003=2013-06-1713:29:21.0,col_004=1372169037,col_005=PENTAX+K-r,col_006=Venice+2013,col_007=Cab+of+the+vaporetto+-+Grand+Canal.,col_008=italia,italy,veneto,venezia,venice,venise,col_009=, col_020=c349232a59}
IMAGE INDEX: 1
  500×344
format: 1
meta:{col_010=,col_011=,col_012=,col_013=http://www.flickr.com/photos/71606984@N00/8736554963/,col_014=http://farm8.staticflickr.com/7288/8736554963_5691f7cfcd.jpg,col_015=Attribution-NonCommercial-NoDerivsLicense,col_016=http://creativecommons.org/licenses/by-nc-nd/2.0/,col_017=7288,col_018=8,col_019=5691f7cfcd,source=http://farm8.staticflickr.com/7288/8736554963_5691f7cfcd.jpg,col_021=jpg,col_000=8736554963,col_022=0,col_001=71606984@N00,col_002=Yelp.com,col_003=2013-05-1220:27:25.0,col_004=1368504993,col_005=Canon+EOS+5D+Mark+II,col_006=Yelp%27s+Poppin%27+Tags+-+The+Thrift+Shop+Party%21,col_007=Photo+by+Douglas+Reynolds+-+Candid+shots+from+Yelp%27s+Poppin%27+Tags+-+The+Thrift+Shop+Party+on+Saturday%2C+May+11th%2C+2013+in+Portland%2C+OR.,col_008=bossanova,bossanova+ballroom,elite,yelp,yelp%27s+poppin%27+tags,yelp+elite,yelp.com,col_009=,col_020=0aacfe183d}

...

IMAGE INDEX: 192
  375×500
format: 1
meta:{col_010=,col_011=,col_012=,col_013=http://www.flickr.com/photos/68928263@N00/5361578330/,col_014=http://farm6.staticflickr.com/5045/5361578330_c219cd30fa.jpg,col_015=Attribution-NonCommercial-NoDerivsLicense,col_016=http://creativecommons.org/licenses/by-nc-nd/2.0/,col_017=5045,col_018=6,col_019=c219cd30fa,source=http://farm6.staticflickr.com/5045/5361578330_c219cd30fa.jpg,col_021=jpg,col_000=5361578330,col_022=0,col_001=68928263@N00,col_002=insidethemagic,col_003=2011-01-1616:07:51.0,col_004=1295212182,col_005=NIKON+COOLPIX+S8000,col_006=Sully+from+Monsters+Inc+-+Pixar+Pals+Countdown+to+Fun+parade,col_007=,col_008=a+bug%27s+life,boo,bullsye,buzz+lightyear,carl,disney%27s+hollywood+studios,dug,flik,frozone,heimlich,jessie,lots-o%27-huggin%27+bear,lotso,mike,monsters+inc,mr.+incredible,mrs.+incredible,parade,pixar,pixar+pals+countdown+to+fun,princess+ada,ratatouille,remy,russell,sully,the+incredibles,toy+story,up,woody,
col_009=, col_020=7164349717}
IMAGE INDEX: 193
  500×375
format: 1
```

```
meta:{col_010=,col_011=,col_012=,col_013=http://www.flickr.com/photos/53005683@N00
/7852823550/,col_014=http://farm9.staticflickr.com/8301/7852823550_0535b9c372.jpg,col_
015=AttributionLicense,col_016=http://creativecommons.org/licenses/by/2.0/,col_017=830
1,col_018=9,col_019=0535b9c372,source=http://farm9.staticflickr.com/8301/7852823550_05
35b9c372.jpg,col_021=png,col_000=7852823550,col_022=0,col_001=53005683@N00,col_002=You
r+photos,col_003=2012-08-2504:54:10.0,col_004=1345838050,col_005=,col_006=image,col_00
7=,col_008=,col_009=,col_020=26331374ac}
Found [194] images.
```

输入文件中的所有字段（例如，描述、作者、标签、地理空间坐标等）都存储在 HIB 中的每个图像元数据中。200 个图像中只有 194 个已成功下载，这是由于图像随时间变得不可用或其他间歇性网络问题。

第 3 步，再次使用 hibInfo 来检查这个集合中的命名为 99.png 的一个图像（如图 10-13 所示），命令如下：

```
./hibInfo.sh download-yf.hib 99 --extract 99.png --meta source
Input HIB: download-yf.hib
Display meta data: false
Display EXIF data: false
Image index: 99
Extract image path: test.png
Meta data key: none
  500×382
  format: 1
Using image encoder: com.sun.imageio.plugins.png.PNGImageWriter@6179e425
Wrote [99.png]
source: http://farm4.staticflickr.com/3171/2297552664_1ee0e8855d.jpg
```

图 10-13　使用 hibInfo 检查图像

习　题

1. 安装 HIPI，并测试本章所述案例。

2. 收集有代表性的春、夏、秋、冬四季景物图像，并尝试打包为 HIB 数据集，通过 HIPI 实现景物图像的季节分类。

参考文献

[1] White T. Hadoop 权威指南: 大数据的存储与分析[M]. 王海, 华东, 刘喻, 等译. 4 版. 北京: 清华大学出版社, 2017.

[2] 林子雨. 大数据技术原理与应用[M]. 2 版. 北京: 人民邮电出版社, 2017.

[3] 董西成. Hadoop 技术内幕: 深入解析 MapReduce 架构设计与实现原理[M]. 北京: 机械工业出版社, 2013.

[4] 吉根林, 孙志挥. 数据挖掘技术[J]. 中国图象图形学报 A 辑, 2001, 6（8）: 715-721.

[5] George L. HBase 权威指南[M]. 代志远, 刘佳, 蒋杰, 译. 北京: 人民邮电出版社, 2013.

[6] Hoffman S. Flume 日志收集与 MapReduce 模式[M]. 张龙, 译. 北京: 机械工业出版社, 2015.

[7] Perera S.Hadoop MapReduce 实战手册[M]. 杨卓莹, 译. 北京: 人民邮电出版社, 2015.

[8] Parsian M. 数据算法: Hadoop/Spark 大数据处理技巧[M]. 苏全国, 杨健康, 译. 北京: 中国电力出版社, 2016.

[9] 范东来. Hadoop 海量数据处理: 技术详解与技术实战[M]. 北京: 人民邮电出版社, 2016.

[10] 黄东军. Hadoop 大数据实战权威指南[M]. 北京: 电子工业出版社, 2017.

[11] 陈长清. 数据仓库与联机分析处理技术研究[D]. 武汉: 华中科技大学. 2002.

[12] 郭景瞻. 图解 Spark: 核心技术与案例实战[M]. 北京: 电子工业出版社, 2017.

[13] Capriolo E, Wampler D, Rutherglen J. Hive 编程指南[M]. 曹坤, 译. 北京: 人民邮电出版社, 2013.

[14] Han Jiawei, Kamber M, PEI J. 数据挖掘概念与技术[M]. 范明, 孟小峰, 译. 北京: 机械工业出版社, 2007.

[15] 王家林. Spark 核心源码分析与开发实战[M]. 北京: 机械工业出版社, 2016.